R.H. Tuttle

R.H. TUTTLE

OLD WORLD MONKEYS

Evolution, Systematics, and Behavior

PROCEEDINGS OF THE CONFERENCE "SYSTEMATICS OF THE OLD WORLD MONKEYS" SPONSORED BY THE WENNER–GREN FOUNDATION FOR ANTHROPOLOGICAL RESEARCH, HELD AT BURG WARTENSTEIN, AUSTRIA, JULY 5–14, 1969.

OLD WORLD MONKEYS

Evolution, Systematics, and Behavior

Edited by

J. R. NAPIER

UNIT OF PRIMATE BIOLOGY
QUEEN ELIZABETH COLLEGE
LONDON, ENGLAND

and

P.H. NAPIER

BRITISH MUSEUM (NATURAL HISTORY)
LONDON, ENGLAND

Academic Press 1970 New York and London

ACADEMIC PRESS, INC.
111 Fifth Avenue, New York, New York 10003

United Kingdom Edition published by
ACADEMIC PRESS, INC. (LONDON) LTD.
Berkeley Square House, London W1X 6BA

LIBRARY OF CONGRESS CATALOG CARD NUMBER: 75-117087

PRINTED IN THE UNITED STATES OF AMERICA

CONTENTS

PART V. SYSTEMATIC CONCLUSIONS

PARTICIPANTS

N. A. Barnicot, Department of Anthropology, University College London, London, England

Irwin S. Bernstein, Emory University, Yerkes Field Station, Lawrenceville, Georgia

Mireille Bertrand, Département de Physiologie, Faculté de Médecine de Paris, Paris, France

F. Bourlière, Département de Physiologie, Faculté de Médecine de Paris, Paris, France

J. S. Gartlan, Department of Psychology, University of Bristol, Bristol, England

C. P. Groves, Duckworth Laboratory, Faculty of Archaeology and Anthropology, Cambridge, England

C. J. Jolly, Department of Anthropology, New York University, New York, New York

H. Kummer, Zoologisches Institut der Universität, Zürich, Switzerland

Lord Medway, Great Glenham House, Great Glenham, Saxmundham, Suffolk, England

J. R. Napier (Organizer), Unit of Primate Biology, Queen Elizabeth College, London, England

P. H. Napier (Rapporteuse), British Museum (Natural History), London, England

Urs H. Rahm, Muséum d'Histoire Naturelle Bâle, Basle, Switzerland

PARTICIPANTS

Suzanne Ripley, Smithsonian Primatological Survey, c/o American Embassy, Colombo, Ceylon

Vincent M. Sarich, Department of Anthropology, University of California, Berkeley, California

Adolph H. Schultz, Anthropologisches Institut der Universität, Zürich, Switzerland

Elwyn L. Simons, Peabody Museum, Yale University, New Haven, Connecticut

Thomas T. Struhsaker, The Rockefeller University, New York, New York

Richard W. Thorington, Jr., Primate Biology Program, Smithsonian Institution, Washington, D. C.

INTRODUCTION

> "The aim of the Conference is to develop a basis for revising the systematics of the Cercopithecoidea, the Old World monkeys. Inasmuch as modern systematics is founded on the total biology of animal populations, revision of major groups must take into account as many of the diverse aspects of primatology as possible. The Conference has been devised to bring together primate biologists with a special knowledge of all the different aspects of the biology of Cercopithecoidea – including their anatomy, physiology, behavior, ecology, zoogeography, paleontology, molecular biology, and genetics. Such a multidisciplinary approach must of necessity be an outcome of team-work."

The foregoing paragraph was precirculated to the participants of the 1969 Wenner-Gren Conference on "The Systematics of Old World Monkeys" held at Burg Wartenstein, July 5–14.

It must be apparent that the aims of the Conference were substantially achieved, otherwise, as organizers, we should hardly have had the face to reprint such an expression of blighted hopes as the above. The paper by Thorington and Groves, which constitutes the last chapter of this book provides an updating of the classification of Old World monkeys as comprehensively as present data permit. Most of the conclusions presented in their paper incorporate information presented at the Conference by taxonomists, zoologists, and ethologists. As Thorington and Groves will be the first to admit, the probabilities are high that this classification will be extensively revised in the next decade as more and more information becomes available through the work of scientists from many disciplines.

Burg Wartenstein, an 11th century Austrian fortress perched on a hill that commands the southern bottleneck of the Hungarian Plains leading into the Semmering Pass, was, metaphorically speaking, a singularly inappropriate venue for this Conference, bearing in mind its aims. The function of such medieval redoubts was to defend the *status quo*. The intent of this Conference was to rebut it, anyway as far as certain aspects of catarrhine taxonomy are concerned. Primate biologists today are the inheritors of a classification which the pioneers of the subject (with none of the technical advantages and travel opportunities of their successors) constructed out of the available data of their time. It is to their everlasting credit that the job was so well done that it is only at the levels of genus and below that major revision needs to be undertaken. Even at this level, much that was obviously built up on the basis of morphology alone remains perfectly acceptable. Of course we may well think differently in ten years time, but one suspects that just as much of Linnaean taxonomy stands firm today so classifications developed in the post-Darwin but pre-population-genetics era will withstand the winds of change for quite some time to come. The solution to this paradox is of course that animals that *look* alike usually *are* alike. For most of the procedures involved in alpha taxonomy, morphology is still the most useful criterion of phylogenetic relationships (but see Jolly, this volume). At other levels, however, the need for supplementary taxonomic characters becomes more pressing, and ethological, geographical, ecological, and serological characters may be the only means by which specific and intraspecific variation can be analyzed. Essentially it is with this level that primate systematists are currently concerned.

It is axiomatic that if anatomists, taxonomists, and ethologists are to work together with a common purpose they must be able to speak and understand each other's language. Surprisingly this problem, for all the diversity of interest and the varied experience of the participants, was no stumbling block. The Lowells and Cabots sharing, in this instance, the common ancestry of zoology could speak to each other as well as to God. Specifically, however, the need for understanding the relevant criteria upon which ethologists and anatomists base their systematic judgments became paramount. The Conference split up temporarily into its principal components to meet this need, and the first and third chapters of this volume summarize the independent deliberations of the two stocks. These sections written by the few express the beliefs of the many; they also speak for themselves and require little further development. Sarich's chapter in Part I outlines his own personal philosophy and is therefore not a "manifesto" emanating from the Conference as a whole. It is perhaps worth emphasizing, however, that Kummer's "manifesto," in which he speaks for all his colleagues at the Conference, represents the first attempt to assess the taxonomic weighting of the various types of behavioral responses of primates.

Bird and insect biologists are well used to adducing behavioral parameters for taxonomic purposes, but primate taxonomists (and particularly those concerned with the classification of the Hominidae) have been slow to follow suit. There appear to be at least two reasons for this: first, many primate taxonomists are anatomists and anthropologists for whom the science of ethology is not only "new" but, worse, "newfangled." Second, and more reasonably, the problems of primate behavior are somewhat different from those of birds and insects. In the latter groups, fixed action patterns of innate behavior provide a sort of one-to-one relationship with the phenotype of the species as a whole and can be so treated, almost typologically in fact, for taxonomic purposes. In the primates, the important addition of a large element of learned behavior renders the analytical process much more complex. So complex is it, that taxonomically oriented ethologists are needed to interpret the data. Perhaps one of the most striking revelations of the Conference (to us at any rate) was that such unlikely beings as taxonomically oriented ethologists actually existed. The chapters by Bernstein, Kummer *et al.*, and Struhsaker, for instance, unmistakably demonstrate the systematic orientation of their research.

Kummer has indicated that fixed action patterns, particularly those concerned with intraspecific communication, are the most promising responses for taxonomic purposes. Struhsaker in his paper comments that "vocalizations seem the most stable and thus the single most reliable indicator of phylogenetic affinities within this genus *(Cercopithecus)*." Fixed action patterns, generally speaking, are stable elements intimately concerned with the ancestral habitat of the species and relatively immune from the confusions of parallelism. Fixed action patterns are the equivalent of the anatomists' principal taxonomic standbys – the teeth, the genitalia, and the pelage characters which serve much the same function for systematists. Learning is not a significant feature of vocal communication in nonhuman primates, thus this form of communication provides a reasonably unadulterated parameter for taxonomy. Added to which, of course, vocal records are not simply value judgments but are comprised of numerical data that can be handled by multivariate analysis.

Social organization, which the participants rightly agreed was altogether too plastic a phenomenon to provide a relevant taxonomic character, has its equivalent in locomotor morphology which is wide open to the pressures of ecological selection, and is therefore one of the least valuable criteria for the systematists at any level lower than that of Family.

The basic issues of this Conference were first played out in 1962 – in the same theater but with a different cast. On this occasion, Bernard Campbell, acting as spokesman, expressed the feeling of the majority in his contribution to "Classification and Human Evolution" (1963), the proceedings of the Symposium at Burg Wartenstein. In this paper he outlined the concept of functional complexes which are sets of interrelated characters composed of equal parts of morphology and behavior. Campbell argued that "Behavior is a connecting link between the environment and the genotype ... . Environment ... acts on the phenotype not only directly, but through behavior and so patterns of behavior operate in conjunction with patterns of morphological features." Specific functional complexes are discussed in detail in the same volume by Biegert, Washburn, and Napier in their individual contributions. The 1963 publication had a tremendous influence on anthropological thinking; many of its tentative proposals have become the articles of faith of the last decade, so much so that the issues settled then were hardly even mentioned at this Conference. Jolly's paper, for instance, is largely structured on the concept of functional complexes.

It is therefore of particular interest to look at the list of participants at the 1962 Conference and reflect on the influence they have had on primate biology since that time:

Josef Biegert
Bernard Campbell
Irven DeVore
Theodosius Dobzhansky
Morris Goodman
K. R. L. Hall
G. A. Harrison
Harold P. Klinger
L. S. B. Leakey
Ernst Mayr
John Napier
J. J. Petter
A. Petter-Rousseaux
Anne Roe
Adolph H. Schultz
George Gaylord Simpson
William L. Straus, Jr.
S. L. Washburn
Emile Zuckerlandl

Since primate and human evolutionary biology achieved scientific respectability in the latter part of the 19th century it has been perpetually bedevilled by the bogey of personal vendettas. These scientific nonevents have affected the growth of primate biology more than they should have done. Acrimonious public controversies, however genuine and noble the aims of the protagonists, invariably redound disadvantageously upon the subject itself. There is a world of difference between scientific controversy and personality confrontation; the former illuminates the scene and the latter inevitably obfuscates it. Scientists are less sophisticated than professional politicians for whom controversy and confrontation are matters of daily routine but who seldom confuse the private image with the professional one and allow their feuds to impinge upon their private lives. Polemicals between scientists are usually derived from an inner certainty of rightness (and even righteousness) and are, thus, more deeply set, more passionate affairs in which the personality element unduly intrudes. Primate biology is a young subject; being young, it is also highly vulnerable and cannot afford to be turned inside out by the sort of feuds that have, seemingly, always accompanied the controversies of human ancestry for example.

In the finest spirit of depersonalized controversy, the 1969 Conference was rich in disagreement as will become apparent on reading these pages. Sarich inevitably comes in for considerable criticism from the paleontologists; Napier is heartily berated for his zoogeographical hypothesis, tacitly in some quarters, outspokenly in others. Groves is challenged for some of his taxonomic conclusions on classification, and various other participants receive their share of obloquy. However, this surely is the stuff of scientific progress.

Although this book covers a wide field of catarrhine biology it is clear that there are still many, many gaps. It is not without reason, for instance, that Groves entitles his paper "The Forgotten Leaf-Eaters." There are many hiatuses in the record of Old World monkeys at all levels of study. For multiple reasons, including biomedical ones, the need for field and laboratory research of this group is of paramount importance. What do we know of the ethology and ecology of *Simias*, for example, or its anatomy for that matter? There is not, as far as we are aware, even a good photograph of *Simias concolor* in existence. *Pygathrix* is still, virtually, an unknown species, as is *Rhinopithecus.* Nearer home, what of the many species of *Cercopithecus* such as *C. hamlyni* and *C. diana* that are still to be studied, even superficially? Gartlan has made a start on the genus *Mandrillus* with *M. leucophaeus*, but what of *M. sphinx*? The Gautiers have provided us with some insights into *Miopithecus talapoin* but what of *Allenopithecus*? John Deag and Michael MacRoberts have independently lifted the edge of the veil over *Macaca sylvanus* in North Africa and Gibraltar as John Crook has done with the gelada in Ethiopia and Kummer and his colleagues with the hamadryas in the same region. But what do we know of *P. hamadryas* in Arabia? Does the species really exist there? Or is it simply a hatched area on distribution maps? If it does how does it differ from the continental race? The African colobines are still a beautiful mystery in spite of the pioneering work of the Schenkels and Peter Marler. The skin of our knowledge of the catarrhines has been scratched but we must not be content until the whole organism has been fully dissected. It is a matter for responsible biological research organizations, for conservation agencies, and for medically oriented pressure groups to see that the increasing body of young graduates who *want* to do this sort of work, and who are academically equipped for it, are enabled to take time out, to sacrifice secure and economically rewarding teaching appointments in order to fill these gaps. Primate biology needs this information if it is to fulfill its considerable promise as the basic science of human biology, and it needs it soon, before the many little-known species and populations of primates join the passenger pigeon and the quagga as examples of man's prodigality with natural resources.

In spite of these limitations we venture to think that the ideas, data, and conclusions presented will have considerable impact on the course of primate biology during the next decade or so, and as editors we take the greatest pleasure that we shall have played some part. However, it is the contributors whose ideas and industry will have ensured the relevance, and, hopefully, the lasting influence of this book.

We have attempted to reflect the closely cooperative spirit of the 1969 Burg Wartenstein Conference by marshalling a naturally heterogeneous collection of papers into something that, we hope, looks more like a book than a bound compilation of offprints. To do this we have had to impose a certain degree of editorial dominance in terms of uniformity of spelling, punctuation, makeup and composition, bibliographical references, hyphenization, etc. We hope that we have not unduly wounded our authors in the process.

In our role as organizers of the Conference we would like to express to the Wenner-Gren Foundation for Anthropological Research our appreciation on behalf of all participants for the opportunity to discuss our favorite subject in the tranquil surroundings of Burg Wartenstein. These summer Conferences, of which the present is the forty-third, are an institution for which primate biology, in particular, has every reason to be deeply grateful.

J. R. Napier and P. H. Napier

PART I

MORPHOLOGICAL AND BEHAVIORAL CHARACTERS IN PRIMATE TAXONOMY

THE INTERPRETATION OF DATA IN SYSTEMATICS

Richard W. Thorington, Jr.

INTRODUCTION

This symposium on the Systematics of Old World Monkeys was organized for the purpose of reviewing our knowledge of the biology of the Cercopithecoidea. However, it is the role of this chapter to linger on the first word, "Systematics", and to discuss the importance of classifications and their derivation.

THE DERIVATION OF CLASSIFICATIONS

Simpson (1961) defined systematics as "...the scientific study of the kinds and diversity of organisms and of any and all relationships among them." It is the emphasis on comparative biology which forms the backbone of the systematic approach. Different animals are compared and judgements are made of their similarities and differences. Zoological classifications are usually based on these similarities, dissimilarities, and on inferred phylogenies of groups of animals. In practice, the blend of these differs depending both on the classifier and the group of animals being classified. Cases have been made for basing classifications solely on similarities and dissimilarities (e.g. Sokal and Sneath, 1963) or, on the other hand, for basing them solely on cladisitic phylogeny (e.g. Hennig, 1966). Neither approach is wrong. However, such classifications are designed to be useful, and a classification may be better or worse than another in being more or less useful. Simpson's (1961: p.110-114) discussion of "classification as a useful art" is a lucid statement of this issue. A classification which is a skillful weighting of similarities and inferred phylogeny becomes a useful tool, both for predicting other similarities and differences between species, and also for interpreting them (Mayr, 1969: 79-80). A classification which is inconsistent with the phylogeny of the group or one which ignores degrees of similarity and dissimilarity tends to be less predictive and hence less useful than one which combines both approaches.

Phylogenies are inferred from diverse data collected from fossil and recent forms. Even if we knew the phylogeny of the Cercopithecidae with a high degree of certainty, we could offer several consistent classifications. One possibility would be to include all or most cercopithecines in one genus. Another would be to place each species in its own genus. Neither classification is wrong but each makes inefficient use of the category of genus, which fact reduces its usefulness. In effect, such classifications don't classify well.

Classifications are hypotheses and as such, they are testable. As new data are obtained, they can be employed to test the usefulness of a classification and can lead to its revision. Before proceeding to test classifications, however, we should understand the logical steps leading to a zoological classification. These have been discussed by Simpson(1961, 1963), Mayr (1969), and many others, but it is valuable briefly to review them. This discussion follows Simpson (1963).

Biologists work with a sample of animals (a hypodigm) and from it obtain their data. From observations of these animals they draw inferences about the population from which the sample was derived. This population is part of a taxon, such as a subspecies or species, and it is this taxon which is given a rank (or placed in a category) and classified. This seems obvious but there are numerous reports in the literature which demonstrate a lack of understanding of the process. Three points in particular seem to require re-emphasis.

First, biologists collect data from a sample of animals. These data may be karyotypes, analyses of serum proteins, measurements of skulls, etc. Each organism from which the skin, blood, or skull was taken is unique and should be uniquely identifiable by the investigator. Where there is any possibility of misidentification of the animal or misrepresentation of it (which is the case for all work with nonhuman primates), the investigator shoud make sure he has some way of providing his identification to a sceptic. For persons working with museum material, fossil or recent, re-evaluation of the identification presents no problem. For persons using live animals, it does. Karyologists and biochemists are particularly blatant offenders and _Cercopithecus_ is a disturbing example. For most published karyotypes or biochemical studies of _Cercopithecus_, there is no way to check the identification of the animals used. At the least, good photographs of the animals should be saved and their place of deposition cited in publications. When practical, voucher specimens should be deposited in museums and references made to their permanent catologue numbers.

A second point was made by Simpson: from the individuals studied, we make inferences about a population. This is generally a statistical inference, and the strength of the inference is based mostly on sample size. The smaller the sample, the less we know about the population. Enough polymorphisms have been demonstrated in populations among characters ranging from gross morphology to protein structure, that their presence or absence should be tested. It may be sobering to some investigators to realize that a homomorphic sample of eight individuals is necessary to demonstrate that a polymorphism with a frequency of 50 percent does not exist in the population ($p<.01$). In other words, sample sizes frequently cited in the literature are not sufficiently large to demonstrate consistently that there are two sexes in the population. Polymorphisms present at lower frequencies can only be consistently detected with larger samples. This is only part of the problem of characterizing a species or even a subspecies, because dealers usually obtain animals from very limited areas and these are the animals we see in our zoos and laboratories. Thus, it is usually possible only to sample a few local populations from a small part of the species range. In studies using museum material, biologists can obtain a somewhat broader picture by studying the variation within populations, between populations and between subspecies. However, the material still falls far short of being an ideal sample in most cases. The magnitude of the problem of characterizing any primate species is emphasized by the fact that anthropologists are still busily characterizing *Homo sapiens*.

A third point mady by Simpson is that is is the populations comprising subspecies and species which are classified. It is not the individuals nor the characters. Here again karyological studies are open to criticism. Too frequently karyologists seem to classify karyotypes, while purporting to be classifying species. Until we can better identify homologies and recognize parallelism, convergences, and similarities due to chance in chromosome morphology, it would seem that present primate classifications have at least as much to offer in elucidating chromosomal evolution as vice versa. This would seem true of many biochemical studies as well.

NEONTOLOGICAL DATA IN SYSTEMATICS

Data collected on recent mammals provide measures of similarities and differences between species. Phylogenies may be inferred from these data, but the inferences are necessarily less direct than those drawn from paleontological data. The strength of inferences from neontologic data lies in the diversity of disciplines which can be employed in comparisons from behavior to microanatomy.

The fascination and interest of systematics increase rapidly as data are integrated from more and more disciplines. Likewise the value of its conclusions increases. The more we know about the many facets of the biology of two species, the more accurately we can state their similarities and differences, infer their phylogenetic relationships, and express these in our classification. These conclusions highlight characters which do not agree and ennable us to single them out for further study. When the evidence of gross morphology and protein biochemistry agree in their assessments of degrees of similarity and difference between species, it is satisfying for the classifier. When they disagree, however, the subject may become more fascinating. It is not just a matter of deciding which body of evidence better reflects the phylogeny but also a matter of discovering why the differences exist. Mosaic evolution is probably universal and has been much described. Different parts of organisms evolve at different rates and therefore different species will be more similar in some characters than in others. Verheyen (1962) studied skull morphology in the cercopithecines and concluded that the patas monkey should be included in the genus *Cercopithecus*. Workers more struck by the body proportions of the animal, place it in a separate genus *Erythrocebus*. Hill (1966) places the talapoin in a separate genus *Miopithecus*, principally on the basis of its sexual skin and small size, whereas Verheyen (loc. cit.) does not consider its skull sufficiently distinctive to rate even subgeneric distinction. A number of questions are raised by these cases of mosaic evolution. How basic are the differences between these two species and species generally included in Cercopithecus? Do they represent considerable genetic differences? If they do, are they the only characters by which the patas and talapoin really differ from Cercopithecus? What are the reasons for more rapid evolution of these characters in comparison with the presumed conservatism of the skull? Considering the proposed classifications of these animals, is it more useful for a neurophysiologist or biochemist to consider these as just other Cercopithecus, or is there evidence that they differ from other Cercopithecus in many characters?

Barnicot et al (1967 and this volume) raise some fascinating possibilities of mosaic evolution among protein molecules. Some proteins such as cytochrome c seem to be very conservative (also Nakagawa et al, 1968), whereas the hemoglobins show a complex evolution involving much polymorphism. The elucidation of the latter with its probable interrelationship with the malarias will make a fascinating chapter in biology. It should also help to emphasize the fact that mosaic evolution and all the evolutionary trends of parallelism, convergence, etc., elucidated by studies of gross morphology will be found in molecular biology. Arguments to the contrary, suggesting that there is non-Darwinian evolution at the molecular level (King and Jukes, 1969), have been lucidly criticized (Richmond, 1970; Clarke, 1970). Therefore all the data of molecular biology, even though they are close representations of the configuration of DNA in the species, will have to be interpreted as carefully as the data of other disciplines and integrated with these other data in systematics.

Long ago most biologists ceased to search for nonadaptive characters to use as keys to phylogeny. However the importance of understanding the adaptiveness or even just the function of taxonomic characters has not been sufficiently emphasized. This is not surprising, for it is a long step from noting the morphological differences between species to understanding the functional differences. In fact, a great deal of good systematic research has been done without any recourse to considerations of function. However, as we incorparate data from more disciplines and as we become more critical in our interspecific comparisons, it becomes increasingly necessary for us to study the functional aspects of the data.

An understanding of function can contribute to systematics by helping us understand the data and how to interpret them. For example, the comparisons of auditory ossicles of Old World primates by Masali and Chiarelli (1967) would be rendered much more informative if accompanies by functional analyses such as that summarized by Davis (1959) for the human ossicles. Functional analyses would allow clearer distinctions between differences related to size of the animal, differences related perhaps to posture, and differences related to auditory acuity at various frequencies. After the mathematical and functional correlations have been recognized it will be easier to recognize the phylogenetic components of the system.

Knowledge of function can lead to the recognition of correlations between bodies of data. In comparisons involving quantitative data it is common practice to calculate correlations between measurements. Highly correlated measurements are not used independently in the comparisons of similarity and difference. Probably they are different measures of a single functional or developmental complex. Among other bodies of data it is not possible to calculate correlations, and yet two sets of data may just as clearly be measures of a single functional complex. There are complex functional interrelations between behavior and morphology, morphology and physiology, general physiology and biochemistry, etc. Data collected in different disciplines are not necessarily independent measures of species differences, and they should not be treated as such. Differences in tooth morphology and digestive enzymes could be part of one functional difference between two forms, associated with different diets. Differences in external morphology associated with behavior patterns present other obviously correlated characters. Some of the characters used by Dandelot (1959) in his classification of *Cercopithecus aethiops* are used in the "red, white, and blue" display peculiar to this species (Struhsaker, 1967). A good example illustrating the ramifications of one basic functional modification is Spatz' (1968) study of the number of features of the skull of *Tarsius* which are correlated with the development of its large eyes. When such related characters are recognized, they can and should be treated as functional complexes in considerations of interspecific similarities and differences.

Another obvious reason for emphasizing function in systematic biology is the increased understanding it brings to evolutionary phenomena. Mosaic evolution can be explained and probably predicted on the basis of function. Changes in ecology or behavior may effect changes in function and more rapid evolution of some parts of an animal than others. This would seem true of the locomotor anatomy of the patas monkey; perhaps it is true, too, for the small size of the talapoin. It is instructive to consider the mosaic evolution of hemoglobins and cytochromes in this way. As intimated by Buettner-Janusch (1967) and by Barnicot et al (1967) the evolution of hemoglobins will probably be understood as changes selected by the same factors responsible for hemoglobin polymorphism today. There is much to learn about what functional properties of hemoglobins are advantageous in some environments and not in others.

Convergence and parallelism in many characters also can be understood in terms of similarities in function. Some transferrin data will probably have to be explained in these terms. In two interesting reports Goodman, et al (1965) and Prychodko, et al (1968) describe the frequencies of transferrins in the Thailand populations of four species of macaques and propose an explanation in terms of interbreeding. As noted by Kuhn (1967), introgression involving M. arctoides is improbable on other grounds. In the case of M. mulatta, M. fascicularis and M. nemestrina, similar frequencies of transferrin alleles could not be maintained in the three species by a small amount of interbreeding. A large amount of interbreeding seems improbable, since there is little evidence of it in other characters. Therefore it seems more promising to seek an explanation in terms of the implications of polymorphisms (summarized by Ford, 1965). The data suggest to me that these are adaptive polymorphisms in Thailand. The similar frequencies of transferrins of M. mulatta, M. fascicularis, and M. nemestrina and and the heterozygosity of M. arctoides are probably maintained by by selection for heterozygous genotypes, although the mechanism of heterozygote superiority in Thailand is unknown.

The many recent publications on cutaneous biology are full of examples of convergences, some of which are undoubtably the results of unknown similarities of function. Hopefully the elucidation of the functional adaptations or other interest to some investigators as the continued documentation of specific differences. In these and other disciplines systematic comparisons enable us to recognize convergences and parallelism which need explaining. Functional studies can contribute to systematics by clarifying some aspects of the meaning of them.

PALEONTOLOGICAL DATA IN SYSTEMATICS

The major contribution of paleontological data to systematics is the element of geological time. A continuing interplay between stratigraphic and chemical dating permits increasingly better estimation of the ages of fossil material. This time dimension provides a perspective that is difficult or impossible to obtain without data from fossils.

The assignment of fossil material to taxa and the alignment of taxa into phylogenies is based largely on morphological similarities and differences. In this respect there is little difference between the use of paleontological data and that of neontological data in systematics. The obvious advantage of a paleontologist is that he works with species which are actually ancestral to other species. When he works with species in a phyletic line he can clearly document that some character states are more primitive than others, and he can determine rates of evolution and changes in these rates. Fossils of known age and geological history place constraints on the phylogenetic hypotheses of the paleontologist and at the same time provide non-morphological data to bolster his arguments.

The sampling problems of paleontology are well known. Fossil material is usually fragmentary, sample sizes are small, and comparative material may be relatively incomparable. These deficiences in the fossil record are constantly being reduced, however, as more material is found and studied. Additionally, paleontologists rely more and more on relevant neontological data where sample size is sufficient to permit good estimates of variability or where studies of function permit better understanding of morphology.

Examples of the contributions of paleontological data to cercopithecid systematics are provided by Jolly, by Napier, and by Simons (this volume). They provide appropriate citations on potassium-argon dating of fossils and discussion of the constraints and sampling problems provided by the fossil record.

CONCLUSIONS

The systematics of primates is an immensely exciting field, in part due to the diversity of disciplines which are being used to further our knowledge. It is a field which permits and requires the integration of diverse bodies of facts and hypotheses in an attempt to understand the comparative biology of primates. Primates are being used widely in medically oriented studies, and therefore there is a wealth of data being collected on all aspects of their biology. However, many of these studies are not taxonomically oriented and the data require critical interpretation for use in systematics. There may be numerous pitfalls - such as incomparable experimental designs - for the unsuspecting systematist who attempts to compare the results of different studies on different species. There are also many potential problems unseen by the specialist who tries to interpret his data more broadly in systematic terms. The obvious way to circumvent these pitfalls and overcome the problems is to increase the exchange of information between disciplines and to encourage constructive criticism. I am indebted to the Wenner-Gren Foundation for encouraging just such activity.

ACKNOWLEDGEMENTS

I am grateful for the suggestions of other persons at the conference. Drs. E. Mayr and G.G. Simpson provided perceptive criticism of the manuscript. I also wish to acknowledge N.I.H. grant FR 00168 to Harvard University.

REFERENCES

Barnicot, N.A., C.J. Jolly and Patricia T. Wade (1967). Protein variations and primatology. Amer.J.Phys.Anthrop. 27: 343-355.

Buettner-Janusch, J. (1967). Hemoglobin and primate evolution. In Progress in Primatology.(Eds. Starck, Schneider, and Kuhn). Stuttgart: Fischer-Verlag.

Clarke, B. (1970). Darwinian evolution of proteins. Science 168: 1009-1011.

Dandelot, P. (1959). Note sur la classification des Cercopithèques du groupe aethiops. Mammalia 23: 357-368.

Davis, Hallowell (1959). Excitation of auditory receptors. In Handbook of Physiology, Sect.1., Vol.1 (Eds. Field, Magoun, and Hall). Baltimore: Williams & Wilkins Co.

Ford, E.B. (1965). Genetic polymorphism. Cambridge Mass.: M.I.T. Press.

Goodman, M., A. Kulkarni, E. Poulik, and E. Reklys (1965). Species differences in the transferrin polymorphism of macaques. Science 147: 884-886.

Hennig, W. (1966). Phylogenetic systematics. Urbana: University of Illinois Press.

Hill, W.C.O. (1966). Primates: Comparative anatomy and taxonomy. Vol VI. New York: Interscience.

King, J.L. and T.H. Jukes (1969). Non-Darwinian evolution. Science 164: 788-798.

Kuhn, H.-J. (1967). Zur Systematik der Cercopithecidae. In Progress in Primatology (Eds. Starck, Schneider, and Kuhn). Stuttgart: Fischer-Verlag.

Masali, M. and B. Chiarelli (1967). The ear bones of the Old World primates. In Progress in Primatology (Eds. Starck, Schneider, and Kuhn). Stuttgart: Fischer-Verlag.

Mayr, E. (1969). Principles of systematic zoology. New York: McGraw-Hill.

Nakagawa, J., P.C. Hudgins, and C.M. Whorton (1968). Molecular evolution of primate myoglobin and cytochrome. Paper given at 2nd International Congress of Primatology.

Prychodko, W., M. Goodman, E. Poulik, T. Miki, and T. Tanaka (1968). Geographic variations of transferrin allelic frequencies in continental and insular macaque populations. Paper given at the 2nd International Congress of Primatology.

Richmond, R.C. (1970). Non-Darwinian evolution: a critique. Nature, Lond. 225: 1025-2038.

Simpson, G.G. (1961). Principles of animal taxonomy. Columbia University Press.

Simpson, G.G. (1963). The meaning of taxonomic statements. In Classification and Human Evolution. (Ed. S.L. Washburn). Chicago: Aldine.

Sokal, R.R. and P.H.A. Sneath (1963). Principles of numerical taxonomy. San Francisco: W.H. Freeman and Co.

Spatz, W.B. (1968). Die Bedeutung der Augan für die sagittale Gestaltung des Schädels von Tarsius (Prosimiae, Tarsiiformes). Folia primatologica 9: 22-40.

Struhsaker, T.T. (1967). Behavior of vervet monkeys and other Cercopithecines. Science 156: 1197-1203.

Verheyen, W.N. (1962). Contribution à la craniologie comparée des Primates. Mus. roy. Afr. centr., Tervuren: Ann. 8, Sci. Zool. 105: 1-255.

MOLECULAR DATA IN SYSTEMATICS

Vincent M. Sarich

In recent years it has become evident that a great deal of evolutionary information of enormous potential value to the systematist is present in the amino acid sequences of proteins and nucleotide sequences of D.N.A. and R.N.A. The complete scope of that potential can hardly be delineated at this very early stage in comparative molecular studies, but even at present two areas where such comparative data are making a significant contribution are apparent.+ One of these is exemplified by Sarich (this volume, Part II) where it is shown that such an approach can, with sufficient care, provide us with phylogenies of both cladistic and temporal dimensions; that is to say that we can develop both the order of branching among the lineages leading to the various modern species being studied, and the time at which such branchings occurred. The following quotation is from Sarich (loc. cit.):

> "The molecular data aid us in developing the history of the modern organisms under study. This history is then a framework upon which we can properly assess the meaning of the evidence we glean from other areas of research. The molecules at this early stage tell us nothing about their own adaptations* but they do allow us to make more meaningful statements about the adaptations of the organisms containing them. This history is essential to my evolutionary study. For the student of catarrhine evolution it allows a coherent picture to be developed that both accounts for the available facts and points toward rewarding new areas of research."

I like to think that the above is a fulfillment of what Nuttall saw when he wrote in 1902:

> "I do not wish these numbers to be taken as final, nevertheless they show the essential correctness of the previous

---

+Dr. Sarich's contribution does not necessarily reflect the views of the participants of this Conference, unlike the other two chapters of Part I. Eds.

*Upon further reflection I see this isn't quite true. V.M.S.

> crude results. To obtain a constant it will be necessary to make repeated tests with the blood of each species with different dilutions and different proportions of antiserum. I am inclined to believe that with care we shall perhaps be able to 'measure species' with this method, for it appears that there are measurable differences in the reactions obtained with related blood, in other words, to determine degrees of relationship which we may be able to formulate."

The question of why it has taken so long for systematists, not to mention the biochemically oriented investigators themselves, to begin seriously to consider such data as a source of useful information would form an interesting study in the history and philosophy of science. Even today most systematists still opt for the primacy of comparative anatomical and paleontological data for developing and understanding phylogenies. I feel the time has come to recognize that while no single technique or source of information has the potential to contribute to all facets of the science of systematics, a division of labor is clearly indicated. There are phylogenies to develop as a first, basic step to evolutionary understanding; this is where the molecular approach has the most to contribute. There are then the functional anatomical, physiological, and behavioral analyses to be made within that phylogenetic framework - and this is a province in which the biochemist or molecular biologist has as yet little to say directly.

The basic problem of how best to organize the biological information and conclusions provided by students in all areas of the life sciences has long occupied the attention of systematists. The presently acceptable solution which takes the form of a hierarchical set has a long developmental history spanning more than two centuries. Its pre-evolutionary roots have led to conflicts not yet resolved - and probably not resolvable - within the strictures that modern taxonomists have tended to place on the system. It can be generally agreed that taxonomies are necessary as means of ordering information; that is, they should have integrative value. In addition, they should possess stability and predictive properties.

Stability requires that the taxonomy represent an objective

reality. The evolutionary process requires that any such reality must have a cladistic base. The integrative aspect requires that the taxonomy provide an appropriate framework within which to study the taxa with which it deals. Finally the taxonomy should allow one to have some reasonable expectation of the pattern of the results to be obtained in a study of a new comparative system. It is my contention, and hardly an original one, that the nearest approach to such taxonomies within the framework of modern systematics is to make them represent phylogenies (in both a cladistic and temporal sense) as closely as possible. It might well be argued that this stops short of the next logical step of using phylogenies only, but to do so would pose many more problems than it would solve. In particular it is useful to have names for monophyletic taxa where the length of the period of monophyly or shared common history legitimizes the taxon as a unit worthy of study in an evolutionary context. A recent effort in this direction dealing with vertebrates has been made (Nelson, 1969). I contend that two of the major objections which have been raised concerning such taxonomies can now be answered.

The first of these deals with the fact that most phylogenies are not known and in fact unlikely to ever be known with a sufficient degree of certainly. I lodge a protest both in principle and in practice against the latter claim. Granted that the available molecular studies are still relatively crude in resolving power, they have provided remarkably valuable information derived from sequence and immunoligical comparisons of but a single protein, as the albumin and cytochrome c (Fitch, Margoliash, and Gould, 1969) studies have demonstrated. The cladistic and temporal information gained leads to a phylogeny whose cladistic resolving power is a function of the internal consistency of the system and whose temporal resolution is limited mainly by the stochastic nature of much of the molecular evolutionary process. Both dimensions can be readily improved through studies of other proteins and, when more rigorously tested, direct DNA comparisons. The phylogenies are recorded in the modern molecules; we do know how to read the message; what remains is more detailed experimental work.

The second, and to most taxonomists the more important, objection considers the fact that, in contrast to the general pattern of protein evolution, morphological evolution is often a most irregular process. The practicing taxonomist therefore feels that a taxon which has undergone a rapid shift into a new grade of evolutionary development should be raised in the rank at which it is differentiated from its closest phyletic relatives. Thus the hominids are given at least family status within the primates, and the pinnipeds at least subordinal rank among the carnivores. In a purely phylogenetic classification these rankings would probably be no higher than generic and familial, respectively. For the pinnipeds, at least, R.H. Telford, of the American Museum of Natural History, would appear to be coming very close to this in his recent taxonomic revision of the carnivores. How, one would then ask, are we to indicate the marked changes in grade which characterize the hominids and pinnipeds? I would suggest that this is not the function of the taxonomist -- mainly because of its extremely subjective nature. Once one leaves an objective system, then the final rank assigned is a matter of personal opinion and thus necessarily unstable. But beyond this are the integrative and predictive aspects. Many systematists have confidently predicted that the marked change in grade of the hominids would be characterized by marked genetic changes as well. This has not turned out to be the case. In fact, as our genetic knowledge concerning man and the apes increases, the initial impression of close genetic similarity is simply reinforced. In this, I feel, lies the solution to the apparent impasse discussed above -- to avoid having to balance cladogenesis versus anagenesis.

What we, as evolutionary biologists and systematists, are ultimately interested in is not a taxonomy -- which is, after all, but a useful tool summarizing what is known, and pointing the way of further research into the causation and mechanisms of specific evolutionary events. As taxonomy becomes more nearly phylogenetic, nontaxonomists using this information will have a stable evolutionary base upon which to orient their research and, specifically, to focus upon the selective significance of their results. In view of the general pattern of regularity thus far observed for genetic change, it is likely to be <u>only those loci which deviate from</u>

this pattern (i.e. those results which do not fit expectations based on a phylogenetic taxonomy) that will have evolutionary significance. It is these data which will indicate to the evolutionary biologist where the functional and selectively important changes have occurred and thereby contribute to our understanding of the taxa under study.

It is recognised that, a phylogenetic (and therefore continuous) system is inevitably distorted when forced into a hierarchical mode. Given our linguistic system, which emphasises discontinuity rather than continuity, it is difficult to see how to avoid such distortion and still construct classifications. It is the events in common ancestry of a monophyletic taxon which allow us to recognize that taxon and therefore legitimize it. This evolutionary reality needs discussion and such discussion is greatly facilitated by giving names to monophyletic taxa. Thus the Linnean hierarchy has evolutionary significance; and even with its liabilities it allows us to coordinate evolutionary events within its semantics (and within no other semantic system yet suggested). The genesis of this chapter was the growing realization that Linneaus gave us a more or less workable system but that attempts to balance cladogenesis and anagenesis often places intolerable strains upon it. Its justification became apparent when the particular utility of a phylogenetic system became clear to me.

There is one other important conclusion to emerge from recent studies of protein evolution. Almost all available protein sequence and immunological data are consistent with the hypothesis that for any given protein the probability of an amino acid substitution occurring over a given length of time is the same for each lineage. It is difficult to explain this apparently stochastic regularity without involving the concept of neutral mutations - whose rate of fixation depends only on their frequency of occurrence. If then, as seems very likely, this regularity of protein evolution reflects the inexorable accretion of neutral mutations, then the large majority of all genetic changes are functionally irrelevant. Accepting that genes evolve over time, we should see in this genetic profile of any species allele frequencies at polymorphic loci varying from 0 to 1.0 depending on where in time and space our randomly evolving system has been intercepted. One had thus best recognize that much of the observed intra- and interpopulational polymorphism seen at a given point in time represents only stages in the inexorable accretion of neutral mutations. Thus, and most importantly, most of this allele frequency variation between populations is simply there - and has no more profound evolutionary significance than that. In light of these considerations, I can only conclude that almost all investigations of such allele frequency variation must inevitably represent pure data-collecting fishing expeditions with the extremely low probability that functional (and therefore selectively) significant fish will be found. Is there a solution to this impasse?

Not, I would suggest, by continuing present survey techniques. The number of possible selective agencies is so large that evolutionarily meaningless correlations will inevitably be found between some allele frequency variations and some selective agency. Masked among these many spurious associations will be real ones - but they will remain invisible to our eyes. If, on the other hand, one takes the suggested tack of first developing the cladistic and, particularly important, the temporal aspects of phylogenies for the taxa under study, then the influence of natural selection should stand out as irregularity in a pattern of regularity. Hopefully the careful and thoughtful observer may then be shown the way toward more meaningful research.

REFERENCES

Fitch, W.M., E. Margoliash, and K.S. Gould (1969). The construction of phylogenetic trees. II. How well do they reflect past history? Brookhaven Symposia in Biology 21: 217-242.

Nelson, G.J. (1969). Gill arches and the phylogeny of fishes, with notes on the classification of vertebrates. Bull.Am.Mus.Hist. 141: 4.

BEHAVIORAL CHARACTERS IN PRIMATE TAXONOMY

H. Kummer

This chapter is the result of a group discussion among the students of primate behavior attending the 1969 Symposium at Burg Wartenstein. The group's task was to judge the taxonomic value of various types of behavioral characters on the basis of their experience with primates. The group agreed with the main conclusions put forward in this chapter but are not responsible for the manner in which I have presented their views.

It has long been recognized that the comparative study of behavioral characters can contribute to systematics. The founders of ethology, in fact, were trained morphologists and taxonomists (Heinroth, 1911; Whitman, 1919) who, in their search for new systematic characters among ducks and pigeons, discovered certain motor patterns characteristic of species, genera and even families. Since these beginnings, the phylogeny of behavioral characters has remained a major concern of ethology; the problem has been recently reviewed and discussed by Wickler (1967). Nevertheless, the direct contributions that ethology has made to systematics are infrequent and are mostly concerned with birds and insects. Selected examples are found in Mayr's paper on Behavior and Systematics (1958). The investigations usually led to new phylogenetic insights and sometimes to revisions of systematic positions. Many were later confirmed by morphological studies.

The reason why behavior in general has played a minor role in taxonomic assessments is easy to understand. A specimen displays all its morphological characters at any one time but its behavior patterns appear only at intervals, under particular internal and external conditions and only a few at a time. Morphological characters survive death; behavior does not, and therefore is unavailable to the museum taxonomist unless he is prepared to go into the field. In addition there are practical difficulties facing the taxonomist, specific to primate behavior research. Comparisons based on published descriptions made by different investigators are always questionable. Significantly, the taxonomic contributions of ethology were made by authors who were familiar with all the relevant taxa from their own observations. Investigators in primatology are usually familiar with one or two species only, firstly because prolonged field work or adequate housing for many species in captivity is beyond the financial means of individual researchers, and secondly because so far very few primate ethologists have developed a primary interest in taxonomy. Notable exceptions are the comparative studies by van Hooff (1962) and Kaufman and Rosenblum (1966). Progress can be expected only under two conditions. Either field ethologists agree on certain standard quantitative methods and each investigator applies them to several species, or taxonomists decide to study animals in the field or in zoos. In these circumstances it would obviously be desirable for them to have a knowledge of both morphological and ethological methods. A recent, systematically oriented study on Tupaia (Martin, 1966a, b) emphasizes the necessity of integrating data from both morphology and ethology.

Even now, on the basis of individual experience with individual species, it is possible to identify the types of behavioral characters most suitable for taxonomic purposes. The participants in our group were almost unanimous as to which behavioral parameters constituted good taxonomic characters, and their conclusions agree with those of vertebrate ethologists in general. The methods required for the study of behavioral evolution and the use of behavioral characters in taxonomy cannot be discussed here. For an introduction, the reader is referred to the chapter on "Evolution and Behavior" in Hinde (1966).

FIXED ACTION PATTERNS

The best and, apparently, the only promising candidate for taxonomic usefulness is the so-called "fixed-action pattern". The flow of an animal's behavior is a series of recurring, recognizable units, each consisting of a characteristic sequence of muscular contractions structured in space and time; terms like "the eyelid display" or "presenting" are labels for such patterns. Independent observers usually produce almost identical lists of such units; and confusion only occurs when different labelling is combined in the literature with an inadequate description of the bahavior pattern.

The length of stereotyped behavioral sequence tends to decrease from the lower to the higher vertebrates. In contrast to the behavior of fishes or squirrels, which often show long rigid sequences of fixed action patterns, primates easily vary the sequence of their patterns. The single pattern, however, is invariable enough to permit definition even in primates, although its duration may average little more than a second.

These short, stereotyped patterns vary little among conspecifics of the same age and sex. In addition, they are not easily modified by changes of the environment. In Bernstein's "mixed taxa group", composed of several primate species, members learned to respond "correctly" to the social action patterns of the other species, but they did not modify their own, species-specific patterns. The fixed action patterns of wild hamadryas baboons are not recognizably different from those found in a captive hamadryas colony after years of life in the Zurich Zoo (Kummer and Kurt, 1965). Differences were found only in the number of patterns. Of 68 patterns displayed by the zoo baboons, 59 were also found in the wild; in addition, the wild troops used two patterns not observed in the captive colony. Environment, thus, can suppress the performance of a behavior pattern, but if the pattern occurs, it does so with the form which is typical of the species. Such observations show that taxonomically relevant behavioral characters can be gathered on captive groups if one bears in mind that the absence of a behavior pattern in a particular group does not mean it is absent from the potential repertoire of the species. As we shall see, frequencies of many behavior patterns are highly plastic and in zoo animals certain behaviors may disappear entirely from the repertoire.

Not all types of action patterns are equally useful in systematics. Some of them, such as feeding or digging movements, are direct functional adaptations to features of the extra-specific environment. Such patterns may often appear in unrelated taxa as effects of convergent or parallel evolution or modification. In contrast, behavior patterns of intraspecific communication are mainly adapted to the intraspecific environment, i.e. to the response mechanisms of the species itself. Acts of communication must merely fulfil a few ecological conditions; in dense vegetation, for example, signals that must be perceived at distances beyond a few meters are preferably acoustic signals in almost any form. The environment to which the signal is primarily adapted is the response of a conspecific, and thus can evolve along with the signal. Analogous signal actions in non-related taxa are therefore less likely to develop than analogous feeding-techniques. As a general rule, the action patterns of intraspecific communication may be expected to provide the most reliable taxonomic characters. In adaptation to their communicative function they generally vary little in form, which facilitates their classification.

Certain social action patterns display morphological structures which apparently have no function other than enhancing the effect of the display. Examples are colored skin patches and crests or tufts of hair. Such structures strongly suggest that the action pattern which displays them is an all-pervading, stable character of the species; the behavior must have been in use long enough to act as a selective agent for the evolution of the displayed structure. It is improbably that geladas would have evolved conspicuous white skin patches above the eye if the threat pattern of "brow raising", which reveals these patches, was not already a well-established characteristic of the species.

Social signals functioning in species recognition, such as territorial calls, evolve partly under the influence of similar signals in sympatric species. Such species may be expected to differ strongly in their respective signal forms. In taxonomy, the main function of these patterns may be to disclose cases of incipient speciation.

Hybridization experiments on birds and insects have shown that action patterns can be genetically determined in at least some of their components or parameters. For instance, hybrids may display intermediate forms (e.g. Hinde, 1956). Primate hybrids have been observed in nature, but their action patterns have not yet been investigated in detail (Bernstein, 1968; Nagel, in preparation).

Like morphological characters, action patterns may suggest

phylogenetic relationships only if they can be homologized. Homology between behaviors of different taxa can be established by using the same criteria that morphologists apply: (1) similarity in a large number of structural elements and (2) a similar relationship of the character in a wider structural context. For behavior, the second criterion becomes applicable if the concept of structural relationship is replaced by one of temporal association. The "context" relates to those behaviors that most frequently precede and follow the one under investigation. Eyelid-displays occurring in similar form in two species are more likely to be homologous if both species perform it while fighting, than if one species associates it with fighting and the other with grooming. The problems of homologizing behavior are discussed in detail by Wickler (1967). This author also provides examples showing that in behavior, similarity of characters is not necessarily a consequence of either common descent or convergent evolution. Tradition can also transfer an action pattern from one generation to the next, and even from one species to another if the two are sympatric. Although there seems to be no evidence for inter-specific tradition transferring action patterns from one primate species to another, the possibility that homology could be simulated by tradition cannot be denied.

Fixed action patterns constitute only one type of behavioral character. The following remarks on other parameters of behavior serve to show why they are likely to be of limited value in systematics.

FREQUENCY

The same action pattern which is rare in one species, can occur with high frequency in another. For example, ground-living monkeys seem to show more dominance-related behavior than arboreal species (DeVore, 1963). Systematics can hardly profit from frequency differences of this kind, even when such differences become sufficiently well documented to be confirmed quantitatively. Frequencies seem to be subject to a high degree of modification by the environment. In the Zurich colony of hamadryas baboons, for example, aggressive behavior was four time as frequent and sexual behavior three times as frequent as the homologous patterns in wild hamadryas troops. Therefore, the study of genetic differences in frequency requires the use of standardized environments. Relevant studies on species differences in primates are available (e.g. Rosenblum, Kaufman and Stynes, 1964), but not in sufficient number to be of immediate significance for systematics.

SEQUENCES

In lower vertebrates and insects, action patterns are frequently integrated into longer sequences which may be considered as behavioral characters of a higher order. Primate behavior, however, is not organized into long rigid sequences, although certain action patterns tend to occur immediately before or after each other. Such temporal associations indicate common internal causation of the associated patterns and, thus, may often assist in homologizing them.

Sequence analysis is a relatively recent method in ethology. It determines the probabilities with which the various fixed action patterns of a species follow each other in the behavioral sequences. For example, it may be found that pattern A is frequently followed or preceded by patterns B or C, whereas it rarely appears in temporal association with pattern D. It is conceivable that two closely related species, sharing a great number of homologous action patterns, would nevertheless differ in the temporal association among these patterns. There is as yet no study in which the behavioral sequences of two or more primate species are analyzed comparatively. The value of the approach cannot therefore be judged at present.

SOCIAL ORGANIZATION

Social organization is a product of social behavior. If the social organizations of two taxa differ for genetical reasons, such differences must be evident already at the more proximal level of social behavior. The resulting group structures yield no additional information on genetic differences; they can only make these differences more conspicuous. For example, a one-male group is a lasting testimony of exclusive behavior among males, which occurs rarely and is therefore more difficult to study than the group composition that results from such behavior. Thus, social organization can only help to find more proximal and thus more useful characters of behavior. By and large, a given type of social structure, such as a harem or a monogamous system, is poorly correlated with systematic position. Taxa as distinct as gibbons and titi monkeys, zebras and seals, can achieve the same or similar group types by different behavioral means. Convergent evolution seems to be more frequent in social organization than in social behavior. There are sufficient reasons to discard much of the available data on primate social organization as being irrelevant for systematics.

LEARNING ABILITIES AND DISCRIMINATION TASKS

The opinion of the group was that the existing attempts to distinguish taxa on the basis of their task abilities have not developed far enough to judge their usefulness in systematics.

SUMMARY. POTENTIAL ROLE OF BEHAVIORAL CHARACTERS IN PRIMATE SYSTEMATICS

The main result of our discussions in the recommendation that vocalizations, facial expressions and gestures serving intraspecific communication are the best taxonomic characters that behavior can provide at the present time. To date there are very few comparative studies on action patterns. The task requires investigators with a major motivation for systematics and for the subtle details of motor patterns, a combination which is not very popular among present students of primate behavior.

Adding behavior to the working material of systematics is not only an attempt to increase the number of characters on which judgements may be based, but is very likely to be a means of acquiring information of a special type. "The significance of behavior research for systematics is that it deals with the living phenotype, which is the one on which evolution works" (Wickler, 1967). Evolutionary changes are often spearheaded by changes in behavior. Reproductive isolation, for example, can be established through behavior long before hybridization becomes physiologically or morphologically impossible; hybrids probably occur more often in captivity than in the wild, because behavioral isolation mechanisms easily break down among confined animals. Two sympatric species may develop traditions of food selection before their teeth and digestive tracts become adapted to different foods. Behavior studies in the wild, thus, may reveal incipient speciation which is not yet evident on morphological grounds. Comparative field studies of closely related species, in areas where they meet or occupy the same habitats, can teach taxonomists not only where but how speciation is proceeding.

REFERENCES

Bernstein, I.S. (1968). Social status of two hybrids in a wild troop of Macaca irus. Folia primat. 8: 121-131.

DeVore, I. (1963). A comparison of the ecology and behavior of monkeys and apes. In Classification and Human Evolution. (Ed. Washburn, S.L.) Chicago: Aldine Publishing Co.

Heinroth, O. (1911). Beiträge zur Biologie, namentlich Ethologie und Psychologie der Anatiden. Verh. 5 Int.Orn.Kong.: 589-702.

Hinde, R. (1956). The behavior of certain Cardueline FI-interspecific hybrids. Behavior 9: 202-213.

Hinde, R. (1966). Animal Behavior. New York: McGraw-Hill.

Kaufman, I.C. and L.A. Rosenblum. (1966). A behavioral taxonomy for Macaca nemestrina and Macaca radiata: based on longitudinal observation of family groups in the laboratory. Primates 7: 205-258.

Kummer, H. and F. Kurt. (1965). A comparison of social behavior in captive and wild hamadryas baboons. In The Baboon in Medical Research I (Ed. Vagtborg, H.) Austin: Univ. of Texas Press.

Martin, R.D. (1966a). Tree shrews: Unique reproductive mechanism of systematic importance. Science 152: 1402-1404.

Martin, R.D. (1966b). Spitzhörnchen und die Evolution der Affen. Die Umschau 66: 437-438.

Mayr, E. (1958). Behavior and Systematics. In Behavior and Evolution. (Eds. Roe, A. and Simpson, G.G.) Yale: Univ. Press.

Remane, A. (1956). Die Grundlagen des natürlichen Systems, der Vergleichenden Anatomie und der Phylogenetik 2nd Ed. Leipzig: Acad. Verlagsgesellschaft Leipzig.

Rosenblum, L.A. I.C. Kaufman, and A.J. Stynes. (1964). Individual distances in two species of macaque. Animal Behavior **12**: 338-342.

van Hooff, J.A.R.A.M. (1962). Facial expressions in higher primates. Symp. zool. Soc. Lond. **8**: 97-125.

Whitman, C.O. (1919). The behavior of pigeons. Publ. Carnegie Inst. **257**: 1-161.

Wickler, W. (1967). Vergleichende Verhaltenforschung und Phylogenetik. In Die Evolution der Organismen I. (Ed. Heberer, G.).

PART II

MORPHOLOGY, MOLECULES, AND EVOLUTION

THE COMPARATIVE UNIFORMITY OF THE CERCOPITHECOIDEA

Adolph H. Schultz

OLD WORLD MONKEYS

The superfamily Cercopithecoidea is composed of a remarkable number of species and genera of widely differing outer appearance. In most of their basic morphological characters, however, Old World monkeys are much more uniform than the other major groups of primates, having acquired fewer diverging specializations than have the prosimians, the platyrrhines or the hominoids. It is mainly due to this fact that the detailed taxonomic subdivision of the cercopithecoids is particularly difficult to justify in all respects.

In the symposium Classification and Human Evolution held at Burg Wartenstein in 1962 and subsequently published in 1963, the writer dwelt chiefly on the great intraspecific variability of the apes. In the present symposium he will emphasize instead the comparatively small intergeneric variability of the Old World monkeys by reviewing briefly some of the relevant facts.

In their body weight fully adult normal primates vary between extremes which among prosimians range from about 0.06 kg in Microcebus to at least 140 kg in some subfossil Megaladapis: the maximum being 2300 times greater than the minimum. Body weight of New World monkeys ranges from 0.14 kg in the pygmy marmoset to 11 kg in some female spider monkeys, the latter being 79 times heavier than the former. In hominoids, male gorillas of at least 180 kg surpass some gibbons of only 4.5 kg by a factor of 40 times. Finally in Old World monkeys, the greatest weight of about 30 kg for wild mandrills is only 25 times the smallest weight of 1.2 kg for talapoins.

From the writer's (1961) data for the numbers of vertebrae of altogether nearly 2000 specimens show clearly the very moderate variability of the numbers of precaudal segments in cercopithecoids in contrast to other primates. For instance, the combined number of thoracic and lumbar vertebrae averages anywhere between 18.7 and 19.0 in the 11 genera of Old World monkeys studied, and varies individually only between 18 and 20, with 20 being exceedingly rare. In the hominoids the combined figure for these vertebrae ranges from 15 - 19, in platyrrhines from 17 - 21 and in prosimians from 18 - 24. While the pelvic ring has shifted in a caudal direction in various prosimians and in a cephalic direction in the highest primates, with manifestations of both trends in some platyrrhines, the superfamily of cercopithecoids is the only one which has retained the conservative number 19 in all living genera with comparatively few and very modest individual exceptions. The number of sacral vertebrae also shows this remarkable constancy in the Old World monkeys since it varies individually only from 2 - 4 with averages near the unspecialized figure of 3. New World monkeys have between 2 - 5 sacral vertebrae, prosimians 2 - 9 and the hominoids 3 - 8, both the latter two groups having independently developed a trend to fuse more than 3 segments into a solid sacrum and this with a high intrageneric variability.

The sternum of all cercopithecoids is very slender and its sternebrae normally do not fuse as they do with advancing age in the far broader sterna of hominoids. While 7 or 8 pairs of sternal ribs are found in all major groups of primates and are typical for the catarrhine monkeys, prosimians may have as many as 12 pairs in some genera; platyrrhines quite frequently have 9 pairs and hominoids often as few as 6 pairs. No cercopithecoids have acquired any such sternal specializations as the bifurcation of the manubrium of howler monkeys, the extension of the manubrium to the third pair of ribs in nearly all hylobatids, or the not infrequent perforation of the corpus sterni in the Asiatic apes and man. The homogeneity of the Old World monkeys is also evident from the shape of the thoracic cavity which is essentially alike in all; the spinal column of cercopithecoids never protrudes into the thoracic cavity to the extent seen in hominoids in whom the transverse thoracic diameter surpasses the sagittal one and the vertebrae have shifted far ventrally in generically differing degrees. The scapulae of all catarrhine monkeys lie at the sides of their comparatively narrow chests regardless of their mode of locomotion, whereas in the apes and man they have migrated to the back of the broad trunk. In connection with these profound distinctions, the clavicles have remained proportionately much shorter in all cercopithecoids than in the hominoids, measuring only around 15 percent of the trunk length in the former, but anywhere from about 26 - 36 in the latter (Schultz, 1968).

The pelves of all cercopithecoids are built in a remarkably

uniform manner with only few generic differences which appear insignificant in comparison with the striking pelvic differences among the other primates, particularly the hominoids. For instance, the length of the hip bone in its percentage relation to the trunk fluctuates in its averages merely from 34 - 41 among the cercopithecoids, but from 39 - 57 in the hominoids. Many other pelvic proportions also show far smaller differences among the catarrhine monkeys than among the apes and man (Schultz, 1969).

The relative lengths of the limbs and their segments are generally regarded as being closely correlated with locomotion, but this rule seems to appertain mainly to those proportions which undergo extensive ontogenetic changes. Table I gives briefly the ranges of the generic averages of some of the most important limb proportions in the great many fully adult primates which the writer has been able to measure in the flesh or as skeletons with strictly comparable techniques. The intermembral index, which intraspecifically is one of the least variable proportions, differs intergenerically far less in Old World monkeys than in any other major group of primates. This remarkable uniformity is even more pronounced if the index is based on total limb length rather than on the summed length of the main long bones. The third and fourth indices in the Table express the total limb lengths in percentage of the trunk length. These data, again, show the comparatively small total range of the generic averages in catarrhine monkeys. The most extreme relative arm lengths have been acquired by the best brachiators, i.e. the Asiatic apes and spider monkeys, and the greatest relative leg lengths by bipedal man and the leaping tarsiers, sifakas, etc. Among the cercopithecoids, however, the greatest relative total limb lengths belong to the terrestrial baboon and arboreal proboscis monkey, and the smallest averages to some guenons and guerezas. Relatively long upper limbs are often, but not always, accompanied by a special elongation of the forearms, resulting in very high brachial indices. This proportion between the lengths of the radius and the humerus is also more uniform among catarrhine monkeys (range of generic averages: 92 - 106) than among platyrrhines (83 - 103) or, particularly, prosimians (92 - 126) and hominoids (75 - 111). The highest averages of each of these four groups pertain to the brachiating gibbons, orangs and

and spider monkeys, to the leaping tarsiers and indrises, to the terrestrial baboons, geladas and patas monkeys as well as some langurs. (See detailed data and discussion of the brachial index in Kussmann, 1967).

The total length of the hand and of the foot in percentage of the trunk length can differ widely among primates as shown by the last two indices of Table I, but this is not nearly as marked in catarrhine monkeys as in the other main groups. Indeed, the total relative foot length of catarrhine monkeys is extraordinarily constant and shows no consistent relationship to locomotion such as is apparent in the extremely long feet of orangs, gibbons, spider monkeys, tarsiers, etc. or the very short feet of man, gorilla, potto and other specialized non-cercopithecoid species. Only in the relative length of the phalangeal parts of the lateral toes can one recognize a moderate locomotor adaptation among the Old World monkeys, since these lengths average slightly less than 30 percent of the total foot length in geladas and patas monkeys, whereas in langurs and guerezas they average 39 percent. These differences, however, seem again very conservative in comparison with the corresponding differences in the highest primates among which the length of the Phalanges III equal over 40 percent of the foot length in gibbons and orangs and merely 19 percent in man (Schultz, 1963). Besides the reduction of the thumb in guerezas (equal to that in Ateles), no catarrhine monkeys show any of those manifold and profound digital specializations as have developed among other primates, notably the reductions of the second digits and the extreme length, strength and abductability of the first toes in various prosimians, the loss of the distal phalanx in the first toe of most orangs, the unique shortening of the lateral toes of man and the grotesque thinning of the long middle finger of the aye-aye.

With a very comprehensive and thorough investigation of the touch pads, flexure folds and dermatoglyphics on the hands and feet of all primates, Biegert (1961, 1963) has demonstrated that these significant features have remained remarkably generalized and have not become sharply differentiated among the cercopithecoid genera. He concludes that "This relative uniformity of the chiridia is an outstanding trait of the Cercopithecoidea, as compared with the Ceboidea and the Hominoidea."

The comparative homogeneity of catarrhine monkeys is very evident in their neurocranium. For instance, the allometric changes in the size of the brain cavity in relation to body weight fit closely in one single curve for all Old World monkeys, while corresponding data for platyrrhines follow separate curves for marmosets, howler monkeys and spider monkeys respectively and the three families of hominoids also differ widely in their relative cranial capacities (Schultz, 1956, 1958). In the formation of the brain itself there also exist far greater differences between the platyrrhines *Cebuella* and *Ateles*, for example, or the hominoids in *Hylobates* and *Homo* than between any two cercopithecoids.

The relative position of the cranial-vertebral joint on the skull-base changes ontogenetically in varying degrees in all simian primates, so that it can differ strikingly among adults. This important character is accurately expressed by the percentage relation between the distance from the nasion to the center of the occipital condyle and the nasion - opisthocranion length, which equals practically 100 in all mammals besides the simian primates. Among the latter it varies between 65 - 75 about the time of birth and changes post-natally to the following range of averages in adults: among platyrrhines from 76 in *Saimiri* to 96 in *Alouatta*; among hominoids from 66 in man (and 83 in *Pan paniscus*) to 95 in orang-utan, but among cercopithecoids only from 87 in *Macaca* to about 93 in most of the colobine as well as cercopithecine genera (Schultz, 1955).

A remarkable uniformity of the entire group of Old World monkeys is also shown by their universal lack of frontal sinuses in contrast to the development of such in some, but not all, species of prosimians, platyrrhines and hominoids (Weinert, 1926; and many other reports).

The facial part of the skull has generally acquired much more marked and significant taxonomic differences than the neurocranium; it not only changes more profoundly in the post-natal period, but also varies individually more intensively (e.g. Schultz, 1962, Vogel, 1966). The most striking generic specializations of the splanchnocranium of Old World monkeys have developed in the relative length and protrusion of the masticatory apparatus of baboons, but these distinguish really only adult males. In spite of the great variety of other and more detailed features differentiating some cercopithecoids in their facial skulls, none equals some of the diverging facial specializations which have developed among the prosimians, platyrrhines and hominoids in connection with especially their in some instances widely differing relative sizes and positions of the orbits.

A further example of the comparative uniformity of the catarrhine monkeys is found in the relative size of the outer ears (ear length x breadth in percentage of head length x total head height) for which the writer (1956 and unpublished data) has collected a great many measurements. The generic averages of this index in adults range among prosimians between 14 in <u>Nycticebus</u> and 89 in <u>Daubentonia</u>; among platyrrhines from 8 in <u>Ateles</u> to 37 in <u>Callithrix</u>; among hominoids from 3 in <u>Pongo</u> to 18 in <u>Pan</u>; but among cercopithecoids the index varies from only 8 in <u>Nasalis</u> to to 19 in <u>Erythrocebus</u>.

In a very complete and detailed description of the teeth of all primates, Remane (1969) concludes that, in the Cercopithecoidea, any specific or generic differences, besides those in size, are very moderate and that even distinctions between the teeth of the two subfamilies exist merely on a general average but are not "sharp diagnostic characters." In all types of Old World monkeys the sequence of eruption of the teeth of the first and second dentition is also extraordinarily uniform in contrast to some significant differences in this process found within each of the other major groups of primates (Schultz, 1956, 1960).

All cercopithecoids for which reliable observations could be obtained so far, are invariably more mature at birth than any hominoids - as judged by their skeletal development as well as some other morphological and physiological conditions. Their permanent dentition is completed long before all of the chief epiphyses have become closed in a remarkably uniform order. The obliteration of the main cranial sutures generally occurs sooner in catarrhine monkeys than in nearly all platyrrhines, but at a comparatively later stage than in the pongids (Chopra, 1957). The ischial callosities, typical of all cercopithecoids, appear very early in prenatal development in contrast to those of the hylobatids. This general lack of really significant taxonomic differences in the main ontogenetic processes of catarrhine monkeys is to be expected from their very limited phylogenetic differentiations and suggests that this limitation is not due to convergence.

The above random examples of data (to which many more could easily be added) must suffice to support the conclusion that the Cercopithecidae form an unusually homogeneous major group of primates. A very evident exception to this generalization exists in the relative length of the tail which differs widely among some Old World monkeys; by far the most variation occurs in the closely related baboons and macaques. While the tail can be more than twice the length of the trunk in some adult langurs and guerezas, it equals only 2 - 18 percent of the trunk length of the stump-tailed macaques (including all species of "*Cynopithecus*" which Fooden, 1969, has convincingly shown to belong to the genus *Macaca*), and the drills and mandrills whose tails comprise only from 4 - 10 vestigial vertebrae instead of the 30 segments of some langurs and guenons. Among the many other Old World monkeys the tail has become considerably reduced in a few macaques, but only in the one colobine species, *Simias*. This reduction has not progressed further than it has in *Cacajao*, which is the only really short-tailed platyrrhine. The apparent ease and rapidity with which the tail can become greatly reduced is also exemplified by the Lorisidae and, among the Lemuriformes, by *Indri*. These prosimians differ as radically in their relative tail-length from the other prosimians as do the stump-tailed macaques from the various long-tailed ones.

Another exception to the rule of the extraordinary uniformity of the catarrhine monkeys is found in the wide variety they display in the degree of secondary sex-differentiation. With regard to general body size, very impressive sex differences exist not only in the largely terrestrial baboons, geladas and patas monkeys, but also in the decidedly arboreal proboscis monkeys and a variety of other colobine and cercopithecine species. The length of the canine teeth and the manifold specializations in the shape of the callosities, the size of the cheek pad of drills and of the nose of *Nasalis*, the extent and location of the sex skin, etc. show also much greater sex differences in Old World monkeys than anything occurring in prosimians or platyrrhines. In no platyrrhine species are the males twice the size of females as is the case among catarrhines. In many instances among platyrrhine monkeys the sexes are of equal size while other secondary sex characters are only moderately developed, although the color or pattern of the coat provides an exception in some species. Among the hominoids, only gorillas and orangs have acquired such extreme sexual differences as those that characterize so many cercopithecoid species.

The most easily seen exception to the general similarity of all Old World monkeys is represented by the amazing variety of specific and subspecific differences in the coloration and patterns of their coats. These differences are fully as striking as those which have developed among the Callitrichidae which in most other respects form also a quite homogeneous family. That the colors and pattern can provide the first, if not the only distinguishing features of local subspecies has been abundantly demonstrated. They clearly show that these characters are more readily subject to change than the more tenacious conditions of the skeleton, teeth, etc.

Catarrhine monkeys have remained a more tightly-knit clan than have the platyrrhines or the hominoids because they have acquired fewer and more moderate specializations in most of their bodily characters as well as in their basic mode of locomotion. This is of particular interest in view of the fact that the catarrhine monkeys have probably existed for at least as long as the hominoids, have a wider distribution and contain types which are more adaptable to different environments and climates than any other non-human primates.

REFERENCES

Biegert, J. (1961). Volarhaut der Hände und Füsse. Primatologia II, P.1, Liefer. 3: 1-326.

Biegert, J. (1963). The evaluation of characteristics of the skull, hands and feet for primate taxonomy. Classification and Human Evolution. (Ed. Washburn, S.L.). Chicago: Aldine.

Chopra, S.R.K. (1957). The cranial suture closure in monkeys. Proc.zool.Soc.Lond. 128: 67-112.

Fooden, J. (1969). Taxonomy and evolution of the monkeys of Celebes. Bibliotheca primatologica Fasc. 10 (148 pp.)

Knussmann, R. (1967). Humerus, Ulna and Radius der Simiae. Bibliotheca primatologica Fasc. 5 (399 pp.)

Remane, A. (1960). Zähne und Gebiss. Primatologia III, P.2: 637-846.

Schultz, A.H. (1942). Growth and development of the proboscis monkey. Bull.Mus.Comp.Zool.Harv. 89: 277-314.

Schultz, A.H. (1944). Age changes and variability in gibbons. Amer.J.Phys.Anthrop. n.s. 2: 1-129.

Schultz, A.H. (1954). Studien über die Wirbelzahlen und die Körperproportionen von Halbaffen. Vierteljahrsschr.Naturf. Ges.Zurich 99: 39-75.

Schultz, A.H. (1955). The position of the occipital condyles and of the face relative to the skull base in primates. Amer.J. Phys.Anthrop. n.s. 13: 97-120.

Schultz, A.H. (1956). Post-embryonic age changes. Primatologia, I: 886-964.

Schultz, A.H. (1960). Age changes in primates and their modification in man. In Human Growth (Ed. Tanner, J.M.) Oxford: University Press.

Schultz, A.H. (1961). Vertebral column and thorax. Primatologia IV, Liefer. 5: 1-66.

Schultz, A.H. (1962). Metric age changes and sex differences in primate skulls. A. Morphol. u. Anthropol. 52: 230-255.

Schultz, A.H. (1963). Relations between the lengths of the main part of the foot skeleton in primates. Folia primatologica 1: 150-171.

Schultz, A.H. (1963). Age changes, sex differences and variability as factors in the classification of primates. In Classification and Human Evolution. (Ed. Washburn, S.L.) Chicago: Aldine.

Schultz, A.H. (1968). The recent hominoid primates. Perspectives on Human Evolution. (Eds. Washburn, S.L. and Jay, P.) New York: Holt, Rinehart & Winston.

Schultz, A.H. (1969). Observations on the acetabulum of primates. Folia primatologica 11: 183-210

Schultz, A.H. (1969). The Life of Primates. London: Weidenfeld & Nicholson.

Vogel, C. (1966). Morphologische Studien am Gesichtsschädel catarrhiner Primaten. Bibliotheca primatologica. Fasc. 4 (226 pp.)

Weinert, H. (1926). Die Ausbildung der Stirnhöhlen als stammesgeschichtliches Merkmal. Z. Morphol. u. Anthropol. 25: 243-357; 365-418.

TABLE I

	Total upper limb 1 / Total lower limb 1	RANGE	Humerus + Radius / Femur + Tibia	RANGE
Prosimians	78 - 115	37	55 - 91	36
Platyrrhines	100 - 137	37	74 - 108	34
Cercopithecoids	106 - 121	15	78 - 96	18
Hominoids	88 - 178	90	69 - 149	80
	Total upper limb 1 / Trunk 1	**RANGE**	**Total lower limb 1 / Trunk 1**	**RANGE**
Prosimians	79 - 163	84	81 - 178	97
Platyrrhines	96 - 180	84	93 - 132	39
Cercopithecoids	105 - 137	32	99 - 120	21
Hominoids	148 - 243	95	116 - 169	53
	Hand length / Trunk 1	**RANGE**	**Foot length / Trunk 1**	**RANGE**
Prosimians	23 - 55	32	35 - 91	56
Platyrrhines	28 - 47	19	44 - 60	16
Cercopithecoids	26 - 36	10	43 - 48	5
Hominoids	37 - 59	22	47 - 62	15

Ranges of generic averages of some limb proportions in adult primates measured by the author (1954, 1956, and some other old and new data).

PALEOECOLOGY AND CATARRHINE EVOLUTION

J. R. Napier

OLD WORLD MONKEYS

"Geology has to choose between the rashness of using imperfect evidence or the sterility of uncorrelated, unexplained facts." (J.W. Gregory).

INTRODUCTION

It seems appropriate at this time to attempt a synthesis of paleoecological data based on evidence derived from sedimentary, marine invertebrate and paleobotanical studies, in order to provide an ecological backdrop against which to study the adaptive radiations of catarrhine primates.

Perhaps to attempt such a synthesis is unduly rash, for the available evidence, such as it is, is palpably richest in just these geographical areas in which primates are not. More, for instance, is known about the past geography of the Bering Strait than the important (for catarrhine primates that is) Afro-Asian land bridge. The past climatology of the present-day temperate latitudes of the Northern Hemisphere are relatively well worked out, while that of the Tertiary of the tropics is still largely unstudied. However, when one bears in mind that the temperate latitudes of today are the tropical latitudes of the early and early-middle Tertiary, the prospect becomes rather more hopeful.

Perhaps, too, it is rash for a primate biologist to attempt to synthesize the available information derived from several unfamiliar disciplines. However, a primate-oriented paleoclimatologist or paleobotanist is as an unlikely a phenomenon as a climatologically and botanically trained primatologist, so it is perhaps unrealistic to await the coming of either. As usual, the information needed to study the interface between disciplines is hard to come by; this is principally because of the difficulty of asking the right questions, but also because these particular areas of information are a sort of no man's land that no one cares to cultivate.

The sort of information gathered together here is necessarily rather broad and imprecise and concerns continents rather than countries and adaptive zones rather than habitats; it takes an equally broad view of geological time. Thus, the questions related to primate evolution that paleoecology may help to elucidate must be broad ones; they must relate to the major shifts of the environmental pattern that have resulted in the emergence of the higher taxa such as superfamilies, families and subfamilies. But, nevertheless, they are very important questions because it is at these levels that symplesiomorphic characters (or characters of common inheritance) are established. Paleoecology provides the opportunity for asking questions about primate evolution, but in the absence of appropriate fossils, the questions often have to be turned inside out. Instead of asking "What was the adaptive significance of this structure?" one can ask "Under the conditions operating at the time, what structures would have been of adaptive significance?" One works inwards from the environmental situation rather than outwards from the structure or behavior. In this way it is possible to confirm the inferences obtained from comparative studies of living animals regarding the probable characters of common inheritance with the paleoecological situation in which they are believed to have evolved. Many primate biologists with an historical perspective have been asking this question, in one form or another, for a long time (e.g. see: Washburn, 1950; Avis, 1962; Simons, 1967; Napier, 1967; Fox, 1967; Walker, 1968).

There are, for instance, many strange locomotor behavioral habits among the living primates; the adaptive value of the leaping of langurs, the slow climbing of the lorises and pottos, the vertical-clinging of the tarsier and certain Madagascan lemurs, and the knuckle-walking of the chimpanzees and gorillas. There are also a number of physical characters that do not quite fit the present day ecology of the animals concerned; the sexual skin of talapoin monkeys, the cheek pouches of guenons, the thumbless hand of Colobus monkeys, the ground-adapted limb proportions of the highly arboreal proboscis monkey, and the modified "dental comb" of the Pitheciinae. This is surely in part the result of ignorance, but it may also be that these structural and behavioral patterns have persisted long after the selection pressures that brought them about have ceased to operate. They have become atavisms. Only by assessing structure and behavior in the light of paleoecology can one hope to separate the structures of common inheritance from those of separate acquisition, and to distinguish between the behavior that is currently adaptive and that which represents an atavistic survival from past times.

The purpose of this communication is to collect and collate as much information as is presently available, on past climates, early vegetational patterns and ancient land bridges and so on, that could conceivably have a bearing on primate evolution. Partly because of the multidisciplinary nature of the information, and partly as a result of the difficulties of interpreting unfamiliar data, this synthesis is far from comprehensive. It is presented as an interim account in a spirit of hopefulness that it may act as a stimulus to those who have not yet considered their own particular problems in the light of past ecology.

The Cenozoic era traditionally started with a "revolution" - the Laramide Revolution, and "revolutions" of one sort or another have not ceased throughout the 65 million years of its duration. Tectonic forces, that have lifted up mountain ranges, submerged continental land masses, raised the sea-bed to form new tracts of dry land and produced faults of the earth's crust, thousands of miles in length, are still familiar, though comparatively muted, phenomena of life today. The Cenozoic era not only brought about profound alterations in the climate and topography of the earth but in the nature of animal life as well.

The rapidly expanding placental stock of the Cenozoic consisted principally of insectivores, primates and condylarths. Although evidence is still fragmentary, the insectivore-primate transition can be reasonably well recognized (Simpson, 1955; Szalay, 1967; McKenna, 1966) in such families as the picrodontids and the paramomyds. It seems reasonable to assume that the ancestors of living primates are to be found in the assemblage of certain primates and near-primates in the Cretaceous and early Paleocene. The story of primate evolution has thus been played out against the background of the Turbulent Tertiary with its constant succession of environmental changes; it is in the nature and tempo of this change that the principal determinants of primate evolution are to be found.

Rather surprisingly, one feature of the earth - the general outline of the continental land masses (including their continental shelves) has undergone only minor change in shape since the Cenozoic began; it is the topography of the surface that has been so radically transformed. This belief in the stability of continental outlines and relationships has made it possible to reconstruct the major events in the zoogeographical history of mammals throughout their Tertiary radiation. But this comfortable belief must be tested in the light of Continental Drift.

CONTINENTAL DRIFT

During the last five or six years the attitude of geophysicists towards continental drift has changed radically. The original widespread scepticism, which has been rife since Wegener put forward his comprehensive theory in 1912, has largely disappeared in the face of major advances in the fields of rock magnetism and radio-active dating, seismology, and, particularly, in determinations of drift based on sea floor spreading. Oceanographic research has produced detailed mapping of ocean floors. The principal systems of submarine ridges and rifts are well known, and their significance for continental drift well appreciated. Sedimentary studies of sea floors have indicated that the deposits are relatively young compared with the age of the earth and date back no earlier than Mesozoic times. Helpful summaries of recent advances in the Continental Drift theory are those of Wilson (1963), Hurley (1968) and Fraser (1967).

While there seems little reason to doubt in the face of overwhelming evidence that in the early Mesozoic land masses of the world consisted of one supercontinent (Gondwanaland and Laurasia), the important questions for primate biologists are: when did it start to break up? How rapidly did the daughter continents separate? Sedimentary evidence places the initial break-up, or rifting, in the Middle Mesozoic, about 160 million years ago but the evidence of paleozoogeography based on both land vertebrates and molluscs suggests that land connections between the continents of Africa and South America were widespread "well into the Cretaceous period" (Smith and Hallam, 1970). More recent evidence from deep sea drilling (Maxwell et al, 1970) strongly supports a Late Jurassic or Early Cretaceous date, some 130 million years ago. Final separation according to Briden (1967) took place in Middle Cretaceous times, some 100 million years ago, although the separation paths of the two continents, drifting about a rotation axis, might have been such that West Africa remained close to Brazil (while South Africa was swinging away from South America) during much of the opening up of the South Atlantic (loc cit). However, eventually "horizontal" as opposed to "rotational" drift must have taken place. The estimated speed of drift is at the rate of 2-6 cm a year, is based on the theoretical velocity of convection currents in the earth's mantle layer, and upon the age of the oldest rocks on Atlantic islands. The islands such as Tristan da Cunha, Falklands, Ascension and Cape Verde etc, are postulated as being of volcanic origin derived from the mid-ocean ridges (which themselves define the line of schism between continents). The islands that have drifted farthest from the mid-ocean ridge, show the oldest rocks, and those nearest, the youngest. Tristan da Cunha, for instance, lies on the South Atlantic mid-ocean ridge and is of recent volcanic origin, the site of active volcanism. The volcanic rocks of Bermuda on the other hand are of Late Eocene age. Plotting the distances of these islands from the mid-ocean ridge against the age of their oldest rocks provides an estimate of the speed of drift. Recent deep sea drilling in the South Atlantic (Maxwell et al, loc cit) confirms that the sea floor of the South Atlantic has been spreading apart at an average half-rate of 2 cm per year since the Cretaceous. These authors estimate that the South Atlantic was some 3000 km (1875 miles) wide by the Campanian (Late Cretaceous). This is less than half of the present distance across that area of the South Atlantic in which Maxwell et al were operating their deep sea drills. Thus at the beginning of the Tertiary the South Atlantic could have had a width of approximately 2600 miles. The nearest points in Brazil and West Africa, which are now separated by only 1800 miles of ocean, might have been rather nearer one another than this, but even 50 miles of ocean constitutes a formidable (though not impassable) barrier to faunal exchanges.

The separation of Africa and India it is suggested (Hurley,

1968) commenced somewhat earlier in the Permian although Mc-Elhinney (1968) has postulated an Early Tertiary date for this event. India's arrival in its present position as part of Asia was heralded by a massive crunch that buckled up the sediments of the Tethys Sea, an upthrust that gave rise to the Himalayan range. This event is usually regarded as of Miocene date.

The implications of Continental Drift for primate evolution are considerable, but its effect during the Cenozoic was local rather than global. The mountain-building episodes of the Tertiary, for instance, can be regarded as the "bow waves" of drifting continents (Hurley, 1967). However it is still very difficult to invoke drift in order to explain the similarity of Old and New World monkeys for instance as Sarich has done (this volume). Zoogeography has not yet found it necessary to invoke theories of Continental Drift to explain faunal relationships as they exist today. (Darlington, 1957; Simpson, 1960; George, 1962). Mammalogists, and particularly primatologists should for the present probably avoid developing hypotheses that rest on the premise that land connections of the Gondwanaland supercontinent persisted well into the Tertiary. Their time may come, but at present Occam's Razor still prevails.

CENOZOIC CLIMATES

The Laramide Revolution at the end of the Cretaceous raised the low-lying lands and shallow Mesozoic seas and produced the initial folding which delineated the major mountain systems of modern times, the Alps, the Himalayas and the Cordilleras. This was the first stage of a process of orogeny that reached its apogee in the Miocene, Pliocene and Pleistocene, and is still in operation today.

Perhaps of more significance to animal life were the climatic changes that accompanied this prolonged period of mountain-building. During the Cretaceous, temperatures of the world were uniformly high and the climate was humid with little in the way of seasonality. The subtropical and tropical province extended to the high latitudes; the 18°C February marine isotherm was in the vicinity of 53°N latitude (Durham, 1950) and cool, temperate forests, which have yielded fossil remains of spruce, hazel, birch and poplar were found at the northern tip of Grinnel Land only 8° from the North Pole (Dorf, 1957). The Bering Strait, at this time a land bridge, was just north of the limits of the subtropics as was also the northern end of the Atlantic bridge.

There are a number of methods available for studying paleoclimates: (1) corals,(2) fossil plants, (3) climatically-sensitive animals such as alligators, musk-ox and reindeer, (4) sedimentary rocks with special reference to glacial deposits, (5) dendrochronology, (6) fossil insects, red coloration, and varving. For the early Cenozoic, only the first two methods have been extensively developed. Among the most valuable indicators of climatic conditions are the reef corals. The average minimum temperature in February at which corals will grow is $18.5^{o}C$, so that discovery of these invertebrates in marine faunas, common throughout the Cenozoic, provides a valuable guide to northern and southernmost limits of tropical zones. Fossil plants are generally the most useful method in paleoclimatology for there is no doubt of the close dependence of their physiology and anatomy on climate but, like all correlative measures, there are notable inherent limitations. The presence of cacti does not inevitably indicate a desert any more than palms indicate a tropical facies. Certain plants, too, such as bracken ferns and willows, are adapted to a wide range of climatic conditions (Kräusel, 1961). Plants can be used very instructively as paleoclimatic keys but it must be borne in mind that plants can be adapted to changes in the environment just as animals can, and that therefore the greater the time span between the paleoplant and present time, the greater is the likelihood of divergent adaptation. Kräusel (1961) who has reviewed the subject with this special problem in mind, concludes that fossil plants <u>can</u> be used as paleontological guides, particularly if the tree and plant specimens represent a large mixed community or flora. Palms are especially valuable as climatic indicators. Among these, <u>Nipa</u> gives the most precise information.

PALEOCENE AND EOCENE

Throughout the Tertiary, except for minor fluctuations in the late Eocene and Middle Miocene, there was a general trend towards cooler conditions with the occurrence of seasonality in high and middle latitudes with cold winters and warm summers (Fig.1). By the beginning of the Eocene the subtropical zone in N. America had reached as far north as Oregon but was starting its southerly retreat. In the Middle Eocene, the Goshen Flora (Chaney and Sanborn, 1933) of west-central Oregon still had a tropical facies. The forests of Oregon were intermediate in type between rain forest and modern warm temperate forests; they were rather similar, in fact, to the vegetational community of Barro Colorado Island, Panama, today. The fossil flora of the London Clay at 50°N latitude indicates that tropical or subtropical conditions similar to that found currently in Indo-Malaysians tropics (Dorf, 1957), still prevailed in western Europe in the Lower Eocene (Reid and Chandler, 1933). In the Ukraine, evergreen tropical and subtropical woodlands prevailed (Schwarzbach, 1961).

The Upper Eocene floras of La Porte, California indicate that western North America still had a subtropical aspect at this time. Magnolias and the thick-leaved Lauracea were dominant plan forms and cycads, the characteristic gymnosperms of the Mesozoic, were still represented in the floras. The flora of Fushun, Manchuria was intermediate between temperate and subtropical. Cool temperate floras are found on the New Siberian Islands (Lat. 75°N) (Kobayashi and Shikama, 1961).

The 18°C February marine isotherm was in the region of 48°N latitude. Molluscan faunas were distributed within a broad tropical belt that extended for 45° of latitude on either side of the equator, and crocodiles are found in Eocene deposits as far south as Victoria, Australia (38°S) and as far north as Halle in Germany, southern England (52°N) and New Jersey (40°N).

OLIGOCENE

The climate of the Lower Oligocene indicates a sharp acceleration of the cooling process. The Florissant floras of Western Colorado (39°N), which resemble those found today in Mexico and south-central China on 25°N latitude (MacGinitie, 1958), suggest a cooler, dryer climate. Tropical or subtropical forest did not extend much further north than southern California; the floras comprising 84 genera of angiosperms were analogous to modern drought-resistant forms. Little is known about climatic and vegetational conditions in the Oligocene but the evidence from marine fauna indicates a steady southerly retreat of the warm seas. In western North America and Britain, the principal sites of Cenozoic paleobotanical studies, the subtropical floras were being replaced by a temperate profile of oaks, ashes, beeches and giant conifers of the genus Sequoia. In western Europe warm temperate forests had replaced subtropical forest, and yearly mean temperatures were in the neighborhood of 18°C (64°F) a climate that one might experience for example, in the Southern States of the U.S.A. (e.g. Charleston, S.C.). In North Africa, a key area in primate evolution at this time, the Fayum of Egypt (30°N) now a featureless desert, was covered with rain forest fringing the numerous tidal creeks of deltaic origin draining into the Tethys Sea. In the late Eocene and early Oligocene arid conditions were fairly widespread.

MIOCENE

By the Miocene the 18°C February marine isotherm had reached 34°N and the climate of California was warm but not subtropical. In the Middle Miocene a brief return to warmer conditions is believed to have occurred (Chaney, quoted in Dorf, 1955). The presence of varving in sedimentary deposits in Sicily show that there the climate was seasonal with rain in autumn and winter. The indications from evaporite deposits were that water temperatures were higher than at present, and that Sicily lay in the lower latitude position of present Mediterranean type of climate but rather than its center as it does at present. In the Miocene of the Ukraine, deciduous trees were succeeded by coniferous-deciduous woodlands. In the Upper Miocene, a period of marked aridity in Europe, desert floras established themselves in the Ukraine (Schwarzbach, 1961).

There is little in the way of paleoclimatological evidence from South America. The Miocene and early Pliocene beds of Chile contain tropical molluscs indicating a tropical or subtropical climate extending as far as 50°S (Gill, 1961). The upthrust of the

Andean section of the Cordilleras in the Miocene produced "local weather" effects on South American vegetation.

By the Upper Miocene average temperatures of the world were much as they are today but summer maxima were reduced and winter maxima were raised, in other words the temperature gradients were less steep and the climate was more equable. The climatic deterioration which began at the end of the Cretaceous, improved briefly during the Paleocene, entered a new cold phase during the late Miocene, a phase which progressed until it culminated in the glaciations of the Pleistocene epoch. The Miocene climate was greatly influenced by enlargement of the polar icecap, and the arctic vegetational zone spread at the expense of areas occupied by temperate and tropical vegetation. During the Miocene, the zone of polar-front weather - characterized by a succession of high and low-pressure areas - was moving gradually southwards bringing with it the conditions for glacier formation - heavy winter snowfalls and cool, cloudy summers.

PLIOCENE-PLEISTOCENE

By the Pliocene, permanent polar icecaps had formed and conditions broadly similar to those of today had become established, although temperatures in eastern Europe were clearly higher according to Schwarzbach (1961). The 18°C February marine isotherm in North American waters had nearly reached its present latitudinal limits of 20-25°N.

By the Pleistocene, the polar ice had extended for southwards and brought glacial conditions into regions of the Northern Hemisphere that are now temperate in character. The ice sheet, unable to melt in the cooler summers, crept down from the north preempting uncountable tons of water, which inevitably lowered sea levels throughout the world and raised previously submerged continental shelves, linking Iceland to the British Isles, the British Isles to Europe, and Europe to Africa. Tundra conditions, exemplified by fossils of musk-ox and walrus, extend into California, Texas and Mississippi. The tropical vegetation belt, the home of primates, advanced and receded as the glaciers fluctuated. It must be remembered too that the snow line on the high East African mountains, such as the Ruwenzoris and Mt. Elgon, fluctuated by several thousands of feet at the same time, although exact correlations are unknown. This relatively unimportant, parochial event of minor significance compared with glacier formations elsewhere must, nevertheless, have had a considerable, if local, effect on African montane forest primate distribution (see Rahm, this volume).

Between the four great Alpine Ice Ages the Gunz, the Mindel, the Riss and the Würm were long intervals or interglacials when warmer conditions prevailed. Primates, including of course man by this time followed the ebb and flow of the glaciers.

It is of particular importance for primate evolution to appreciate that the humid equatorial zone has shown no appreciable change in climate during the Cenozoic. The overall pattern of climatic change has been one of steady cooling, increasing aridity and evolving seasonality. The principal effects of climatic changes in the Cenozoic were in high and middle latitudes, while the low latitudes, (23°N - 23°S) underwent relatively little alteration. The effect of Cenozoic climates in tropical vegetation however has been profound; migration of temperate floras in both hemispheres has had the effect of compressing the tropical vegetational zone to less than half its width at the end of the Cretaceous. Tropical rain forest, today, has even narrower limits. It has been eroded by natural means; in the remote past, by cooling and aridity and, in the Middle Tertiary, by mountain-building and its local consequences resulting from rain-shadow effects; today it is being further eroded by human agency. It might be estimated that tropical evergreen forest at the present time occupies less than 5% of the area it occupied at the dawn of the Cenozoic. To summarize the climatic changes of the pre-Pleistocene epochs of the Cenozoic the following mean annual temperature figures from Schwarzbach (1961) for N.W. Europe are fairly representative: (see also Fig. 1.)

Epoch	Region	Climate Type	Tertiary	Present
Eocene	London Clay	Tropical to subtropical	21° C	10°C

Epoch	Region	Climate Type	Tertiary	Present
Upper Oligocene	Rott near Cologne	Subtropical to warm temperate	$18^{o}C$	c. $9^{o}C$
Miocene	Lake Constance	Warm temperate	$16^{o}C$	c. $9^{o}C$
Pliocene	Frankfurt-am-Main	Warm temperate to temperate	$14^{o}C$	c. $9^{o}C$

EVOLUTION OF FLOWERING PLANTS

The Cenozoic was the age of mammals but it was also the age of flowering plants. Angiosperms evolved in the Triassic or even earlier as tropical plants occupying the upland areas (Axelrod, 1951); by the Jurassic they were spreading to lowland areas where they were in competition with the gymnophytic flora. During the Cretaceous they gradually replaced the gymnosperms. By the start of the Cenozoic the angiosperms had evolved many modern species such as maples, sycamore, figs, birch, magnolias and oaks. Many of these trees that are now considered as temperate were evergreen in their primitive ancestral form.

The forests of the Mesozoic were sparse by comparison with those of the Cenozoic, lacking their richness, variety, and animal life. The grasses, monocotyledonous angiosperms, evolved at the beginning of the Cenozoic, but it was not until the Middle Tertiary (Oligocene-Miocene epochs) with the widespread colonization of open areas by grasses and herbs, that monocotyledons began to dominate the vegetational scene in many parts of the world. This process was accelerated during the Miocene as contraction of the humid tropical forests,with the increasing aridity, allowed grasslands to expand.

Both subclasses - mono- and dicotyledons - have played a critical role in the evolution of the primates. In the late Cretaceous and early Tertiary the evolutionary radiation of flowering plants produced an arboreal ecosystem in which plants, insects, birds and primates were involved in a complex food net. Plants found the trees immensely attractive places to live in, on account of the abundant insect life attracted there by the rich variety of available tree-products including fruits, buds, leaves, and seeds. Later, in the Upper Eocene or early Oligocene, certain primates that had acquired the physical adaptations to arboreal life (notably affecting extremities, the brain, and the eyes) were able to range freely within the canopy of the forest and, for reasons perhaps related to increasing body size, converted from an insectivorous to a largely fructivorous and foliivorous diet. Today arboreal primates, with one or two notable exceptions, are omnivorous in their diet, eating fruit, leaves, buds, bark - in fact all the trees have to offer including insects, birds, small mammals and reptiles.

The great expansion of grasslands during the Miocene brought about the evolution of herbivores, including primate herbivores, the ancestral macaques and baboons. With the addition of grasses, herbs and roots to their menu, baboons have become even more generalized in diet that the arboreal monkeys. Man has proceeded even further towards omnivorousness and added another item - meat - to his diet, which he largely obtains from herbivores. Thus, both directly and indirectly, man is almost wholly angiosperm-dependent for his survival.

The grasslands of the Miocene provided a new major ecological niche for the expanding populations of primates which lived in the rapidly shrinking forests. It seems undeniable that several Old World primate stocks accepted the challenge of this new environment for which, paradoxically, their arboreal preadaptations ideally suited them.

GEOMORPHOLOGY

It is generally understood that the geographical changes during the Cenozoic, affected the surfaces of continents rather than their shape. The principal exception to this generalization is southern Europe and the northern parts of peninsula India, areas that were once covered by sea. The Tethys Sea, a great stretch of tropical water that linked the Atlantic with the Indian Ocean and the Bay of Bengal, lay throughout the early part of the Tertiary in a vast geosyncline that had been in existence since the Cambrian. The Tethys Sea has in its history undergone extraordinary fluctuations in shape and size. Today all that is left of this ancient ocean are the Mediterranean, the Aegean, the Black and Caspian seas. But during the first half of the Cenozoic, the Tethys formed a major barrier between Europe and Africa. The Uralian trough, an arm of the Tethys sweeping northwards, cut Eurasia into two parts. It was not until the Miocene that this sea corridor disappeared and Eurasia was, once again, united. The Uralian trough must have acted as a major barrier to east-west faunal movements in the Paleocene, Eocene and Oligocene although there is some scientific evidence that a land bridge existed linking the two halves of Eurasia in the Upper Eocene (Kurten, 1966). By the Upper Miocene or early Pliocene, the Tethys Sea was disturbed by the abutment of peninsula India to Eurasia. The buckling effect that this north-eastward continent had upon the Eurasian land mass, gave rise to the mountain ranges that now form a west-east chain from the Atlas to the Himalayas.

During the Miocene epoch the world-wide mountain-building activity that characterized the Cenozoic was in full swing. Of particular importance to primate zoogeography was the Alpine-Himalayan system which stretches from western Europe to the East Indies. Folding along this axis began in the Jurassic, but most of the mountains remained submerged beneath the Tethys Sea until the beginning of the Cenozoic; there was further uplift early in the Cenozoic but the first great paroxysms of alpine orogeny occurred in the Lower Oligocene during which the submarine folds arose out of the Tethys Sea and thrust northwards; the mountain-building reached its apogee in Middle to late Miocene times. Towards the end of the Oligocene there was considerable orogenic activity at the eastern end of the Tethys Sea where the Atlas range appeared.

Rift faulting at the site of the present Red Sea had begun and volcanoes were active in East Africa. The final elevation of the Alps and the Himalayas took place in the Pliocene and Pleistocene.*

*The extent of this uplift is dramatically emphasised by the presence of Eocene marine deposits at 20, 000 ft. altitude in the southern range of the Himalayas.

In North and South America the Cascadian Revolution of the Miocene established the general form and extent of the Cordilleras, but the final thrusts, that produced the present day Andean peaks, came later. In western North America the "block mountain" chain of the Sierra Nevadas and the Coastal range were uplifted in the late Pliocene.

The principal effects for primate ecology of the mountain-building orgasms of the Miocene and Pliocene were (1) the spread of grasslands, (2) the opening up of a land corridor between Asia and Africa which resulted in exchange of African and Asian primate fauna which up to this time had evolved largely in continental isolation.

LAND BRIDGES

Simpson has summarized the problem of intercontinental faunal travel in a classic paper (1940). He observes, by way of introduction, that the inference to be drawn from the occurrence of an animal on one continent but not on another is, simply, that there is no possible way for it to get to the second continent. The corollary to this inference is that _if_ the same animal species should, at a later date, be found on the second continent, then something has happened in the meantime to make inter-continental travel possible. The alternative explanation that the animal species has finally "come round" to using a route that has existed all along, is unacceptable. The paleontologist, in other words, assumes that what _can_ happen _will_ happen.

Fluctuations in the level of epicontinental seas may result in the formation of land bridges or corridors that bring into contact animals from separate faunal realms. Land bridges must always to some extent be filter bridges which, as their name implies, hold back certain animals and allow others to pass. The "filter effect" may be climatic in nature. Tropical animals for instance cannot make use of a bridge in a temperate or cold climate. Thus, the Bering Strait land bridge during the Pleistocene was open to bears but closed to hyaenas. On the other hand the filter effect may be vegetational. A semi-desert bridge such as that between Africa and Asia in the Pliocene, or between India and Ceylon in the Pleistocene, would present an insurmountable barrier to forest-living animals, but would allow free passage to gazelles and other desert-adapted fauna. The colonization of the East Indies is a story of the successive rise and disappearance of filter-bridges between Sumatra and Borneo, and Borneo and the Philippines.

Simpson favors the use of the word "corridor" for bridges which are wide open and non-selective. On the whole, true intercontinental corridors are rather rare. The Bering Strait connection in the early Cenozoic, however, was one such example. Before leaving the subject of the principles of intercontinental movements of animals, it is as well to remember that they are not simply "expeditions" or positive migrations involving actual, purposeful taxes of individuals, but, rather are extensions of the species-range spreading outwards from a center of origin. It is the genes that migrate, not the animals.

SWEEPSTAKE ROUTES

Oceans are formidable barriers for land animals but the inference must be made that they have been crossed many times. The likely mechanism is by "rafting" or "waif immigration" (Patterson and Pascual, 1968) which implies a chance crossing of rivers or open water by means of floating vegetation, such as tree-trunks or blocks or matted grass-roots. Many well authenticated instances of this method of transportation in modern times are known; "islands" of vegetation have been seen many miles out to sea (e.g. Powers, 1911; Chasen, 1940). It is assumed that rafting accounts for the presence of cricetid rodents on the Galapagos, cebid monkeys in South America and lemurs in Madagascar. Simpson has felicitously named this method of migration, the "sweepstake route" epitomising in this phrase the essential ingredient of chance. In contrast to movements through corridors and across filter-bridges in which animals can pass in both directions, the sweepstake route is essentially a one-way system dictated by the directions of currents and prevailing winds. It is fairly obvious, too, that rafting is possible only for small animals.

Filter-bridges, corridors and sweepstake routes must have played a large part in the zoogeographical history of the primates and will now be considered in some detail.

THE BERING STRAIT

This northern link between Asia and North America was open intermittently throughout the Cenozoic (Hopkins, 1967). In the Paleocene the climate of the Strait was warm-temperate and the land surface was forested. During this time the assumption is that early Tertiary mammals (primates and insectivores, for instance) were able to pass freely between the two continents. Later in the Paleocene, rodents first which appear in North American fossil faunas, used this corridor to spread to Eurasia. By the early Eocene the Bering Strait was becoming a "cold bridge" permitting the passage of animals with large climatic ranges, physiologically speaking. But it effectively slowed, and eventually put a stop to, the movements of subtropical animals such as primates. The Bering Bridge was open in the Upper Miocene when the first wave of the great Asian-American faunal exchanges of open country animals, horses (from America) and elephants (from Asia) occurred. The Bering Strait acted as a filter-bridge throughout the Pliocene and Pleistocene. In the late Pleistocene, man was a notable filtrate.

NORTH ATLANTIC

Most paleogeographic atlases show the presence of a North Atlantic land bridge stretching in a south-easterly arc between modern Europe, the British Isles, Iceland, Greenland and North America (e.g. Termier et Termier, 1960). This route is thought to have been open in the Paleocene and early Eocene, some 60 - 70 million years ago. Recent geomagnetic studies of ocean floors (Vine, 1966) has strengthened the belief that Europe and N.E. America were for a brief period in the early Tertiary joined by a land bridge across the present North Atlantic. The closeness of the North Atlantic bridge to the "hinge" of continental drift (postulated as being sited approximately in the area of the New Siberian Islands) would mean that the opening of the North Atlantic occurred late in the Tertiary. During the same period, geological evidence points to the complete separation of Asia and Europe by the Uralian trough, sometimes referred to as the Turgai Strait. Zoogeographical evidence tends to support these conclusions (Kurten, 1966). The European element in Asian faunas is insignificant in the Paleocene and early Eocene while the North American element is strong. At the same time European and North American faunas are extremely similar and included the primate families Plesiadapidae and Paromomyidae; in fact, during the Paleocene, Europe and America shared the same primate genera, *Plesiadapis* and *Phenacolemur*. Some faunal exahange between Europe and Asia occurred in the Upper Eocene or early Oligocene (Termier and Termier, 1960) due possibly to the existence of a land bridge across the middle of the Uralian trough at the southern limit of the Ural Mountains that lasted into the Miocene, at which time Asia and Europe were fully re-united.

CENTRAL AMERICA

It is thought that North and South America were connected by land in the early Cenozoic but by the middle of the Eocene the connection had been broken. Thereafter, until the Pliocene when a union took place in the area of the Bolivar syncline, South America was separated from North America by a broad seaway (Patterson and Pascual, 1969). These authors disagree with Hershkovitz (1966) who stated that the water barrier would not have provided any formidable obstruction to the penetration of "herds" of immigrants. Movement of animals during the late Eocene, Oligocene and Miocene would have been by a process of island-hopping which could have been achieved by swimming, fording or rafting. The only mammals known to have reached South America between Middle Eocene and late Pliocene were primates and rodents. The evolution in South America of the caviomorph rodents seems to be entirely comparable with that of the New World monkeys. Both have close relatives in the Old World but there is no conceivable way (short of a hypothetical South Atlantic land bridge or a much delayed separation of Africa and South America) by which one group of rodents or primates could have given rise to the other. To explain both rodent and monkey evolution in South America it is necessary - but not difficult in the case of primates - to accept an extensive degree of parallel evolution from ancient basal (Middle Eocene) stocks. This reconstruction, Sarich (this volume) rejects on biochemical grounds. Sarich places the New World - Old World dichotomy at some 35 million years ago at end-Eocene or in early Oligocene times, and seems on balance to favor an African origin for New World primates.

AFRICA

The earliest fossil primates known in Africa are found in the Lower Oligocene of the Fayum of Egypt. The earliest mammals are in Upper Eocene of the same region. This suggests that Africa and Eurasia must have been connected at some time during the Eocene, possibly at the western end of the Tethys Sea where the Straits of Gibraltar now stand or between Tunis and Sicily (Moreau, 1952), areas which are today linked by an isthmus of shallow seas. Intermittent land connections are not improbable in view of the mountain-building activity in the bed of the Tethys from this time onwards.

For most of the first half of the Cenozoic, however, Africa was probably isolated and developed its own fauna superimposed on a older Eurasian fauna. The terminal phases of drift of the Indian peninsula produced the uplift of the Alpine-Himalayan system in the Oligocene and Miocene which both obliterated much of the Tethys Sea and ended the isolation of Africa. A wide corridor spanned the site of the present Red Sea and was intermittently connected through Arabia and the Persian Gulf with the general area of the Siwalik Hills in northern India, which was part of the Eurasian land mass (Laurasia) and not part of Gondwanaland. At its inception this corridor was probably forested but during the late Miocene and early Pliocene it became progressively more and more arid. In terms of faunal movement, in fact, it became a filter-bridge. During its forested period, the Afro-Asian corridor was the site of extensive two-way faunal movement, in fact, it became a filter-bridge. Carnivores, rhinoceroses, giraffes and rodents were westward-bound, and elephants were heading east. Nothing is known of primate movements at this time but there can be little doubt that such movements took place. Moreau (1952) has expressed the contrary view that this corridor was unlikely to have provided, at any time, a passage for forest animals. The presence of Cercopithecines in Africa, India and the Far East, on the other hand, would not be impossible, granting Moreau's premises; for this subfamily is essentially a terrestrially-adapted one and theoretically capable of utilizing an arid land bridge. A land connection with western Europe probably existed throughout the medium of the Betic Massif during the Middle and Upper Oligocene and in the Upper Miocene. Brinkmann (1960) suggests that this was in the region of the Straits of Gibraltar.

Cooke's (1968) comprehensive review of the paleogeography of Africa is particularly valuable for the new information it contains on the duration and extent of the Afro-Asian land bridge. Cooke points out that the Arabian block has been essentially a part of Africa throughout Cenozoic times (and earlier).

In Paleocene times the Tethys was drastically reduced and there were broad land connections through Arabia to India, and via the Balkans to Europe. By the Eocene, the Tethys had reclaimed much of the North African coastal plain and southeast Arabia and Africa were virtually isolated.

In the Oligocene once again the Tethys Sea withdrew as the mountain-building process of the Alpine chain commenced. At the end of the Oligocene it seems likely that there was a land connection between Africa and Western Europe, and between Africa and India via Arabia.

The rifting which produced the Red Sea trench began in the Oligocene but it was not until the Miocene that the Tethys overflowed the Suez peninsular and filled the rift-valley trough to provide a sea barrier between Africa and Arabia. Even then the barrier was not complete for there existed a considerable tract of land (300-350 miles) between the Farasan Islands and the Bab el Mandeb or Aden Straits which separate Ethiopia from the Yemen. This land bridge survived until the late Miocene or early Pliocene when the Gulf of Aden and the Red Sea became continuous.

In the late Miocene further uplift of the Eurasian land mass caused the Tethys to withdraw once more, creating the Isthmus of Suez. While Africa was finally isolated from Arabia (except for the Suez bridge), Arabia was probably still connected to India across the Persian Gulf which became drained at this time; so African fauna isolated in Arabia by the earlier severance of continuity could continue to pass into Asia and vice versa.

Recent studies have abundantly expanded our detailed knowledge of this area but the overall picture remains the same - that between the upper Oligocene and the upper Miocene, or early Pliocene, there was an intermittent continuity between Africa, Arabia, Asia and Europe. The crude 'wide corridor' concept can now be modified, but the potential for faunal exchange is unaltered. Recent studies, however, throw no light upon the really important question for primate paleozoogeography - the vegetational history on this great East-West, West-East commuter belt.

MALAYSIA

During the Cenozoic the sea levels in Southeast Asia have fluctuated considerably and Malaya, Sumatra, Java and Borneo, which lie within the 100-fathom line of the Sunda Shelf, became intermittently connected by dry land. The present coastline was probably determined in Pleistocene or post-Pleistocene times (Ho, 1964). To the east and west of the Sunda Shelf are some of the deepest marine trenches in the world. Alfred Wallace drew his "line" through one such trench, the Straits of Makasser, an ancient and permanent water barrier between Borneo and Celebes. Celebes has both tarsiers and macaques in its present fauna as well as the puzzling _Cynopithecus_ now accepted as a macaque (Fooden, 1969). The colonization of Celebes by primates cannot be easily explained on the basis of land bridges but the spread of primates to the Philippines probably can, since two separate tongues of the Shelf run in a northeasterly direction linking Borneo with the Philippines. Perhaps, however, the colonization of Celebes by primates can be explained by a combination of the Borneo-Philippine route and the interrupted Sangi island chain connecting Mindanao and the northeast peninsular arm of Celebes. One primate, _Tarsius spectrum sangirensis_, a subspecies of _T. spectrum_ of Celebes, is found on these linking islands. By Pleistocene times, India and Southeast Asia were cut off from northern Asia and Europe by the Himalayas and from Africa by the Arabian desert, so that faunal isolation of Southeast Asia was complete; it is unlikely that any intercontinental faunal movements have taken place since then.

The significance of these corridors and filter-bridges for the present zoogeographical distribution of primates, is still largely hypothetical. Darlington (1957) has stated " .... the best evidence for the geographical radiation of most mammals is not their present distribution but their fossil records." Unfortunately the fossil record of post-Eocene primates does not compare with that of other mammalian groups such as the horses or the elephants which left behind traces of their zoogeographical wanderings.

SIGNIFICANCE OF CENOZOIC CLIMATIC FLUCTUATIONS FOR CATARRHINE EVOLUTION

There can be no doubt that the climatic oscillations during the Cenozoic and their influence on the distribution of forests were of great significance for primate evolution. Primates, as we believe, are fundamentally a forest-living Order. (Of the 189 species extant at least 88 percent are arboreal; the breakaway exceptions to the rule, man and the macaque-baboon group, are more significant by reason of their population numbers than they are by the richness of their speciation). Thus, we can make the assumption that the key to the zoogeographic history of the primates and to the adaptations that led to the emergence of major radiations, lies in the distribution, the floristic composition and the seasonality of Cenozoic forests.

The major ecological revolution of the Middle Tertiary unquestionably affected the phylogeny of the catarrhine primates but had, strangely enough, little apparent effect on the New World monkeys. Non savanna-living platyrrhine monkeys are known, either living or fossil. The explanation which suggests itself is both a climatic and vegetational one. Old World primates in the Middle and Tertiary had a vast range of territory in Eurasia which are now warm-temperate and temperate, but which in the Miocene were subtropical (Fig. 2). The cooling of late Cenozoic times had its effect primarily on areas outside the present tropics. From these considerations, the inference might be made that savanna-living Old World primates first evolved in parts of Europe and Asia well removed from the equator and, thus, were the first to seize the opportunity of the novel ecological niche offered by the emerging grasslands. Consider now the geographical range of New World monkeys, which from late Eocene times has been restricted to the Latin-American continent and well within climatic zones that, today, we regard as tropical or subtropical. It is generally agreed that there was little, if any, change in climatic conditions during the Miocene or Pliocene that could have affected the distribution of South American forests. Consequently no new ecological opportunity, comparable to that offered to Old World primates, presented itself to the platyrrhines in mid-Tertiary times. Thus they have remained - for want of an environmental alternative - an essentially arboreal stock. The special nature of the tropical rain forest habitat of the tributaries of the Amazon River, has had its own, unique effect on platyrrhine evolution (Hershkovitz, 1966).

THE CERCOPITHECOID-HOMINOID DICHOTOMY

While the Lower Oligocene deposits of the Fayum of Egypt provide undoubted evidence of the early radiations of the hominoids they provide little certain information on the origin of the cercopithecoids. The fossil evidence for an Eocene origin of this superfamily is not particularly strong. It depends on how such forms as *Apidium* and *Parapithecus* of the Fayum Oligocene and the Eocene fossil *Pondaungia* and *Amphipithecus* of the Eocene of Burma are interpreted. Parapithecids are extremely common in Quarry I (late Oligocene) of the Fayum site but also appear in Quarry G (a middle Oligocene horizon) along with many jaws of *Apidium*. Simons (1967) tentatively regards both *Apidium* and *Parapithecus* as having both reached the monkey grade of evolution; the overt similarities of the frontal bones of *Apidium* to a New World monkey is impressive. Even more striking are the dental similarities of *Parapithecus* to the modern talapoins (*Cercopithecus talapoin*). There are interesting similarities too, to the African fossils *Victoriapithecus* and *Prohylobates* of the Lower to Middle Miocene, both likely cercopithecoids. The recently discovered Rooneyia of the Oligocene of Texas, North America (Wilson, 1966), an undoubted omomyid, suggests that this family possessed the potential for evolving a "monkey"-like form.

Molar bilophodonty, the characteristic dental specialization of the Cercopithecoidea, is found for the first time in Parapithecus sp. nov. of Quarry I of the Upper Oligocene of the Egyptian Fayum. This might suggest an early or middle Oligocene date for separation of the hominoid and cercopithecoid stocks.

Whether the Old World catarrhine stock underwent a dichotomy in the Eocene, the Oligocene or the Miocene, the ecological question remains as to the nature of the selection processes which brought about such diversification. The following hypothesis is offered with reservations.

The primary behavioral differences between apes and monkeys are concerned with their locomotion and their diet. In respect of locomotion, the differences in adaptation are obvious and striking. In respect of diet, the differences are less clear-cut due to the wide variety of secondary habitats invaded by Old World primates in relatively recent times. However, among the more conservative members of each group, the predominantly leaf-eating habits (and associated anatomical adaptations) of the Colobinae and the predominantly fruit-eating habits of the arboreal apes, strongly suggest that fruit-eating and leaf-eating represent extremely ancient divergent specializations. Diet, however, is demonstrably variable in the relative proportions of its minor components even within such a homogenous group such as the Colobinae. Ripley, (this volume) for instance emphasizes the eclectic feeding habits of the gray langur (P. entellus thersites) in dry mixed deciduous forest in Ceylon. The diet of these "leaf-eaters" varies from leaves through fruit, to a high occasional dependence on bark in Mysore, India, an area where during the winter the trees are practically leafless. Manley (personal communication) on the other hand stresses the predominantly foliivorous nature of the diet of Presbytis senex monticola which is an inhabitant of high altitude tropical evergreen forest in Ceylon.

Locomotion and diet are so intimately related that it is hardly meaningful to give one or other function an adaptive priority; animals must move to eat and eat to move. Both these primary, or first-order, behaviors are closely linked to the nature of the habitat. Therefore, it is logical to look for an ecological explanation for such a major adaptive shift as that which gave rise to the two existing superfamilies of the Old World catarrhine stock. The habitat shift that is most likely to have produced such a major furcation is from an arboreal life to a ground-living one, but clearly this cannot have been the whole explanation for the following reason. It is generally agreed that the colobine subfamily of the Cercopithecoidae are more primitive than the cercopithecine subfamily. Colobines are essentially arboreally adapted in locomotion and diet and it is well-nigh inconceivable, on anatomical and physiological grounds alone, that they should have evolved from a ground-living stock. Therefore an alternative ecological explanation must be sought.

The major difference between tropical rain forest, and monsoon and dry deciduous forest is seasonality of rainfall. This affects, among other things, the ratio of deciduous/evergreen trees, the physiognomy of the forest, the patterns of fruiting and the cycles of leaf change. In tropical rain forest, fruit is available all year round and one or other fruit-bearing species are in season at all times of the year. But as Medway points out (this volume) although tropical rain forest as a whole may be evergreen, there are many tree species within it that exhibit annual cycles of leaf change and fruiting that tend to be concentrated in certain parts of the year. On the other hand, there are other species that show multiple fruiting cycles such as several species of epiphytic figs. In monsoon and dry deciduous forest, fruiting is seasonal and usually corresponds with the onset of the rainly season or seasons. Primates of strictly seasonal forests must possess the adaptations to enable them to subsist on food-items other than fruit. Among vegetarians, the available alternatives are leaves, shoots, roots, bark, pith, buds, flowers and sap. For arboreal animals, leaves are the most prolific food-item that, in a mature or immature state, are available all the year round. Leaf-eating, then, the principal (though not necessarily the habitual) dietary habit of the present day Colobinae (the most primitive section of the Cercopithecoidea) would have provided a selective advantage for monkeys over apes in seasonal forests. The monkey stock, thus being more generalized in diet, would have been able to spread widely into geographical regions where the more specialized fructivorous apes would have been at a selective disadvantage.

Present-day zoogeography indicates that while monkeys live in tropical forest, monsoon forest, dry deciduous forest and even temperate forest, the fruit-eating apes such as *Pan* and *Pongo* are largely restricted to the tropical rain forest belt (Fig. 3). Gibbons, however, which are largely fruit-eaters (80 percent: Groves and Napier, in press) extend into seasonal forests, where their territorial behavior may be an adaptation to seasonal fruit shortages.

Many cercopithecoids have moved into tropical forests. Gorillas have shifted to a ground-living way of life and, although some fruit is eaten by western, lowland gorillas (about 10 percent), generally speaking they favor a more bulky, fibrous diet. The primitive dietary dichotomy of apes and monkeys is therefore no longer clear-cut, having been blurred in modern catarrhine families by secondary ecological shifts; such adaptability of feeding preferences is characteristic of higher primates.

The anatomical correlates of leaf-eating, the specialized leaf-eating stomachs and salivary apparatus of *Colobus* and *Presbytis*, clearly must have evolved after the cercopithecine stock separated from the Colobines in Miocene times, for the Cercopithecinae show no evidence of these adaptations. By the same token, cheek pouches, a hallmark of cercopithecine monkeys, must have been acquired since their separation from the Colobinae which show no signs of this particular specialization. The presence of cheek pouches in the genus *Cercopithecus* is one of the strongest indications that this genus is secondarily arboreal having evolved from a ground-living cercopithecine stock, as was first suggested by Napier and Davis (1959).

The validity of this hypothesis for cercopithecoid evolution depends on the accumulation of considerably more evidence of the periodicity of fruiting patterns in the forests of the tropics and subtropics; but in spite of the secondary habitat shifts, already referred to, there is still the evidence from the geographical distribution of present day forms that the colobines are capable of surviving in vegetational zones outside the non-seasonal forest belt: whereas apes with their relatively narrow range of diet would appear to be less fortunate. Cercopithecines are capable of surviving in even more arid and seasonal areas than are colobines. The generalized nature of their diet, which includes roots, grass, tubers and bark, has permitted them to range into areas which are markedly seasonal for five or six months in the year (for example see Suzuki, 1965). Limiting factors to the present day northerly migration of primates are concerned partly with mean annual temperatures and rainfall, partly with the shortness of winter daylight for food-finding (Jolly, 1964), partly with the southernmost limit of permafrost in the northern hemisphere, and partly with the distribution of man.

During the critical period of the Oligocene-Miocene, the tropical and subtropical zone extended considerably further north than it does today. Warm temperate zones in Europe probably reached 50°N compared with their present northern limit of 40°. Vast areas of Eurasia, now devoid of nonhuman primates, would then have provided adequate environments for primates to live in. Much of this zone would have been subject to seasonality and, thus, the seasonal and non-seasonal forests that exist today are likely to have existed in the mid-Tertiary, but with a greater latitudinal range. Hence, there would then have been more clear-cut ecological separation of leaf-eating and fruit-eating catarrhines. Today, primates have been driven from the higher latitudes, partly as a result of climatic deterioration and, partly, from the presence of competition in the form of man, into the narrow equatorial band to which they are now virtually restricted. As a result, both monkeys and apes are forced to occupy the same biotic provinces, and the resulting secondary dietary adaptations have obfuscated the primitive, leaf-eating/fruit-eating dichotomy that once existed (and which is still "immortalized" in the dental disparity of the two subfamilies). Due to the persistence of certain living leaf-eating species such as the Hanuman langurs and *Rhinopithecus* in higher latitudes the original pattern of wide dispersal is still apparent.

ORIGIN OF THE CERCOPITHECINAE

The assumption that the colobine-cercopithecine separation took place during the middle Miocene from a primitive stock has been based largely on paleoecological hypotheses (Jolly, 1964; Napier, 1967) and on inferences derived from the *Mesopithecus* fossils of the Pontian, and from the Pliocene genera *Libypithecus* and *Dolichopithecus* all of which show an admixture of colobine and cercopithecine features (Gabis, 1960; Jolly, 1964, 1967; Verheyen, 1962). More direct evidence for this dichotomy comes from the recent discovery of a cercopithecine tooth and a colobine-like frontal bone at Napak, Uganda, a site with an estimated K/Ar date of 19 ± 2 million (Pilbeam and Walker, 1968), and on the reappraisal of the affinities of *Prohylobates tandyi* from Wadi Moghara by Simons (this volume). The implications of these discoveries are that the furcation must probably be put back at least to upper Oligocene or lower Miocene times. Sarich's postulate (this volume) that the cercopithecine-colobine division took place in the middle Miocene, some 18 million years B.P., is somewhat at variance with this conclusion.

The separation of the ancient colobine-like, cercopithecoid stock into the two subfamilies known at present, was essentially an ecological event. Climatic and botanical changes which led to habitat differences would in turn have led to the structural and behavioral distinctions that characterize these two subfamilies today. Paleoecological studies of various sorts indicate that in the early Miocene, temperate, seasonal conditions were replacing tropical conditions in middle latitudes in a fairly abrupt manner. The climax tripical rain forest deteriorated into a more open type of dry deciduous woodland. The global climatic changes combined with the "local" effects of mountain-building, provided the conditions for a widespread colonization of open areas by grasses and herbs (Axelrod, 1951). Principally these conditions were in operation *outside* Africa, the present tropical belt being largely unaffected. An assumption can therefore be made that the ground-living adaptations that led to the radiation of the cercopithecines, took place in Europe and Asia, areas remote from the present tropics. Cercopithecines, in other words, evolved principally in Eurasia, probably in the early Miocene.

This hypothesis is in accord with the prevailing climatic and vegetational conditions of the time, and it is also in accord with the fossil evidence from Napak and Wadi Moghara, and the geological and oceanographic evidence of an Afro-Asian land bridge.

The Afro-Asian land corridor, connecting north east Africa to the general region of the Siwalik Hills in north east India, was established at the end of the Oligocene (George, 1962). During the Miocene, it became progressively more arid and eventually, during the Pliocene, the final convulsions of the developing African Rift Valley system served to sever its wide connection with the Middle and Far East, leaving only the Suez link. Lagomorphs, rodents, canidae, viverridae, chalicotheres, rhinoceroses and many species of artiodactyls invaded Africa during the Miocene (Cooke, 1968) - and so, presumably, did the primates. All these families and groups were responding to the vis-à-tergo - a deteriorating climate in the middle latitudes of Europe and Asia.

It is uncertain whether the Afro-Asian land bridge was forested at any time in its existence. Moreau (1952) regards it as highly unlikely, while George (1962) considers it was forested at least in the early Miocene and thereafter became progressively arid, though not desertic. It would seem that land connections whether forested or not, would favor the arrival of cercopithecines from Eurasia throughout the first half of the Miocene. The spread of the savanna biome in Africa (although, as Moreau (loc cit) contests, it may well have pre-existed there as a minor vegetational feature), in the middle to late Miocene would have prompted a a major cercopithecine radiation towards the end of this epoch. The hypothesis however is not in accord with the fossil record of the catarrhine monkeys. As Simons (this volume) states, there are no fossil monkeys in Eurasia which antedate the Pontian. This is undeniably true always bearing in mind the proviso that negative evidence proves nothing. The Miocene Eurasian sites represent only a very minute fraction of the total areas involved. Africa's Miocene sites are much better known and, while still of course constituting only a fraction of possible sites, their monkey yield during this epoch is very poor as Simons (loc cit) states. Perhaps it is not surprising that during the Miocene, apes and lorisoids were abundant and that monkeys in Africa were scarce. The African apes, prolific during the early to middle Miocene, were functionally and ecologically monkeys; they moved like monkeys (e.g. Limnopithecus and Proconsul) and thus they occupied present-day monkey forest and woodland niches. Lorisoids, presumably nocturnal, then as now, offered little in the way of competition. The Miocene was optimal for hominoids and pessimal for monkeys, a situation that today is radically different in African forests.

It seems likely that the primitive cercopithecines were somewhat macaque-like in their limb proportions (Gabis, 1960) while still retaining the dental and skull features of their primitive colobine ancestry. Jolly (1967) states that the long manal phalanges of Mesopithecus suggest an arboreal habitat but, in fact, the length of the third digit based on Gaudry's (1862) measurements of the hand clearly indicate that, judged by colobine standards, the fingers were not particularly long. While the thumb is on the short side, it is considerably longer than that of any colobus monkey. In the accompanying table the limb proportions of Mesopithecus are compared with those of certain other relevant cercopithecoids.*

It will be seen that the proportions of Mesopithecus fall within the range of Colobinae in two out of five indices (crural index and phalangeal index) and within the range of ground-living Cercopithecinae in five out of five. Within the Cercopithecinae the closest "match" to Mesopithecus is the living genus Macaca. Thus, the behavioral implications derived from limb indices would suggest that the specialized adaptations related to live on the ground are moderately far advanced in M. pentelici. Groves (this volume)

*It should be noted that the author's methods of measuring limb lengths differ slightly from those of other authors (e.g. Schultz). All indices in this Table are derived from measurements taken by the author and so are strictly comparable. For author's methods see Napier and Napier, 1967.

disagrees with this conclusion preferring to regard the "macacoid" characters of Mesopithecus as primitive for the cercopithecoids. The alternative explanation which I favor is that "macacoid" characters are not symplesiomorph for the Cercopithecoidae as a whole but, on the contrary, had been recently acquired by the Cercopithecinae and are therefore symplesiomorph for that subfamily only.

	Brachial Index	Crural Index	Intermemb. Index	Thumb L. Index	Phal. L. Index
Colobus spp. (30)	89-96	83-92	77-83	-	52-54
P. aygula grp. (23)	101-113	86-94	73-81	27-34	50-56
P. cristata grp. (10)	89-98	85-88	79-83	28-33	50-54
Cercopith. spp. (30)	90-102	91-100	77-87	35-42	49-54
C. aethiops grp. (12)	89-102	89-98	80-87	35-40	48-52
Papio spp. (16)	96-106	80-86	91-100	39-48	41-46
Macaca spp. (33)	91-102	83-98	84-98	33-43	47-53
Mesopithecus (1)	98	85	88	38-39*	51-52*

Macaca while historically an Eurasian genus entered North Africa in late Miocene or early Pliocene times according to Arambourg's (1959) evidence for Macaca flandrini. Both von Koenigswald (1969) and Simons (this volume) find M. flandrini an acceptable macaque.

In its total morphological pattern, the true baboon, as Jolly suggests (this volume) is of a late Pliocene origin. Baboons, the gelada and patas monkey, are physically specialized for savanna life to a much greater degree than is Macaca. It would seem that the physical specializations of baboons are responses to life in the

---

*Estimated length of carpus based on generalized O.W. monkey dimensions.

specific habitat of the African tropical grassland, an ecosystem that has no exact equivalent in Asia. Behaviorally, however, baboons and macaques are very close. The inferences that might be made are twofold: firstly, that African grassland is a relatively recent vegetational climax and that baboon-like African cercopithecines did not become 'true' baboons until this particular biome evolved. Secondly, on the behavioral side, the inference is that the multi-male troops and the male dominance characters of the social systems of macaques and baboons are universals that relate to ground-living *per se* rather than specifically to savanna-living.

ORIGIN OF THE COLOBINAE

I believe it is necessary to regard the colobine subfamily of the Cercopithecoidea as having arisen from autochthonous stocks in the continental areas that they now occupy, having had little or no truck with risky things like land bridges. In a conservative order, such as the primates, major structural differences within subfamilies are not common, particularly in the conservative parts of the body such as the hand, where primitive pentadactyly has been tenaciously retained. The absence of the thumb in *Colobus* and its retention (though in a diminished form) in Asian colobines, such as *Presbytis*, argues an ancient separation of these genera as indeed their biochemical constitution also argues. Such a separation must have been more ancient than Sarich (this volume) proposes on biochemical grounds. The same ecological factors that led to the division of the catarrhine stock into hominoids and cercopithecoids undoubtedly operated, though to a lesser extent, in Africa as well as in Europe. *Colobus*, therefore is regarded as an African genus of the primitive colobine stock that may have either: (1) evolved in Africa from an Oligocene anthropoid stock, or (2) entered Africa with the first wave of primate invaders early on in the Oligocene-Miocene when the Afro-Asian land bridge was still forested. After this date the land bridge would surely have become too arid to have permitted the passage of a purely arboreal stock such as the Colobinae.

I am only too well aware of the tenuous nature of some of the hypotheses of catarrhine evolution outlined here, but as J.W.Gregory, whose words introduce this chapter, has said in effect: one has to choose between the audacious and the sterile. If audacity stimulates controversy, and controversy stimulates new research, then audacity is a part of science.

ACKNOWLEDGEMENTS

I would like to thank the Wenner-Gren Foundation for Anthropological Research, Mrs Lita Osmundsen and her colleagues for providing such an agreeable and academically stimulating setting for the Conference for which this paper was prepared.

My thanks are also tendered to all participants at the 1969 Conference who made me see some of the errors of some of my ways. I warmly acknowledge the advice of Dr Lewis Gazin of the Smithsonian Institution on some aspects of this study, and the help of my administrative assistant in London, Miss Barbara Dickson, who worked so hard on the final manuscript, my secretary in Washington Miss Dorothy Osborn and Mrs Audrey Besterman who prepared the diagrams.

REFERENCES

Arambourg, C. (1959) Bertebres continentaux du Miocene superieure de l'Afrique du Nord. Publ.Serv.Geol.Algerie, Pal. 4: 1-59.

Avis, Virginia. (1962). Brachiation: The crucial issue for Man's ancestry. Southwest.J.Anthropol. 18: 119-148.

Axelrod, D.I. (1951). A Theory of Angiosperm Evolution. Evolution 6: 29.

Briden, J.C. (1967). Recurrent Continental Drift of Gondwanaland. Nature Lond. 215: 1334-1339.

Brooks, C.E.P. (1949). Climate through the ages. New York: McGraw-Hill.

Chaney, R.W. and E.I. Sanborn. (1933). Goshen flora of W. Central Oregon. Carnegie Inst.Wash.Publ. 439: 103.

Chasen, F.N. (1940). Handlist of Malaysian mammals. Bull. Raffles Mus. 15: 1-209.

Cooke, H.B.S. (1968). Evolution of mammals on Southern Continents. II. Fossil Mammal Fauna of Africa. Quart.Rev. Biol. 43: 234-264.

Darlington, P.J. (1957). Zoogeography. New York: Wiley.

Dorf, Erling. (1955). Plants and geological time scale. Geol.Soc. Amer. Special paper No.62: 575.

Durham, J.W. (1950). Cenozoic marine climates of the Pacific Coast. Bull.Geol.Soc.Amer. 61: 1243.

Fooden, J. (1969). Taxonomy and evolution of the monkeys of Celebes. Biblioteca Primat. 10: 1-148.

Fox, Robin. (1967). In the beginning: aspects of hominid behavioural evolution. Man 2: 415-433.

Fraser, Ronald. (1967). Understanding the Earth. Penguin Books Ltd.

Gabis, R.V. (1960). Les os des membres des singes cynomorphes. Mammalia 24: 577-607.

Gaudry, A. (1862). Animaux fossiles et Geologie de l'Attique d'apres les recherches faites en 1855-56 et en 1860. Paris: F. Savoy.

George, Wilma. (1962). Animal Geography. London: Heinemann.

Gill, E.D. (1961). The climates of Gondwanaland in the Kainozoic times. In Descriptive Paleoclimatology. (Ed. Nairn, A.E.M.) New York: Inter-Sciences Publ. Inc.

Hershkovitz, P. (1966). Mice, land bridges and Latin America faunal interchange. In Ectoparasites of Panama. (Eds. Wenzel, R.L. and Tipton, V.J.) Chicago: Field Museum.

Ho, R. (1964). The Environment. In Malaysia: a Survey. (Ed. Wang Gungwu). London and New York: F.A. Praeger.

Hopkins, D.M. (1967). The Bering land bridge. Stanford: Univ. Press.

Howell, F. Clark. (1967). Recent advances in human evolutionary studies. Quart.Rev.Biol. 42: 471-513.

Hurley, Patrick, M. (1968). The confirmation of continental drift. Sci.Amer. April: 53.

Jolly, C.J. (1964). Origins and specialisations of the baboons. Univ.Lond. Ph.D. Thesis.

Jolly, C.J. (1966). Introduction to the Cercopithecoidea with notes on their use as laboratory animals. Symp. zool. Soc. Lond. No. 17: 427-457.

Jolly, C.J. (1967). Evolution of baboons. In The baboon in medical research Vol. 2. (Ed. Vagtborg, H.) Austin and London: Univ. of Texas Press.

Kobayashi, Teiichi, and Shikama, Tokio. (1961). The climatic history of the Far East. In Descriptive Paleoclimatology. (Ed. Nairn, A.E.M.) New York: Intersciences Publ. Inc.

Krausel, R. (1961). Palaeobotanical evidence of climate. In Descriptive Paleoclimatology. (Ed. Nairn, A.E.M.) New York: Intersciences Publ. Inc.

Kurten, Björn. (1966). Holarctic land connexions in the Early Tertiary. Comm. Biol. Soc. Sci. Fenn. 29(5): 1.

MacElhinney, M.W. (1968). Northward Drift of India - Examination of Recent Palaeomagnetic Results. Nature 217: 342-344.

MacGinitie, H.D. (1958). Climate since the Cretaceous. In Zoogeography. (Ed. Hubbs, Carl L.) Amer. Assoc. Advance. Sci. No. 51.

Matthew, W.D. (1915). Climate and Evolution. Ann. N. Y. Acad. Sci. 24: 171-318.

Matthew, W.D. (1939). Climate and Evolution. 2nd edit. Ann. N. Y. Acad. Sci. Special Publ. 1: 1-223.

Maxwell, Arthur, E., Richard P. von Herzen, K. Jingwa Hsü, James, E. Andrews, Tsunemasa Saito, Stephen F. Percival Jr., E. Dean Milow, Robert E. Boyce. (1970). Deep Sea Drilling in the South Atlantic. Science 168: 1047-1059.

McKenna, M.C. (1966). Paleontology and the origin of the primates. Folia Primat. 4: 1.

Moreau, R.E. (1952). Africa since the Mesozoic. Proc.zool.soc. Lond. 121: 869-913.

Napier, J.R. (1967). Evolutionary aspects of primate locomotion. Amer.J.Phys.Anthrop. 27: 333-341.

Napier, J.R. and P.R. Davis. (1959). The forelimb skeleton and associated remains of Proconsul africanus. Fossil Mammals of Africa No. 16. London: British Museum (Nat.Hist.)

Napier, J.R. and P.H. Napier. (1967). A Handbook of Living Primates. London and New York: Academic Press.

Patterson, Bryan and Rosendo Pascual. (1968). Evolution of Mammals on southern continents. V. The Fossil Mammal Fauna of South America. Quart.Rev.Biol. 43: 409-451.

Pilbeam, D. and Alan Walker. (1968). Fossil monkeys from the Miocene of Napak, North-East Uganda. Nature Lond. 220: 657-660.

Powers, S. (1911). Floating islands. Popular Science Monthly 79: 303-307.

Reid, E.M. and M.E.J. Chandler. (1933). The London Clay flora. London: British Museum.

Sarich, V.M. and A.C. Wilson. (1966). Quantitative immunochemistry and the evolution of primate albumins. Science 154: 1563-1566.

Sarich, V.M. (1967). Rates of albumin evolution in primates. Proc.Nat.Acad.Sci. 58: 142-148.

Sarich, V.M. (1968). The origins of hominids: an immunological approach. In Perspectives on Human Evolution. (Eds. Washburn, S.L. and Jay, Phyllis, C.) New York: Holt, Rinehart and Winston.

Schwarzbach, M. (1961). The climatic history of Europe and North America. In Descriptive Paleoclimatology. (Ed. Nairn, A.E.M.) New York: Intersciences Publ. Inc.

Simons, E.L. (1967). The significance of primate paleontology for anthropological studies. Amer. J. Phys. Anthrop. 27: 307.

Simpson, G.G. (1940). Mammals and land bridges. J. Wash. Acad. Sci. 30: 137.

Simpson, G.G. (1955). The Phenacolemuridae, new family of early primates. Bull. Amer. Mus. Nat. Hist. 105: 411.

Smith, A. Gilbert and A. Hallam. (1970). The fit of southern continents. Nature Lond. 225: 139-144.

Suzuki, A. (1965). An ecological study of wild Japanese macaques in snowy areas - focused on their food habits. Primates 6: 31-72.

Szalay, Frederick S. (1968). The beginnings of primates. Evolution 22: 19.

Termier, H. and G. Termier. (1960). Atlas de Paleogeographie. Paris: Masson et Cie.

Verheyen, W.N. (1962). Contribution a la craniologie comparee des primates. Musee Roy. Afrique Centrale - Tervuren, Belgique. Ser. 8 Sci. zool. 105: 1-256.

Vine, F.J. (1966). Spreading of the ocean floor - new evidence. Science 154: 1405.

von Koenigswald, G.H.R. (1969). In: Fossil Vertebrates of Africa. Vol. II. London: Adademic Press.

Walker, Alan. (1967). Patterns of extinction among Madagascan subfossil lemuroids. In Pleistocene Extinctions. (Eds. Martin, P.S. and Wright, J.E. Jr.) New Haven and London: Yale Univ. Press.

Washburn, S.L. (1950). The analysis of primate evolution with particular reference to the origin of man. Cold Spring Harbour Symp. Quart. Biol. 15: 67.

Wilson, J.A. (1966). A new primate from the earliest Oligocene. Folia Primat. 4: 227-248.

Wilson, J. Tuzo. (1963). Continental Drift. Sci. Amer. April.

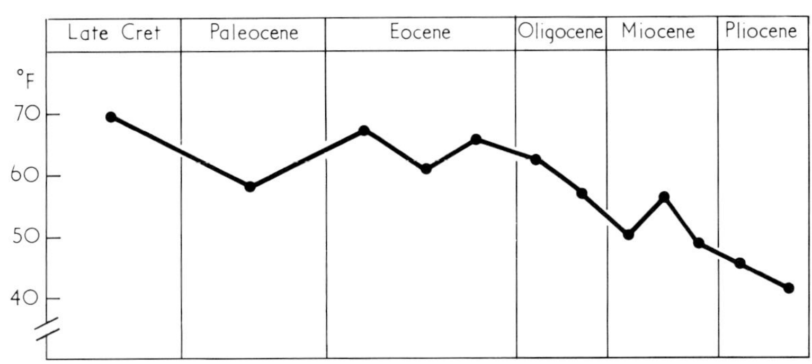

Fig. 1. Mean annual temperature, in degrees Fahrenheit, at latitudes 40° - 50° during the Tertiary (after Dorf, 1957).

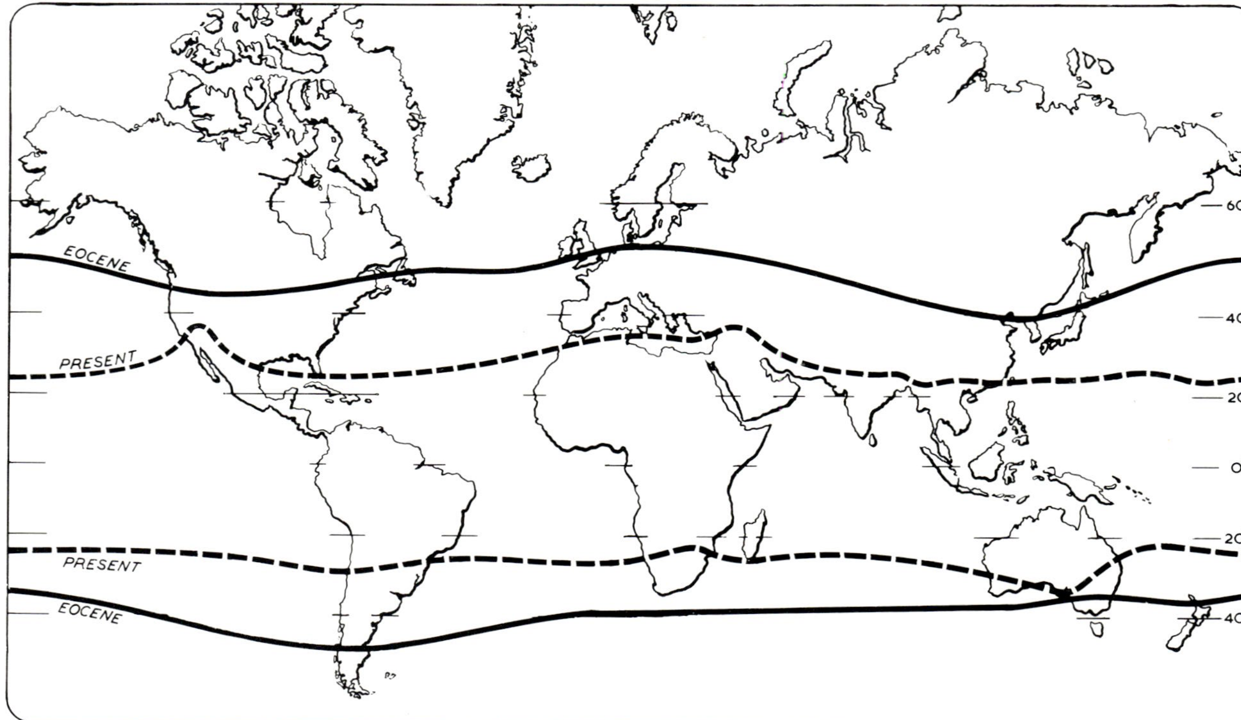

Fig. 2. A map of the world (Mercator projection) indicating northern and southern limits of subtropical and tropical biomes in the Eocene (solid line) and the Present (interrupted line).

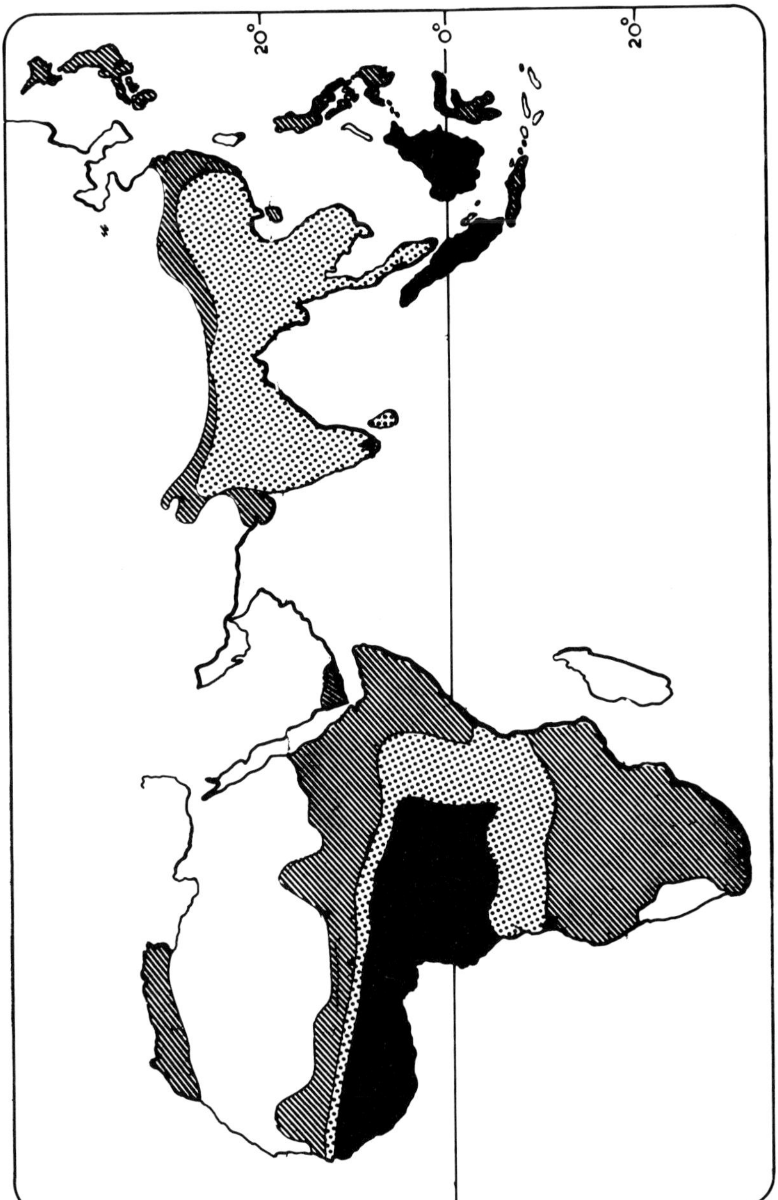

Fig. 3. The maximum range of present day Cercopithecinae (cross-hatching), Colobinae and Hylobatidae (dots) and Pongidae (solid black).

THE DEPLOYMENT AND HISTORY OF OLD WORLD MONKEYS (CERCOPITHECIDAE, PRIMATES)

Elwyn L. Simons

OLD WORLD MONKEYS

INTRODUCTION

An overall review of the fossil monkeys of the Old World has not been published to date. This is primarily because many of the oldest materials of cercopithecoid fossils are either fragmentary or undescribed.* In addition, they come from widely scattered localities and consequently have never been studied comparatively by a single scientist. Delson is currently surveying Old World fossil Cercopithecoidea but pending publication of his study the following review is provided.

The principal Pleistocene finds of Old World monkeys are from numerous sites in South and East Africa and from scattered Eurasian localities. Pliocene occurrences reported to date appear to be restricted to North India, Europe, Wadi Natrun in Egypt, Kanapoi and Lothagam in northern Kenya and (as the result of extensive recent collections) the Omo basin of Ethiopia. Not all of the fossil monkeys of the latter two regions have yet been described. It is possible that some of the fossil monkeys recovered from South Africa, principally in association with *Australopithecus*, are as old as the upper Pliocene, but pending more accurate dating of these sites that question cannot be resolved.

Miocene finds are apparently all restricted to Africa and this, therefore, constitutes a piece of negative, but significant evidence that Cercopithecidae did not emigrate from Africa before the Pliocene. Since the African Miocene sites sample only a very tiny fraction of the areal extent of Africa in Miocene times it is clear that our knowledge of early cercopithecids is as yet exceedingly fragmentary. Although I have myself suggested elsewhere (Simons, 1963) that the cercopithecid radiation may have been relatively late, the bases for this view are still nothing more than extrapolations from biochemical and interspecific hybridization data. It would be premature to suppose that the relatively late

---

* Such species as have been described are brought together in Hill's (1953 et seq.) series on the Primates.

dates of most of the known fossils of this group indicate with certainty that their radiation in Africa was later than that of the hominoids.

African Miocene monkeys have been reported to date from Rusinga and Maboko Islands, Lake Victoria, Kenya; Fort Ternan, Kenya; Napak, Uganda; from the Congo basin; and from Wadi Moghara in Egypt. A species of _Macaca_ has also been described from the latest Miocene (or earliest Pliocene) lignites of Marceau, in the department of Oran, Algeria.* Even including in the total all of the latter finds in Oran these African Miocene cercopithecines constitute only about fifteen jaw fragments, one limb bone and some fifty isolated teeth. Paleontologically this is exceedingly poor representation for a whole continent over an entire epoch.

Our knowledge of African Oligocene monkeys is even more restricted since all known finds come from an area no more than three miles square in the Fayum badlands of Egypt. However, the number of good specimens recovered so far exceeds that known from the African Miocene. Yale Quarry I alone has yielded more than forty partial upper and lower jaws of parapithecines. I have recently proposed (Simons, 1969b) that the Fayum primate species of the genera _Parapithecus_ and _Apidium_ should no longer be considered as representing a distinct primate family, Parapithecidae, but instead should be classified as the Parapithecinae, a primitive subfamily of the Cercopithecidae.

In sum, the Tertiary record of Cercopithecoidea is scanty and (apart from the Fayum Oligocene sample) there is reason to believe that many of the taxa we know are not associated with tropical lowland forest faunas and floras. These taxa include, for example, _Victoriapithecus_, _Prohylobates_, _Mesopithecus_ and perhaps _Dolichopithecus_. As a consequence of the limited number of individuals and limited habitats samples, the rarity of Miocene

---

* These deposits contain _Hipparion_, but a more primitive species than the typical Pontian form from Pikermi. The Marceau lignites interfinger with beds containing a Sarmatian (late Miocene) marine mollusc fauna. Consequently, the fauna is considered Miocene by some, Pliocene by others.

Cercopithecoidea might simply be an artifact. It is possible that were our Tertiary cercopithecoid samples from dense, wet lowland forest sites (in the western Congo area, for instance) we might have a very different picture of the evolution of monkeys in Africa.

A further, relevant piece of evidence, suggesting that the apparent rarity of cercopithecoid monkeys in the East African Miocene sites is in fact real (that is, does not result from "collector's bias") is that provided by the lorisoid prosimians from these same beds. The associated fossil postcranials taken along with the behavior and ecology of the living forms indicate that these Miocene lorisoids were clinging and leaping forms which almost certainly occupied a forest habitat and most probably possessed no semi-terrestrial or open-country adaptations at that time. These Miocene lorisoids of East Africa are much less common than the apes but more common than the monkeys. See Simpson (1967) for a tabulation of the material. Many mandibles, maxillae, postcranials, and several partial or nearly complete skulls have been found, totalling between thirty and forty specimen numbers. These lorisoid species are all smaller in absolute size than the two cercopithecoid species so far named from East Africa. Thus, their better preservation suggests that they were living at or near the several sites of deposition of these East African Miocene faunas and that they were there in greater numbers than the monkeys.

Inasmuch as the later Pleistocene African monkey fossils have been (or will be) discussed elsewhere by Freedman (1957, 1963) and Jolly (in press) they will be dealt with only briefly here. A full consideration of the Parapithecinae from the Fayum is also to be published elsewhere (Simons and Pilbeam, mss.) and consequently will not be reviewed in detail here either. This paper will concentrate principally on a discussion of the relationships and distribution of the Miocene and Pliocene monkeys of the Old World.

OLIGOCENE CERCOPITHECOIDEA

Apidium and *Moeripithecus* from the Egyptian Fayum were

suggested by Gregory (1922) and by others as possible ancestors of monkeys or of Oreopithecus. The many new finds of dryopithecine jaws and teeth from our Egyptian Oligocene quarries G and I, belonging to species of both Propliopithecus Schlosser (1910, 1911) and to Aegyptopithecus Simons (1965), show that these animals should undoubtedly be considered dryopithecine pongids. They further show that the unique type mandible of Moeripithecus markgrafi belongs to a pongid species that is nearly intermediate between Propliopithecus haeckeli and Aegyptopithecus but apparently closer to the former. I have consequently referred the genus "Moeripithecus" to junior synonymy under Propliopithecus (Simons, 1968). P. markgrafi is a distinct species but it clearly can no longer be considered a monkey forerunner. The Fayum dryopithecines are not very similar to their contemporaries among the parapithecines (Simons, 1969b) and consequently a common ancestry would have to have been much earlier in time than the Oligocene. In fact the parapithecines are so different from the Egyptian dryopithecines that several scholars (but see principally Hürzeler, 1968) have even challenged the fact that they are primates.

The many new finds of mandibles, maxillae, isolated teeth and a considerable number of isolated postcranial bones which can be referred to the Fayum Parapithecinae, establish the relatedness of these animals to the cercopithecoid monkeys. What is equally clear is that they are not some otherwise unknown group of prosimians -- even in their most primitive features; judging from the numerous postcranials, parapithecines are at least as "simian" as many platyrrhine monkeys. Cranially also these animals have reached the grade of Anthropoids, for they exhibit pre-juvenile (perhaps pre-natal) fusion of the metopic and symphyseal sutures as well as the development of post-orbital plates. In 1967 I discovered part of an Apidium phiomense frontal associated with teeth of this species at Yale Quarry I and a right maxilla with the orbital margin and post-orbital bar was found at Quarry M. These show that the more complete frontal described by Simons (1959) belongs to A. phiomense. Together with a mandible and isolated upper canines and incisors from Quarry I, these two specimens can be made the basis for the composite reconstruction of Figure 1.

To date, four species of parapithecines have been discovered in the Fayum; two of _Apidium_ and two of _Parapithecus_. Species of these two genera resemble each other closely in morphology of the antemolar teeth. They have relatively small canines compared to primates other than the marmosets, which they approximate in size. Like platyrrhines, species of _Apidium_ and _Parapithecus_ possess 3 pairs of premolars above and below. Consequently loss of $P\frac{2}{2}$ in Cercopithecoidea must have occurred independently, and later than it did among ancestral Hominoidea. In molar and mandibular morphology _Apidium_ and _Parapithecus_ species differ distinctly. Relative to tooth size, _Apidium_ horizontal mandibular rami are typically deeper than in _Parapithecus_. In lower molars of both genera the paraconid is reduced to a crest and the $M_{\overline{1-2}}$ are long and block-like as seen from above (rather than nearly circular in outline as are the Fayum pongid $M_{\overline{1}}$ and $M_{\overline{2}}$'s). This is because, in parapithecines, the trigonid and talonid are slightly separated from one another by median internal and external indentations.

Narrow or "waisted molars", are typical of most later Cercopithecoidea. Such indentations are not typical of the vast majority of hominoids but do occur in _Oreopithecus bambolii_ and _Gigantopithecus blacki_. Species of _Apidium_ differ markedly from those of _Parapithecus_ in that there is a clear-cut size increase in the molar series posteriorly. The upper and lower increase in the molars of _Apidium_ are distinctly polycuspidate for a primate while in _Parapithecus_ species, cuspules beyone the basic five are almost never present, and $M_{\overline{3}}$ is always distinctly smaller than $M_{\overline{2}}$. In many cases (just as in the Talapoin) the $M_{\overline{3}}$ of _Parapithecus_ is considerably smaller. In _Parapithecus_ the foveal height of the lower molars is great, and lower molars have four principal cusps (the paraconid and hypoconulid are very low and reduced in size) arranged in a quadrate pattern which is almost identical in form to that of teeth in the smaller species of _Cercopithecus_, or _Miopithecus talapoin_, see Figure 2.

At Quarry I in the Fayum the two most common mammalian species are _Apidium phiomense_ and _Parapithecus sp. nov_. Since

Quarry I definitely represents a river-lain sediment it is tempting to conjecture that the Oligocene parapithecines lived in the canopy along river banks and over swamps just as many small, arboreal, living cercopithecoids such as the Talapoin and Allen's swamp monkey do today.

MIOCENE CERCOPITHECOIDEA

Until quite recently the Miocene monkeys of Africa have remained largely undescribed, unnamed, or misidentified. The principal earlier papers which dealt with them at least in passing, are: Fourtau, 1918; Remane, 1924; MacInnes, 1943; Hooijer, 1963, and Remane, 1965. The work of Van Koenigswald (1969) constitutes a major advance in that this author has now described the Miocene monkeys of East Africa which were found during the nineteen-forties (and before) in association with fossil Hominoidea at the Rusinga, Ombo, and Maboko sites in Kenya. Previously only a brief discussion of these earliest monkeys had been published by MacInnes (1943) who referred a lower right mandibular ramus with $P_{\overline{4}}$ and $M_{\overline{3}}$ from Rusinga Island to cf. *Mesopithecus* sp. but remarked that the form would probably prove generically distinct. In the opinion of Von Koenigswald (1969) generic distinction for this and other specimens, principally from the younger deposits of Maboko Island, is warranted. He proposes two species of a new genus *Victoriapithecus*, a smaller one, *V. macinnesi*, and a larger one, *V. leakeyi*, see Figure 3. The latter species is said to have more advanced and more quadrate upper molars but the differences between the two, evident in Von Koenigswald's illustration, are very slight.

The proximal part of an ulna of appropriate size to belong with *Victoriapithecus leakeyi* is also discussed by Von Koenigswald, Figure 3. He notes that the olecranon process is comparatively longer than in *Mesopithecus* from the Pontian of Greece. Since this ulna is not retroflexed proximally it would indicate, if correctly referred, that *Victoriapithecus* was not to any great degree terrestrially adapted. Although the upper molars of *Victoriapithecus* are quadritubercular and incipiently bilophodont, some of them show remnants of a *crista obliqua* (a protocone-metacone crest). Von Koenigwald believes that this character demonstrates that the upper molar pattern of cercopithecid monkeys was derived from a form with a distinct trigon and hypocone. This may well be the case but since upper molars in nearly all of the thirty to forty Eocene and Oligocene primate genera had distinct trigons and hypocones the upper molar morphology of *Victoriapithecus* provides little new evidence as to the group from which Old World monkeys originated.

The lower and upper molars of Fayum *Parapithecus* are much closer morphologically to *Victoriapithecus* than are those of *Apidium*. The Fayum dryopithecines show no significant morphological resemblance to *Victoriapithecus* in either upper or lower molars. Recent suggestions that Old World monkeys might be derived from forms like *Propliopithecus* and *Aegyptopithecus* are highly improbable. Species of the latter genera are too *Dryopithecus*-like, and the Fayum parapithecines far too monkey-like, to make this a useful suggestion. In the light of the hundreds of new Fayum primates recently recovered, it should be stated clearly that many of the views about the Fayum primates to be found in the literature can be shown to be incorrect. After all, conjectures as to the phyletic positions of the Fayum primates before 1960 were based on only four individual specimens. These conjectures were summarized in 1961 by Kaelin, but in spite of his careful and well illustrated treatment of these four finds he was mistaken in a number of his major conclusions. In considering the question of the ancestry of *Victoriapithecus* then, the many fossils now known must be studied and not the completely outdated literature on African Oligocene primates.

Two other problems worth raising about *Victoriapithecus* are its age and its possible ecologic competitors. Von Koenigswald in naming *Victoriapithecus macinnesi* failed to discuss the fact that the type mandible from Rusinga is probably much older than the rest of the hypodigm. The hypodigm, as a whole, comes mainly from the Maboko Island fauna, now generally regarded as probably a late Miocene equivalent of that found at Fort Ternan, Leakey (1968). Van Couvering and Miller (1969) have recently published evidence that almost all the Rusinga primates come from about the same time period, K/Ar dated at about 18 million years ago. Maboko, if the same age as Fort Ternan, would be some four million years younger. Few chronospecies of mammals have been shown to have a time spread as long as 4 million years. The possibility exists that the Maboko and Rusinga specimens do not belong to the same species.

Stringent competition may perhaps account for the apparent paucity of monkeys in the East African Miocene sites. Possible competitors might have been the somewhat gibbon-like species of *Limnopithecus*. These small apes, contemporary with *Victoriapithecus* are much more common and much better preserved skeletally than the monkeys but are not significantly larger. From the nearly complete skeletal knowledge we have of *Pliopithecus* (*Epipliopithecus*) from Europe and *Limnopithecus* from East Africa it is clear that species of this genus were proportioned rather like some of the more agile arboreal South American monkeys, particularly *Alouatta* and *Ateles*. This does not, however, obviate the possibility of their being near the ancestry of the Southeast Asian gibbon and siamang. There are a number of special common features shared by *Limnopithecus*, *Pliopithecus* and *Hylobates* which are unlikely to be parallelisms, but it is clear that the Miocene forms had not yet acquired the elongate forelimbs and caudal reduction of the modern hylobatines. The locomotor and feeding preferences of *Pliopithecus* and *Limnopithecus* were thus probably not dissimilar to those of a generalized monkey. Relative to the caliber of the humeral shaft and the narrower distal end the humerus of *Limnopithecus* is more elongate than in its similar-sized contemporary *Dryopithecus africanus*: *Limnopithecus* was clearly more arboreally adapted than the latter. Although they were relatively successful in the Miocene, the absence of lesser apes from subsequent African faunas could perhaps be due to their failure in competition with a more successful radiation of arboreal monkeys such as the cercopithecines, which might just possibly be better fitted than apes to feed on fruits (cf. Napier, this volume).

The question of the antiquity of the separation of the two extant subfamilies of Cercopithecidae was touched on recently by Pilbeam and Walker (1968), who have described a small cercopithecine $M\frac{1}{-}$ or $M\frac{2}{-}$ from Napak, Uganda, which is fully quadritubercular and which does not show a crista obliqua. Also from Napak, they report an isolated frontal which in Walker's view is closest morphologically to those of Colobus species. Napak is apparently a little older than Rusinga, with a K/Ar date of 19 ± 2 million years. A separation of these subfamilies prior to this date is inconsistent with the biochemically estimated age of separation of 17 ± 2 million years recently published by Sarich (1968, and this volume)

Hooijer (1963) described a single lower left M3 of a cercopithecid monkey from the Sinda Valley at Ongoliba in the Congo, collected in 1958 by Dr. X. Misonne. He referred to this as "cf Macaca c.q. Mesopithecus spec.". From map 1 of this paper it can be seen that Ongoliba is not very far to the east of the sites which have yielded Victoriapithecus, and moreover the specimen is closely comparable in size and morphology to V. leakeyi. The Ongoliba tooth should probably in future be considered as cf. Victoriapithecus.

Fortau (1918) described two Miocene primate species from the Wadi Moghara local fauna of northern Egypt. These fossils were recovered from ridges running along the northeastern margin of the Qattara Depression about thirty-eight miles southwest of El Alamein. Fourtau mistakenly considered his primate materials as belonging to apes, -- one was said to be allied to the gibbons, the other to Dryopithecus. Recently Simons (1969a) has shown that these two supposed species, Prohylobates tandyi and Dryopithecus mogharensis, are synonyms; of the two, P. tandyi has page priority. The three Wadi Moghara specimens quite clearly belong to a cercopithecid monkey species. Nearly all the "primitive" or non-cercopithecoid features which Fortau and Remane (1924) attributed to this material are unreal or represent mistaken identification of characters. Fortau and several subsequent workers (see Said, 1962) on the basis of invertebrates recovered from these and interfingering marine sediments have considered the Wadi Moghara fauna a Burdigalian equivalent. If this correlation proves correct then Prohylobates tandyi would be the oldest African Miocene cercopithecoid, being probably more than 20 million years old. The inflated cross-section of the superior transverse torus, coupled with the shallow, anteroposteriorly elongate simian shelf in *Prohylobates tandyi* is clearly of the cercopithecid type and closely resembles the same region in *Macaca libyca* from Wadi Natrun, Egypt, figure 4. Comparisons made by me in 1963 between the types of *Victoriapithecus macinnesi* and *Prohylobates tandyi* indicated a very close correspondence in size and in mandibular-dental morphology. *P. tandyi* differs, however, in showing a relatively smaller $M_{\overline{3}}$ talonid and a rather more robust mandibular horizontal ramus. As far as the present evidence goes it does suggest that a generic distinction between *Victoriapithecus* and *Prohylobates* should be maintained. Since the $M_{\overline{3}}$ hypoconulid is poorly developed in *Prohylobates*, *Parapithecus*, and in the possibly 'primitive' cercopithecine genera such as *Miopithecus* and *Allenopithecus*, it is just possible that small talonids are the primitive condition for Cercopithecidae and that the enlargement of this region of $M_{\overline{3}}$ is a parallel acquisition in various subsequent lines. There is certainly ample corroboration from other groups of mammals that $M_{\overline{3}}$ talonid enlargement including acquisition of a third lobe frequently arises independently.

There is little fossil evidence of late Miocene monkeys in Africa. Leakey (1968) reports that from Fort Ternan, Kenya: "At least two different types of "monkey" are represented by isolated teeth ...". These teeth have yet to be described.

Arambourg (1959) has described a cercopithecine monkey species, *Macaca flandrini*, which he considers of latest Miocene age. The estimated age of this fauna again depends on whether all faunas which contain *Hipparion*, as this one does, should be considered Pontian and consequently of Pliocene age. There are, however, *Hipparion* species which are smaller and more primitive than that which occurs in the classic Pontian Pikermi fauna of Greece. Arambourg's fauna from the lignites of Marceau in the department of Oran, Algeria includes such a small and primitive *Hipparion* together with the materials he refers to *Macaca flandrini*. The type of *M. flandrini* consists of a maxillary fragment with $M^{\underline{1}}$ and $M^{\underline{2}}$ and the hypodigm of a mandibular symphysis with two central incisors and the right lateral incisor, three partial mandibles and forty-two isolated teeth. As fossil samples go, this is a reasonably good one and Arambourg's lengthy comparative analysis with other living and fossil cercopithecoids appears to succeed in establishing that *M. flandrini* is correctly referred to genus *Macaca*, or alternatively that later *Macaca* species are conservative in dental morphology. This would then be the most ancient documen tation of an extant genus of cercopithecoid, since the Marceau lignites are probably 9 - 12 million years old. The existence of a species of an existing genus of cercopithecine monkey at such a remove, suggests -- although it certainly does not prove -- that many of the present African cercopithecoid genera could have differentiated by that time. It is well to remember that we are woefully ignorant of African faunas in late Miocene and early Pliocene times. Von Koenigswald's interpretation of *M. flandrini*, admittedly based on study of Arambourg's plates, is that it represents two species.

PLIOCENE CERCOPITHECOIDEA

One important fact which has emerged from the recent studies of Old World fossil monkeys is that there are no fossil finds from Eurasia which antedate the typical Pontian (as represented at Pikermi). This strongly suggests that the radiation of the cercopithecines (at least) out of Africa long postdated that of the Hominoidea which were well established in Europe during the later Vindobonian at an age of about 15 million years, according to Evernden et al (1964). Thus I have to differ strongly with the conclusions presented by Napier (this volume) that the ancestors of modern African cercopithecines once existed outside of Africa. A mean age for the Pontian might well fall under 10 million, perhaps at about 8 million years. One could argue that the Asian colobines, which seem more distinct biochemically from their African allies than do Asian cercopithecines from African cercopithecines, might have left Africa before the Pontian, but through tropical forest corridors unsuited for the deposition or preservation of fossils. Nevertheless, it is equally possible that the stock of which the Asian colobines are the modern representatives had a long history within Africa separate from that which produced *Colobus* species.

The earliest European monkey, and one of the best known fossil monkeys of any age, is Mesopithecus pentelici. Skulls and postcranial bones of this species have been recovered in great numbers from the Pikermi deposits in the vicinity of Athens, Figure 5.

One of the earliest primate species to be discovered, Mesopithecus pentelici was described by Wagner in 1839. In the 1860's Gaudry reviewed the osteology of this species which is known in considerable detail because numerous postcranial bones have been recovered from the Pikermi sites. Several dozen skulls have been found and the species has been reported from a number of other sites, including Salonika; Maragha, Iran; Baltavar, Hungary; near Tiraspol, Russia and in Czechoslovakia. Mesopithecus has generally been assessed as having affinities with the langurs, e.g. by Abel (1931), Piveteau (1957) and Genet-Varcin (1963). Patterson (1954) questioned the colobine affinities of Mesopithecus on the grounds that its association with an open country fauna was implausible for a colobine. Nevertheless, a full anatomical assessment places it clearly with Colobinae. Cranially the structural similarity with Presbytis and Colobus is unmistakable (see Vogel, 1968). Jolly (1967) has remarked that the long phalanges indicated by the indices reproduced by Gabis (1960), together with the low inclination of the olecranon process suggests that "Mesopithecus was at least as arboreal as the vervets (Cercopithecus aethiops)". Napier (this volume) however regards Mesopithecus as "macacoid" on the basis of its limb proportions, including the phalangeal index remarked on by Jolly (loc. cit.). D.R. Pilbeam has recently pointed out to me that although it is fully arboreally adapted, Presbytis entellus does occur in open woodland or wooded savanna, (see Ripley, this volume). Species of Presbytis might even be ultimately derived from Mesopithecus. In view of its abundance and wide distribution at an early date it seems more likely than not that Mesopithecus directly pertains to the Asian radiation of the Colobinae, although for confirmation of its range the material referred to this species from localities other than Pikermi should be reassessed.

Definite records of middle Pliocene monkeys in Europe are

limited to the several isolated teeth which have been assigned to *Presbytis* from deposits of probable Plaisancian age at Monte Casino.

In Asia, the fossil record is hardly more informative. It seems clear that monkeys did not occur in North India or Pakistan during the time of deposition of the Chinji and Nagri faunal zones there. Scores of mammals smaller than monkeys have recently been found in these Siwalik deposits, particularly by the Yale expeditions at Haritalyangar. Consequently, it can no longer be argued that the monkeys were not found because of collecting techniques favoring recovery of large mammals. Pilgrim (1915) discusses two monkey species described on the basis of jaw fragments and isolated teeth from the Dhok Pathan zone near Hasnot in the Salt Range, West Pakistan. These would be of middle Pliocene age. The first, *? Macaca sivalensis*, was described by Lydekker in 1878; the second, *Semnopithecus* (now *Presbytis*) *asnoti*, by Pilgrim in 1910. Pilgrim (1915) adopted the position that this latter species should be transferred to *Cercopithecus*, a genus otherwise unrecorded in Eurasia. Of Pilgrim's three specimens of "*C. asnoti*", (G.S.I. numbers D.120, D.121 and D.182), very little can be said except that any case utilizing them to establish the presence of *Cercopithecus* in Eurasia is incorrect. D.120 consists of a right maxillary fragment with DP^{4} (P^{4} dissected out) and M^{1}; D.121, a left maxilla with DP^{3} and DP^{4}; and lastly D.182, an isolated M^{1}. Thus from all three specimens only M^{1} and P^{4} represent the adult dentition. Although Hill (1966) has continued to treat this species as belonging to *Cercopithecus* it might just as well be a colobine or macaque. In regard to this species, Pilgrim was at his most accurate when he observed: "I may preface my remarks by suggesting the extreme probability that the fragment is generally indeterminable." In referring the species to *Cercopithecus* Pilgrim (1915) appears to have been overly impressed by the presence of a fairly distinct anterior shelf on M^{1} in D.120 and stated that this does not occur in "any species of *Presbytis* know to me." Had he looked no further than the nicely preserved right maxillary fragment of *Presbytis entellus* from the Karnul Caves, Madras in the G.S.I. collections (number F.201a) he would have seen the same condition in a colobine. There is perhaps no better example than this of the now outmoded practice of basing far-reaching conclusions on little more than a single tooth. Even more surprising than Pilgrim's perplexity in 1915 are the fanciful conclusions drawn by Hill (1966) who would derive a colobine directly from *Cercopithecus*:

> "The existence in the Pontian of western Asia of a species pertaining, or at least closely allied to the African Recent *Cercopithecus*, suggest either that *Cercopithecus* emerged there, later migrating to Africa. Alternatively it might be supposed that a common cercopithecoid ancestral form occurred here whence an African migration developed into *Cercopithecus*, the local population becoming converted into *Semnopithecus* and its allies."

Also in 1915, Pilgrim described a fragment of mandibular horizontal ramus (D.184) with $M_{\overline{1}-\overline{2}}$ and tentatively referred it to Lydekker's 1878 species, *Macaca sivalensis*. Lydekker's material consisted of two maxillary fragments (D.2 and D.2a) and to these can be added Y.P.M.19135, an undescribed mandibular fragment with most of $M_{\overline{3}}$ collected for Yale on August 23, 1932 by G. Edward Lewis, illustrated here (Fig. 6). All four of these specimens, as well as the three of *P. asnoti* discussed above come from the Dhok Pathan zone near Hasnot, Pakistan. Considering their common provenance and the absence of any other Pliocene monkey fossils from elsewhere in India, one should raise the question as to whether or not they could all belong to a single species. With the possible exception of D.120, all these Indian fossil monkey jaw fragments are smaller than comparable parts of modern species of macaques or of *Mesopithecus pentelici* typically are. The slight foveal height of the lower molars and the high, sharply set off cusps of Y.P.M.19135, which is hardly worn, suggest *Presbytis* or one of the other Asian colobines. Moreover, $M_{\overline{3}}$ of Y.P.M.19135 has a distinct hypoconulid, a feature never seen in cercopithecines with teeth as small as those of the Hasnot monkeys. Without studying all the originals little more can be said about these middle Pliocene monkeys of Asia, but it is clear that they need comparative restudy. Should this species or both species belong to *Presbytis*, then it emerges from this study that there is no evidence of cercopithecine monkeys in Eurasia before the late Pliocene. In addition there is no fossil evidence from Eurasia to show that any member of that subfamily other than *Macaca* (*sensu lato*) ever got outside the African-Arabian land mass. The justification for this conclusion is also indicated by the following discussion on *Dolichopithecus* and related forms.

In 1890 Depéret described a species of monkey which he called *Dolichopithecus ruscinensis* from late Pliocene deposits near Perpignan, France. There are nearly complete skulls of both male and female individuals as well as partial or complete humerus, femur, radius, ulna and some phalanges. Depéret considered this species a colobine but showed that it was somewhat atypical for that group in having short limbs and a rather long face. Recently Jolly (1967) has discussed these postcranials and has concluded that they are macaque-like. Presumably these resemblances are parallelisms toward the cercopithecines since the skull and dentition indicate that this species is a colobine. As Jolly remarks:

> "The rather short, broad nasals, the expansion of the gonial angle of the mandible, and the high molar cusps and small incisors support Depéret's conclusion that this was a colobine monkey."

Hornbeck and Swindler (1967) have pointed out the differences w which exist in morphology of $\overline{P_4}$ between *Colobus* and *Papio*, and in this context also it seems that *Dolichopithecus ruscinensis* resembles *Colobus*.

In 1929, Depéret described a nearly complete female skull from Villafranchian deposits near Seneze, France, which he considered a second species of *Dolichopithecus*, *D. arvernensis*. This is the cranium most often figured as *Dolichopithecus* but actually this species does not belong with the type in the same genus or even subfamily, as both Jolly (1967) and Vogel (1967) have pointed out. "*D.*" *arvernensis* is actually a large terrestrially adapted macaque. Figure 7 illustrates its similarity to large species of *Macaca*. The species should be transferred to genus *Procynocephalus* Schlosser 1924, as should also *Cynocephalus subhimalayanus* from "the hills near to the Sutlej" according to Baker and

Durand (1836) and presumably of Pleistocene age (Fig. 8). But this find, an almost entire right maxilla with parts of C, P^3, and P^4-M^3 intact, and another smaller mandible described by Lydekker (1886) as *Cynocephalus falconeri*, have often been referred to as documenting the occurrence of *Papio* in Asia but this is not the case. From figures and the casts at Yale it seems apparent to me that "*D*." *arvenensis* and "*C*." *subhimalaynus* are very similar indeed, as was also noted by Jolly (Fig. 8). Both types of these species are females and both are similar in size of comparable parts to the largest modern *Papio* females. Both species should be associated with and probably referred to *Procynocephalus wimani* Schlosser (1924) the postcranial skeleton of which indicates a terrestrially adapted form. Necrasov et al (1961) have described a similar fossil from Rumania as *Paradolichopithecus geticus*, which like "*D*." *arvernensis* is large and has the comparatively enlarged incisors typical of cercopithecines. This is probably also a *Procynocephalus*.

Jolly (1967) concluded:

> "The available evidence therefore suggests that *Procynocephalus* was an offshoot of the Asiatic Cercocebini, which took to the ground, probably some time during the Pliocene, and became extinct at the end of the Lower Pleistocene."

These various Pleistocene species or races of *Procynocephalus* resemble in turn the Celebes black ape *Macaca* (*Cynopithecus*) *nigra* but whether this resemblance is convergent due to their similar size or shows special relatedness is not clear. On present evidence it seems more likely that *M. nigra* is an independent and local Celebes product in regard to size, standing at the upper end of a clinal distribution of regional species and subspecies (Fooden, 1969).

Several species of *Macaca* (*sensu stricto*) have been described from Eurasian deposits but, in general, these are either very fragmentary or of no great antiquity. Consequently they have little phylogenetic significance. Gervais (1859) gave the name *Macaca prisca* to a mandibular fragment from the late Pliocene of Montpellier, France. *M. prisca* has also been reported from

Astian age deposits at Gundersheim and Csarnota, see Viret (1954). During Early and Middle Pleistocene times, species of Macaca have been reported from far-flung localities in Eurasia and north Africa but not to my knowledge in sub-Saharan Africa. In Europe, the Florentine Macaque, M. florentina, was described by Cocchi (1872) and its occurrence has been reported from Villafranca d'Asti and Val d'Arno, Italy; Tegelen, Holland and Saint-Vallier and Seneze in France. Macaca pliocaenica Owen 1845 has been reported from Middle Pleistocene deposits in England near Greys Thurrock and West Runton in East Anglia.

Another species, described by Hedinger (1892), M. suevicus occurs in Middle Pleistocene deposits at Wurttemburg, Germany. Fossil macaques closely resembling the living magot or Barbary ape have been recovered from caves in Algeria, and in 1892 Harle described the species Macaca tolosana from the cave of Montsaunes near Toulouse. According to Arambourg (1959) this specimen does not appear to differ from Macaca sylvana. Kurten (1968) is of the opinion that Macaca florentina is directly ancestral to M. sylvana and that M. pliocaenica is a synonym of the latter.

A species of Macaca, M. (= Aulaxinuus) libyca was described by Stromer (1920) from the late Pliocene of Wadi Natrun, Egypt, a site from which he had earlier (1913) described a new genus and species of primate, Libypithecus markgrafi, which was based on a nearly complete skull upper jaw fragment. This cranium appears to show most affinities with the colobines and Jolly (1967) has suggested that it may be conspecific or at least congeneric with Depéret's species Dolichopithecus ruscinensis from the late Pliocene of France. The two are compared in Fig. 9.

It thus appears that in Pliocene-Pleistocene times in the Mediterranean region there were only three principal lineages of cercopithecoid monkeys, a macaque-like colobine Dolichopithecus (=? Libypithecus), a lineage leading to the modern Barbary ape, and the large, terrestrially adapted macaque Procynocephalus wimani of which "Dolichopithecus" arvernensis may be a synonym, and which apparently became extinct at the end of the early Pleistocene.

In concluding this consideration of European fossil monkeys one form that is not a monkey deserves mention. Haupt (1935) described a broken first lower molar from the Pontian deposits near Eppelsheim and Wissberg in Rhenish Hessia as *Semnopithecus* (now *Presbytis*) *eppelsheimensis*. Hurzeler (1954) transferred this species to *?Pliopithecus* and, in 1956, von Koenigswald pointed out that this was a dryopithecine and not a monkey tooth and proposed for it the prabably invalid genus *Rhenopithecus*. It seems more probable to me that this tooth should be assigned to *Dryopithecus laietanus*.

Von Koenigswald (1969) is of the opinion that monkeys probably did not reach China and the Far East before the latest Pliocene. He never located a single tooth in the enormous series of Chinese Pontian fossils which he examined in drug stores that sell fossils. Schlosser (1924) named a species *Macaca andersoni* from a probable Villafranchian equivalent in Honan, China and another species *M. robusta*, which does not differ significantly from the former, has been described from the deposits at Chou-Kou-Tien near Pekin, China. Jolly (1967) is of the opinion that Lydekker's *Cynocephalus falconeri* may be conspecific with *M. andersoni*. If so, *M. falconeri* has priority.

AFRICAN PLEISTOCENE CERCOPITHECOIDEA

Inasmuch as the African Pleistocene monkeys occur too late in time to tell us much about the phylogeny of the cercopithecids and are currently under study by Jolly, Howell, Richard Leakey and others, only a brief review can be attempted here.

The species *Simopithecus oswaldi* was described by Andrews (1916) on the basis of a mandible recovered at Homa Mountain, west of L. Victoria. Since that time very extensive further finds of this and related species have been made in East and South Africa and these new discoveries include plentiful postcranial material. These finds are currently being written up by Jolly who now believes (pers. comm.) that species of *Simopithecus* should be referred to the same genus as the extant gelada baboon of Ethiopia: *Theropithecus*. Species of this genus (*sensu lato*) are the most ground-adapted of all the Old World monkeys. In fact most of their present range lies in the treeless high plateau or "high semyen" of Ethiopia. During the Pleistocene several large-bodied species or races of "Simopithecus" evolved. At Olorgesailie in East Africa, and elsewhere, forms as large as chimpanzees are known. Jolly has demonstrated that two correlates of this size-increase are that together with increasing body bulk there is a decrease in the relative size of the anterior dentition and a distinct increase in the length of the forelimbs so as to shift backward the centre of gravity of the larger animal (see Jolly, this volume). Both trends appear to correlate with the marked development of a terrestrial, grass-feeding adaptation. Jolly (1967) remarks:

> "The fact that Simopithecus is so common in lakeside deposits, but one of the rarer cercopithecoids in cave or subaerial sites suggests that it may have been dependent upon the waterside habitat, with its rich ground vegetation, to sustain its large bulk. There are signs in the Olduvai material that humans preyed upon Simopithecus at the time of the Chellean culture, and that by the time of the Acheulian of Olorgesailie it was extremely hunted, most of the animals taken being juveniles. It is possible that the evolution of relatively advanced human hunters, who preyed directly upon Simopithecus and competed with it for the waterside habitat, contributed to its extinction at the end of the Middle Pleistocene."

Recently a lower molar of "Simopithecus", showing fully the distinctive cusp arrangement and enamel foldings has been identified in the Harvard University collections from Lothagam. Since the level of this find dates to more than 4 million years it shows that at least by that time some of the distinctive dental adaptations of Theropithecus had arisen. This in turn permits an assessment of the limitations for determining times of divergence by means of the I.D. distance calculated by Sarich (this volume) who reports "No antisera have as yet shown any difference between the albumens of Papio and Theropithecus ....." although these two are said to differ by about 5 units from the forest dwelling drill and mandrill. If Theropithecus was distinct from Papio which also occurs in the

Pliocene at Lothagam, then the common ancestry of the East and South African savanna baboons and the West African forest baboons must have been much earlier. On anatomical and behavioral grounds, Jolly (1967) would place Papio and Theropithecus in different cercopithecine tribes, Cercocebini and Theropithecini respectively but he has since abandoned this view, because of his belief that the Tribe Cercopithecini is distinctly more divergent from either of the two above groups than they are from each other. Also the reduction of the number of contained genera, facilitates submergence of Theropithecini in the Cercocebini. This, then, leaves a 42-chromosome group, the Cercocebini, and the Cercopithecini with higher chromosome numbers. Nevertheless mandibles and postcranials of the two different genera, Papio/Parapapio and Theropithecus (= Simopithecus) are contemporaneous in the fossil record of East and South Africa throughout the Pleistocene and before indicating a divergence which could hardly be less than 6 - 8 million years. If Sarich can find no I.D. differences between these two genera to reflect their millions of years of measurable separation, then the immunochemical measures arrived at by this method are clearly not precise enough to extrapolate times of phyletic divergence with any accuracy. Naturally, it would be nice to know meaningful dates of divergence arrived at by measuring I.D. differences, and perhaps someday we will. Meanwhile paleontologists will continue to find interest in the actual fossils themselves, their dating, their anatomy, adaptations and ecology. Just as with human history, the most interesting things about the history of fossils are what happened, and how.

The remainder of extinct genera of Pleistocene cercopithecoids have been based on fossils recovered at the South African sites which have yielded Australopithecus such as Sterkfontein, Swartkrans, Taung and Kromdraai or at nearby sites of similar age. These include the type species of Parapapio, Jones 1937; Dinopithecus, Broom 1937; Gorgopithecus, Broom and Robinson 1949; Cercopithecoides Mollett 1947 as well as a number of extinct species of Papio. These fossil monkeys are fully dealt with in a series of papers by Freedman (1957, 1960, 1961 and 1965) and they are also discussed in some detail as they relate to east African finds by Jolly (1967). Of these monkeys, Freedman believes that Dinopithecus and Gorgopithecus are related to Theropithecus, while the fossil species of Papio and Parapapio were clearly closer to the modern south African baboons. Cercopithecoides on the other hand is fairly certainly a colobine of quite large size. Recent geological explorations in the Baringo Basin of central Kenya, under the direction of W.W. Bishop, have succeeded in establishing a succession of deposits of Pliocene age. From the basal part of the Chemeron beds there, at locality J.M. 91, a skull and nearly complete skeleton, which R. Leakey (1969) has recently described as Paracolobus baringensis, has been recovered. The further study of this postcranial skeleton should be of great interest because its skeletal adaptations can be compared with modern monkeys as to locomotor type. In addition to this find at Baringo, the recent collections in the Omo basin of Ethiopia made under the direction of Clark Howell of the University of Chicago and from the Paris Museum by Coppens have yielded numerous monkey skulls and skeletal parts. These finds all date from between a little less than 2 to about 3.5 million years, B.P. In the Omo sites also, Papio-like forms and "Simopithecus are distinct and other smaller cercopithecines occur as well. At both Lothagam and Kanapoi in Northern Kenya monkeys occur, and Patterson has recently (1968) described a specimen of Parapapio from Kanapoi. Kanapoi is probably about four million years old, and (as mentioned above) the Lothagam tooth of Theropithecus probably somewhat older. When all the monkeys from these sites have been fully studied it is clear that we will know much about the final stages of cercopithecoid evolution in Africa.

ACKNOWLEDGEMENTS

Certain aspects of the research reported here were made possible by grants from the Wenner-Gren Foundation for Anthropological Research, New York, by the National Science Foundation (grants G-18102, GP-433, GA-723) and Smithsonian Foreign Currency grants 5 and 23.

I also wish to thank P.C.Ettel, I.M.Tattersall, D.R.Pilbeam, and David Roberts for assistance in the compilation of this paper, and Clifford Jolly for valuable discussion of the topic.

The photographs for the figures, except for figures 3 and 5, were taken by A.Coleman of the Yale Peabody Museum.

REFERENCES

Abel, O. (1931). Die Stellung des Menschen im Rahmen der Wirbeltiere. Jena: Verlag von Gustav Fischer.

Andrews, C.W. (1916). Note on a new baboon (Simopithecus oswaldi; gen. et sp. nov.) from the (?) Pliocene of British East Africa. Ann. Mag. Nat. Hist. 18 (8): 410-419.

Arambourg, C. (1959). Vertébrés continentaux du Miocène Supérieure de l'Afrique du Nord. Publ. Serv. Geol. Algérie, Pal. 4: 1-159.

Baker, W.E. and H.M. Durand. (1836). Sub-himalayn Fossil Remains of the Dadoopoor Collection. Jour. Asiatic Soc. 5: 739.

Broom, R. (1937). On some new Pleistocene mammals from limestone caves of the Transvaal. S. Afr. Jour. Sci. 33: 750-768.

Broom, R. and J.T. Robinson. (1949). A new type of fossil baboon. Gorgopithecus major. Proc. zool. Soc. Lond. 119: 379-386.

Clark, W.E. Le Gros, and L.S.B. Leakey. (1951). The Miocene Hominoidea of East Africa. Fossil Mammals of Africa No. 1. London: British Museum (Nat. Hist.).

Cocchi, I. (1872). Su di due Scimmie fossili italiane. Boll. R. Com. Geol. Ital. 3: 55-71.

Depéret, C. (1897). Les Animaux pliocènes du Roussillon. Mém. Soc. Géol. France (Pal) 3: 1-164.

Depéret, C. (1929). Dolichopithecus arvernensis Depéret. Nouveau Singe du Pliocène supérieur de Sénèze (Haute-Loire). Trav. Lab. Géol. Fac. Sc. Lyon. Mém. 12: 5-12.

Evernden, J.F., D.E. Savage, A.E. Curtis, and J.T. James. (1964). Potassium-argon dates and the Cenozoic mammalian chronology of North America. Amer. Jour. Sci. **262**: 145-189.

Falconer, H. and P.T. Cautley. (1868). Palaeontological Memoirs and Notes. Vol. 1. Fauna antiqua sivalensis. London: Hardwicke.

Fooden, J. (1969). Taxonomy and evolution of the monkeys of Celebes (Primates: Cercopithecidae). Bibliotheca Primat. **10**: 1-148.

Fourtau, R. (1918). Contribution à l'étude des Vertébrés miocènes de l'Egypt. Cairo: Egypt. Surv. Dept.

Freedman, L. (1957). The fossil Cercopithecoidea of South Africa. Ann. Transvaal Mus. **23**: 121-262.

Freedman, L. (1960). New cercopithecoid fossils, including a new species from Taung, Cape Province, South Africa. Ann. S. Afr. Mus. **46**: 1-14.

Freedman, L. (1960). Some new fossil cercopithecoid specimens from Makapansgat, South Africa. Pal. Africana **7**: 7-45.

Freedman, L. (1965). Fossil and subfossil Primates from the limestone deposits at Taung, Bolts Farm and Witkrans, South Africa. Pal. Africana **9**: 19-48.

Gabis, R.V. (1960). Les os des membres des singes cynomorphes. Mammalia **24**: 577-607.

Gaudry, A. (1862). Animaux fossiles et Géologie de l'Attique. Paris: F. Savoy (atlas 1862-1867).

Genet-Varcin, E. (1963). Les singes actuels et fossiles. Paris: Boubée.

Gervais, P. (1859). Zoologie et Paléontologie Français. 2e édit. Paris.

Gregory, W.K. (1922). The Origin and Evolution of the Human Dentition. J.Dent.Res.New York 2: 89-175; 357-415; 607-717; 3: 881.

Harlé, E. (1892). Un mandibule de Singe du repaire d'Hyènes de Montsaunès (Haute Garonne). Bull.Gre.Hist.Nat.de Toulouse. 26: 9-11.

Haupt, O. (1935). Andere Wirbeltiere des Neozoikums. in Salomon-Calvi, Oberrheinischer Fossilkatalog Leifg. 4 Hefte 9: 1-103.

Hedinger, A. (1892). Über fossile Affen. Jaresh.Ver.Naturk. Wutt. 68: 114-115.

Hill, W.C.O. (1953-1966). Primates, Comparative Anatomy and Taxonomy. Vols. I - VI. Edinburgh: University Press. New York: Interscience Publications.

Hooijer, D.A. (1963). Miocene Mammalia of Congo (with a chapter by A.Gautier and J.Lepersonne). Ann.Mus.Royal de l'Afrique Centrale, Tervuren, Belgique (Sci.Géol.) 46: 1-77.

Hornbeck, F.V. and D.R. Swindler. (1967). Morphology of the Lower Fourth Premolar of Certain Cercopithecidae. Jour. Dent.Res. 46: 979-983.

Hürzeler, J. (1954). Contribution à l'odontologie et à la phylogénèse du genre Pliopithecus Gervais. Ann.Pal. 40: 1-48.

Hürzeler, J. (1958). Oreopithecus bambolii Gervais. A preliminary report. Verb.Naturf.Ges.Basel 69: 1-48.

Hürzeler, J. (1968). Questions et reflexions sur l'histoire des Anthromorphes. Annales de Pal. 54: 195-233.

Jolly, C.J. (1967). Evolution of Baboons. In The Baboon in Medical Research. Vol. 2. Austin: Univ. of Texas Press.

Jolly, C.J. (1970). The seed-eaters: a new model of hominid differentiation based on a baboon analogy. Man N.S. 5: 5-26. (in press) The classification and natural history of Simopithecus Andrews (= Theropithecus, Geoffroy). Bull. Brit. Mus. Nat. Hist. (Geol. Ser.)

Jones, T.R. (1937). A new fossil primate from Sterkfontein, Krugersdorp, Transvaal. S. Afr. Jour. Sci. 33: 709-728.

Kaelin, J. (1961). Sur les Primates de L'Oligocène Inférieur d'Egypte. Ann. Pal. 48: 1-8.

Kaelin, J. (1962). Über Moeripithecus markgrafi Schlosser, und die phylogenetischen Vorstufen der Bilophodonties der Cercopithecoidea. Bibl. Primat. 1: 32-42.

von Koenigswald, G.H.R. (1956). Gabissreste von Menschenaffen aus dem unter-Pliozen Rheinhessens, 1. Konmkl. Nederl. Lekad. von Wetenschappen, Amsterdam Proc. Ser. B. 59(4): 318-334.

von Koenigswald, G.H.R. (1969). Miocene Cercopithecoidea and Oreopithecoidea from the Miocene of East Africa. In Fossil vertebrates of Africa. Vol. 2. London: Academic Press.

Kurtén, B. (1968). Pleistocene Mammals of Europe. Chicago: Aldine Publ. Co.

Leakey, L.S.B. (1968). Notes on the mammalian faunas from the Miocene and Pleistocene of E. Africa. In Background to Evolution in Africa. (Eds. W.W. Bishop and Desmond Clark). Chicago: University Press.

Leakey, R.E.F. (1969). New Cercopithecidae from the Chemeron Beds of Lake Baringo, Kenya. In Fossil vertebrates of Africa. Vol. 1. London: Academic Press.

Lydekker, R. (1878). Notices of Siwalik Mammals. Rec.Geol. Surv.India 11: 66.

Lydekker, R. (1886). Indian Tertiary and Post Tertiary Vertebrata. Pal.Indica Ser. 10.

MacInnes, D.G. (1943). Notes on the East African Primates. Jour.East Afr.and Uganda Nat.Hist.Soc. 17: 141-181.

Mollet, O.D. van der Spuy. (1947). Fossil mammals from the Makapan Valley, Potgietersrust. 1. Primates. S.Afr.Jour. Sci. 43: 295-303.

Necrasov, et al. (1961). Sur un nouveau singe catarrhinien fossile découvert dans un mid fossilifère d'Oltenie (R.P.R.). Annalele Stiintifice Ale Universitatii Al.1.Cuza, Iasi (Jassy), sect.11, stiinte nat. 7: 401-416.

Owen, R. (1845). Notice sur la découverte, faite en Angleterre, de restes fossiles d'un quadrumane du genre macaque, dans une formation d'eau douce appartenant au nouveau pliocène. C.R.Acad.Sci. Paris 21: 573-575.

Patterson, B. (1954). The Geologic History of Non-hominid Primates in the Old World. Human Biology 26: 191-209.

Patterson, B. (1968). The extinct Baboon Parapapio jonesi in the early Pleistocene of Northeastern Kenya. Mus.Com.Zool. 282: 1-4.

Pilgrim, G.E. (1910). Notices of new mammalian genera and species from the Tertiaries of India. Rec.Geol.Surv.India 40: 63-71.

Pilgrim, G.E. (1915). New Siwalik Primates and their bearing on the question of the evolution of man and the Anthropoidea. Rec.Geol.Surv.India 45: 1-74.

Pilbeam, D.R. and A.C. Walker. (1968). Fossil Monkeys from the Miocene of Napak, North-East Uganda. Nature Lond. 220: 657-660.

Piveteau, J. (1957). Primates Paléontologie Humaine: Traité de Paléontologie VII. Paris: Masson et Cie.

Remane, A. (1924). "Einige Bemerkungen über Prohylobates tandyi R. Fourtau und Dryopithecus mogharensis R. Fourtau". Cbl. Min. Geol. Pal. 220-224.

Remane, A. (1965). Die Geschichte der Menschenaffen. In Mensliche Abstammungslehre: Fortschritte der Anthropogenie 1863-1964. herausgegeben von Gerhard Heberer, Gottingen.

Said, R. (1962). The Geology of Egypt. Amsterdam: Elsevier.

Sarich, V.M. (1968). Quantitative immunochemistry and the evolution of Old World Primates. In Taxonomy and Phylogeny of Old World Primates with references to the origin of man. (Ed. Chiarelli, B.). Turino: Rosenberg and Sellier.

Schlosser, M. (1910). Über einige fossile Säugetiere aus dem Ologocän von Aegypten. Zool. Anz. 35: 500-508.

Schlosser, M. (1911). Beiträge zur Kenntnis der Oligozänen Landsäugetiere aus dem Fayum. Beiträge Pal. Geol. Oesterr. -Ung. 24: 51-167.

Schlosser, M. (1924). Fossil primates from China. Pal. sinica (C) (2): 1-14.

Simons, E.L. (1959). An anthropoid frontal bone from the Fayum, Oligocene of Egypt: the oldest skull fragment of a higher primate. Amer. Mus. Novitiates 1976: 1-16.

Simons, E.L. (1960) Apidium and Oreopithecus. Nature Lond. 186: 824-826.

Simons, E.L. (1963). A critical reappraisal of Tertiary Primates. In Genetic and Evolutionary Biology of the Primates. (Ed. Buettner-Janusch, J.) New York: Academic Press.

Simons, E.L. (1965). New fossil Apes from Egypt and the initial differentiation of Hominoidea. Nature Lond. 205: 135-139.

Simons, E.L. (1968). Hunting the "Dawn Apes" of Africa Discovery 4: 19-32.

Simons, E.L. (1969a). Miocene Monkey (Prohylobates) from Northern Egypt. Nature Lond. 223: 687-689.

Simons, E.L. (1969b). Origin and radiation of the Primates. Ann.N.Y.Acad.Sci. 167: 319-331.

Simons, E.L. and D.R. Pilbeam. (1965). Preliminary revision of Dryopithecinae (Pongidae, Anthropoidea). Folia primatologica 3: 81-152.

Simons, E.L. and D.R. Pilbeam. (in prep.). Early Cenozoic Mammalian Faunas Fayum Province. Egypt. Part III. The African Oligocene Primates. Bull.Peabody Mus. New Haven: Yale Univ. Press.

Simpson, G.G. (1967). The Tertiary Lorisiform Primates of Africa. Bull.Mus.Comp.Zool.Harvard. 136: 39-61.

Stromer, E. (1913). Mitteilungen über die wirbeltierreste aus dem Mittelpliocän des Natronales (Aegypten). 1 Affen. Zeits deutsch.geol.Ges. 65: 350-372.

Stromer, E. (1920). Miteilungen über Wirbeltierreste aus dem Mittelpliocän des Natronales (Aegypten). 5. Nachtrag zu 1 Affen. Sitz.-Ber.bayer.Akad.Wiss. 15: 345-370.

Van Couvering, J.A. and J.A. Miller. (1969). Miocene stratigraphy and age determinations, Rusinga Island, Kenya. Nature Lond. 221: 268-632.

Verheyen, W.N. (1962). Contribution à la craniologie comparée des Primates. Musée Roy. Afrique Centrale, Tervuren, Belgique. Ser. 8. Sci. Zool. 105: 1-256.

Viret, J. (1954). Le Loess à bancs durcis de Saint-Vallier (Drôme) et sa faune de Mammifères villafranciens. Nouv. Arch. Mus. Hist. Nat. Lyon. 4: 1-200.

Vogel, C. (1968). In Taxonomy and Phylogeny of Old World Primates With References to the Origin of Man. (Ed. Chiarelli, B.). Torino: Rosenberg and Sellier.

Wagner, A. (1839). "Fossile Überrests von einem Affenschädel und andern Säugetieren aus Griechenland"; Gelehrte Anzeigen von dem Bayerisches Akademie der Wissenschaften in Munich 8: 305-312.

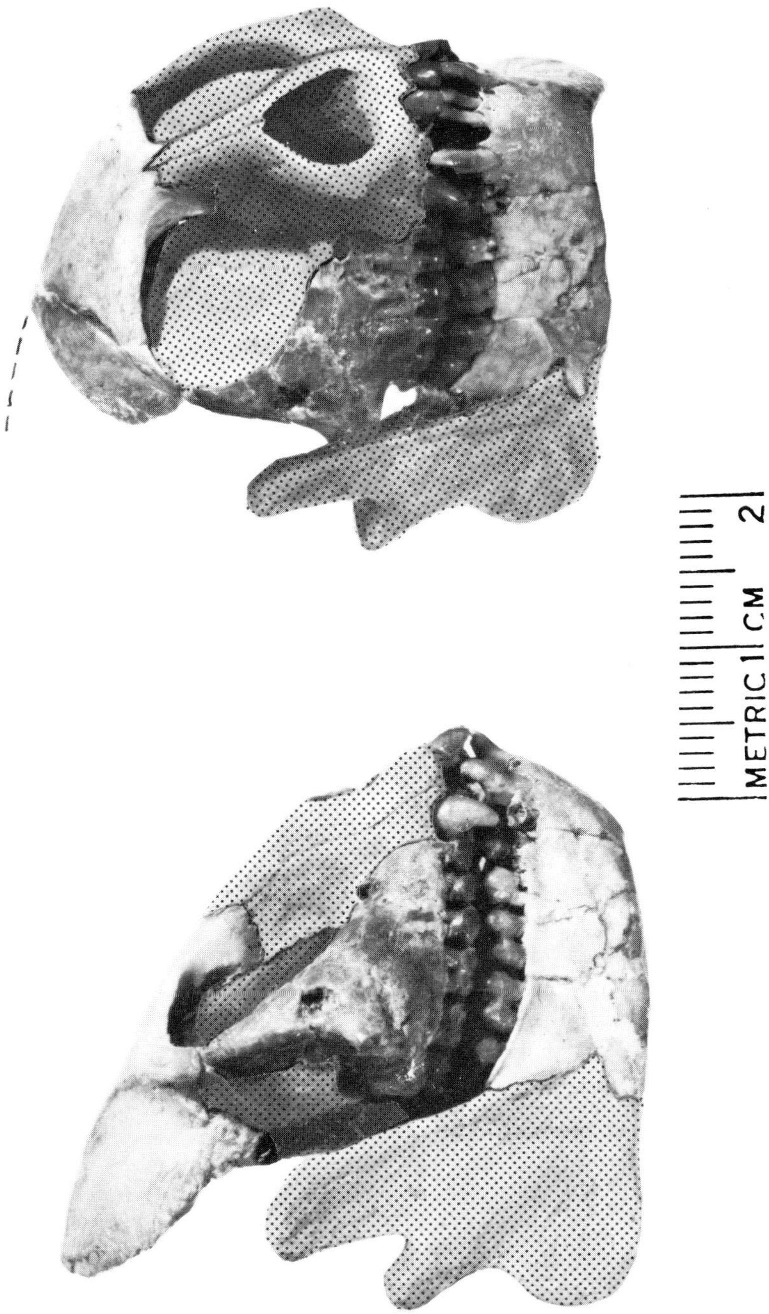

Figure 1. Composite reconstruction of the face of *Apidium phiominse* based on separate finds; A.M.N.H. 14556 (frontal), C.G.M. 26920 (maxilla), Y.P.M. 21018 (mandible), and isolated upper canines and incisors associated in Quarry I. Stippled areas reconstructed from other specimens.

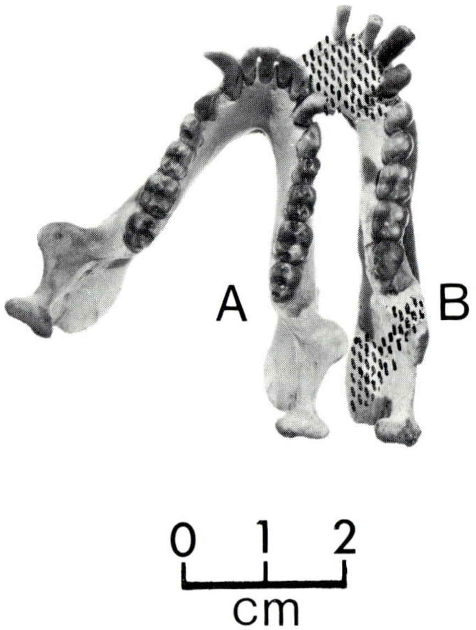

Figure 2. Comparison between mandibles of the modern African swamp monkey, Miopithecus talapoin (A) and a new species of Parapithecus from the Fayum (B). The four anterior teeth of B are from the same site but do not belong to the same individual. In both these catarrhines there are four principal lower molar cusps and in the Oligocene species the hypoconulids are so reduced and flattened that they do not receive significant wear except in extreme dental age.

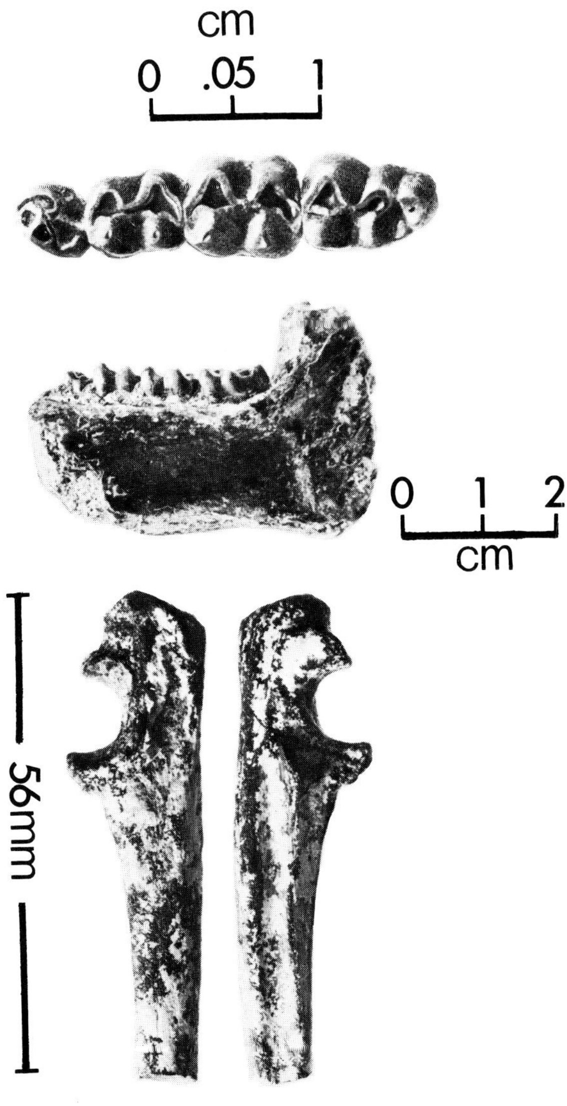

Figure 3. Victoriapithecus from Miocene sites on Rusinga and Maboko Islands, Lake Victoria-Nyanzae, Kenya, East Africa. A-detail of cheek-teeth, B-internal view of the most complete known jaw from Rusinga Island, C-Proximal part of an ulna referred by von Koenigswald to Victoriapithecus. This is the only known post-cranial fragment of a Miocene Old World monkey. Figure from von Koenigswald (1969).

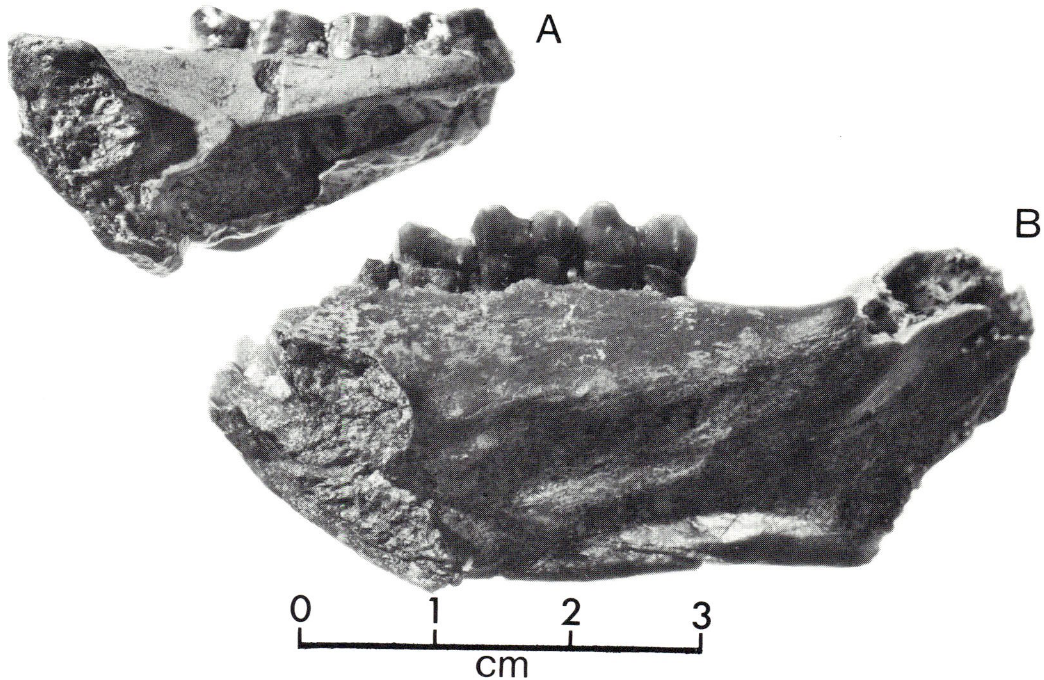

Figure 4. Comparison of mandibular fragments two Egyptian fossil monkeys: A - Prohylobates tandyi from Miocene deposits at Wadi Moghara south of El Alamein; B - Macaca libyca from a lacustrine site of Pliocene age at Wadi Natrun north west of Cairo. Both show the characteristic cercopithecid symphysis in cross-section (a bulbous superior torus, above a shallow, elongate "simian shelf").

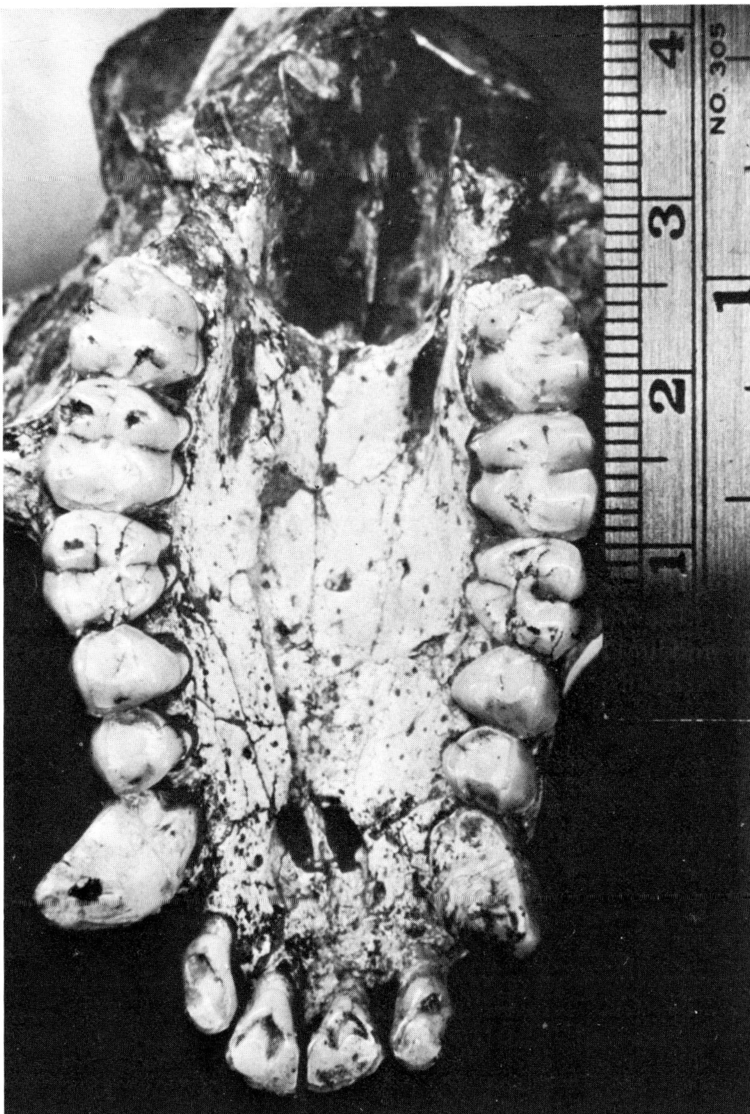

Figure 5. Palate with upper dentition of Mesopithecus pentelici from the Pikermi fissure fills of early Pliocene age, near Athens, Greece (Photo by courtesy of E. Delson).

Figure 6. The Yale specimen of Presbytis? asnoti (left $M_{\overline{3}}$)from early or middle Pliocene deposits near Dhok Pathan deposits near Hasnot, Pakistan.

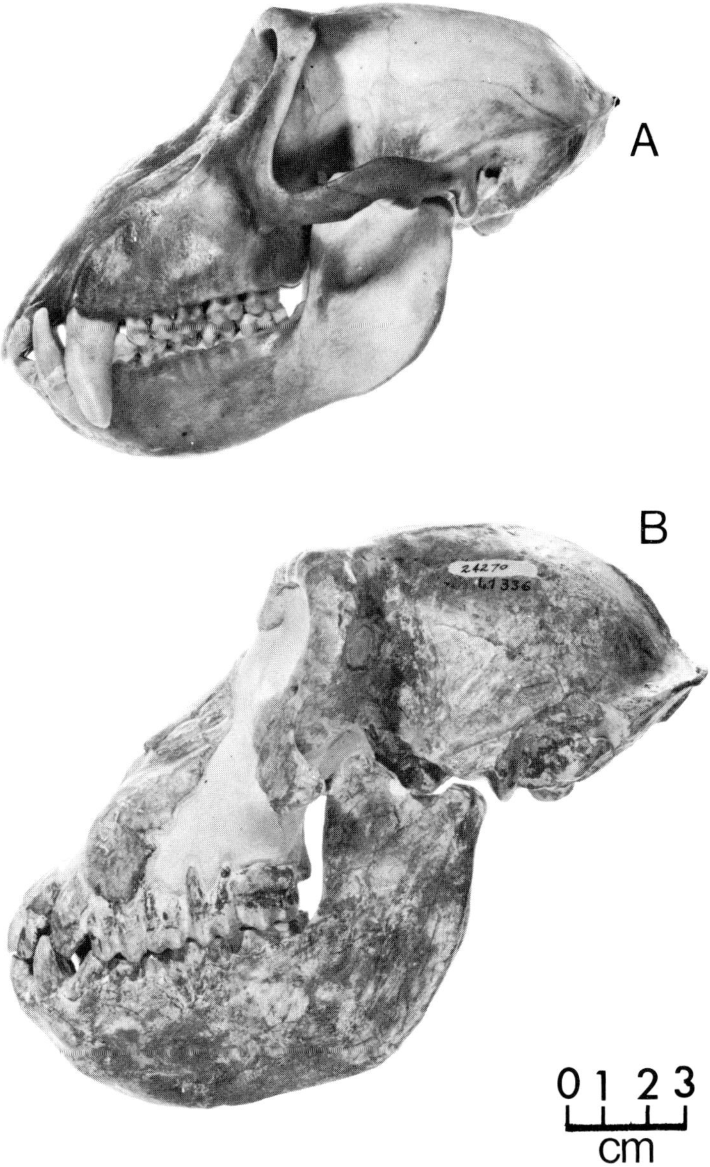

Figure 7. Comparison of (A) *Macaca nemestrina* male, A.M.N.H. 35045 with (B) the most complete skull, probably female, of *Procynocephalus*? *arvernensis* from the vicinity of Senèze, France. Although widely figured as an example of genus *Dolichopithecus* the Auvergne monkey does not at all resemble the type species of that genus which appears to be a colobine, rather the Auvergne monkey is apparently a species of large, terrestrially adapted macaque. Morphology of the snout is not baboon-like, but resembles macaques.

Figure 8. Right maxilla of *Procynocephalus? subhimalayanus* (A) the first fossil monkey ever described, by Baker and Durand (1833), with (B) the Auvergne macaque both of early Pleistocene age.

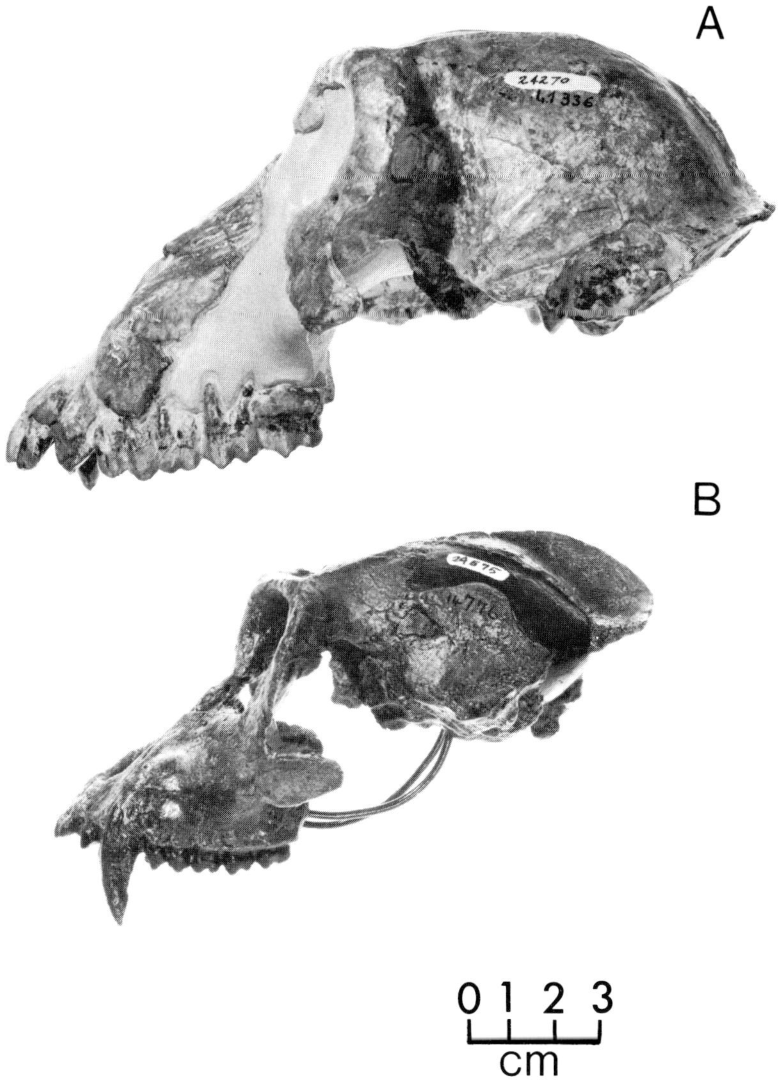

Figure 9. A. The Auvergne macaque *Procynocephalus*? (not *Dolichopithecus*) *arvernensis* compared with B. The type of *Libypithecus markgrafi* from the Pliocene deposits at Wadi Natron north west of Cairo, Egypt. *Libypithecus* may be a junior synonym for *Dolichopithecus* since it closely resembles the type species of the latter *D. ruscinensis*. *Procynocephalus* is a large, terrestrially adapted macaque, perhaps inseparable generically from *Macaca*. Both *Libypithecus markgrafi* and *Dolichopithecus ruscinensis* appear to be ground living colobines. They may not be separable generically.

THE LARGE AFRICAN MONKEYS AS AN ADAPTIVE ARRAY

C. J. Jolly

INTRODUCTION

The three groups of large African Cercopithecine monkeys (the drills, the common baboons, and the gelada) have been variously classified by different authorities. On the one hand, older works such as those of Lydekker (1893), impressed by the superficial similarity of the three, placed them in a single genus *Cynocephalus* (now *Papio*). This view has recently been defended by Buettner-Janusch (1966), who further suggests that each group includes only one species. At the other extreme, generic separation for each group, with separation at the level of tribe between the gelada and the others has been proposed (Jolly, 1966; Hill, 1967). Compromise schemes include that of Fiedler (1956), who classifies common baboons and drills as separate subgenera of a single genus, but retains full generic status for the gelada, and Dandelot (1969) who reduces all three to subgeneric status within *Papio*.

In its concern to produce a classification that is both practically useful and consistent with the facts of evolution, modern taxonomic practice take into account both overall similarity and propinquity of common ancestry of the organisms to be classified (see Fig. 1). The end product of an aberrant and fast-evolving lineage of comparatively recent origin may thus be separated at a higher taxonomic level from its more conservative relatives than these are from one another, even though it may share a common ancestor with one such relative to the exclusion of the others (Simpson, 1963). On the other hand, if morphological similarity were allowed to take complete precedence over community of descent in classification, then taxonomy would lose its phylogenetic referent and return to its pre-Darwinian state. This is recognized in the taxonomic usage which prohibits the formation of a single taxon from units derived phyletically from two or more taxa of equal or higher rank, however morphologically similar the members of the groups have become. The amount of weight to be given to morphological similarity and propinquity of divergence in classifying organisms is, with the limits imposed by this usage, determined by the judgement of the classifier. Thus, as Simpson has emphasized (1961), more than one classificatory scheme may be compatible with a particular conception of the phyletic interrelationships of the organisms to be classified. A case in point among the Cercopithecinae is the classification of the patas monkey and the guenons. Most of the evidence seems to point to a relatively recent divergence of the patas from a species of *Cercopithecus*, probably subsequent to the divergence of the ancestral stocks of the living members of this genus (Sarich, this volume). Since the divergence, the patas has become adapted to open-country life with striking changes in its postcranial anatomy. Those classifiers who are more impressed by its general similarity to, and obvious close common ancestry with, the guenons will put it in *Cercopithecus*, while those wishing to emphasize its divergence will give it full generic status as *Erythrocebus*. Both schemes are quite compatible with the proposed phyletic interrelationships between patas and guenons, and with taxonomic usage.

In making taxonomic classifications, therefore, we are to a large extent expressing opinions about the configuration of the phyletic 'tree' upon which the organisms lie. A phyletic tree is itself essentially a model representing the events of speciation which are believed to have occurred in the history of the group being studied, and the relative positions of such events in time. The essential dimension of such a tree is the 'vertical', chronological one on which time is expressed either in purely arbitrary units, or geological units, or 'absolute' dates. Although the horizontal dimension can be used to represent in a general way the amount of divergence that has occurred since speciation, such divergence is of course multidimensional, with different features evolving at different rates, and therefore cannot be adequately represented as divergence along a single dimension. To construct such a model, we again examine the morphological differences and similarities between taxa, but this time selectively, with a view to extracting and weighing their phyletic significance, rather than to expressing degree of morphological difference between the groups. The two separate pathways by which the data of comparative anatomy may be used to construct taxonomic classifications are shown in Fig. 1.

The dotted horizontal line represents a common but, it is suggested, erroneous pathway by which the topography of phyletic

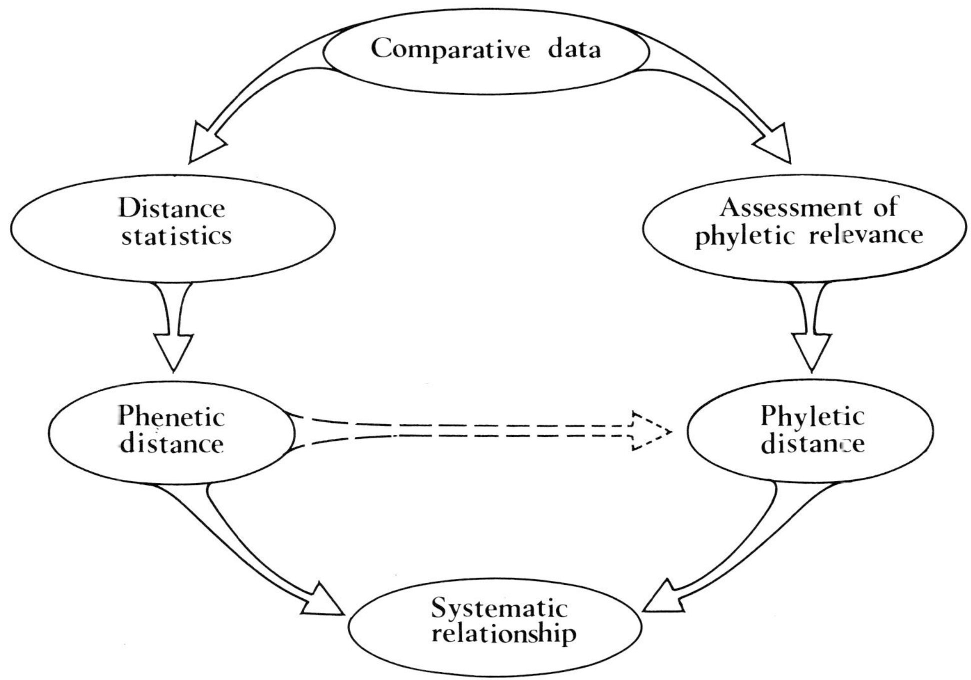

Fig. 1 The use of data from comparative anatomy in systematics.

trees is inferred from data whose phyletic significance has not been assessed (see, for example, some papers in Chiarelli, 1968).

When they are being used to construct phyletic trees, the comparative data are perhaps best organized as 'characters', that is, attributes of homologous structures in the organisms being compared. The structures may range from the whole organism, or even populations of organisms (as when social organization is being discussed) down to single polypeptide chains of protein molecules. Each character is usually most clearly expressed in terms of a contrast with an alternative condition; "upper molars four-, not five-cusped" is a character of comparison between Cercopithecidae and Hylobatidae. In this case, the character is a discordant one, occurring in the first group but not the second (+/-). In the case of a comparison between, say, *Colobus* and *Papio*, the character would have been concordant (+/+). In a comparison between *Pan* and *Hylobates* the character would also be concordant, but this time (-/-). The rules by which the phyletic significance of a character is assessed in a particular case are generally accepted but seldom formulated. Either discordance or concordance with respect to a given character can be of phyletic significance and it would seem that in general terms the phyletic significance is inversely proportional to the extent to which their occurrence can be accounted for in terms of function in the contemporary natural history of the animal concerned. To take the example of four- and five-cusped molars in Catarrhines; there is no factor in the present-day diet of *Papio* and *Colobus* on the one hand, and *Pan* and *Hylobates* on the other which can explain the concordance of the members of each pair in this character, and the discordance between the pairs. The diet of *Papio* is much closer to that of *Pan* than it is to that of *Colobus*. We can therefore infer that the concordance is likely to be due to heritage characters within each group, and hence that *Papio* shares a common ancestor with *Colobus*, and *Pan* with *Hylobates*, more recently than either of the monkeys with either of the apes. Of course, such single concordances might well be due to chance rather than either to function or heritage; the phyletic tree which shows *Papio* and *Colobus* diverging later than the common stocks of monkeys and apes is the product of many parallel observations.

Hence, it would be true to say that the most valid judgements about the shape of phyletic trees are to be made on the basis of non-adaptive characters. This is not, of course, the same as saying that judgements should be based only or even mainly upon structures that are redundant or vestigial (molar teeth are obviously neither). Nor is it to say that such characters as the difference between the number and disposition of the molar cusps in Hominoidea and Cercopithecoidea did not have adaptive value at the time of initial divergence of the species from which the two groups sprang. In fact, it may plausibly be suggested that the Cercopithecoid molar was originally a leaf-eater's modification of a five-cusped Hominoid-like tooth (Jolly, 1965, 1966).

Thus, as a general rule we may state that concordances which cannot be plausibly interpreted on the grounds of function of the structure in question are good indications of common ancestry, while similarly inexplicable discordances are evidence of separate ancestry extending back beyond the entry of one of both of the organisms into their present adaptive zone.

Conversely, either concordances or discordances that can be plausibly explained in terms of the current ecology of the organisms concerned carry little or no phyletic weight. The characters in question may or may not be heritage characters of long standing, but this cannot be inferred from the data of comparative anatomy. As an example, we may take the character of elongation of the forearm which is common to Pongidae, Hylobatidae, *Oreopithecus*, and the lemuroid *Palaeopropithecus*, all of which are, or were, brachiators. If concordances of this type are used to support assignment to the same taxon (as they are in classificatory schemes which unite gibbons and great apes in a single family excluding Hominids) then it is on the strength of general resemblance (that is, the left-hand pathway of Fig. 1), not because they indicate phyletic relationship (the right-hand pathway).

Thus, phyletic models built up on the basis of the comparative anatomy of the terminal products of evolutionary lines, though involving an element of judgement, are far from being dependent solely on the subjective whim of the classifier, as some proponents of numerical taxonomy would maintain.

The study of fossils is often said to yield more direct evidence for the reconstruction of phyletic trees than the use of comparative anatomical techniques. It is certainly true that the fossil record is ultimately the only means of relating cladogenic events to a scale of geological of chronometric time. This is true even should 'protein clocks' prove reliable; these must be calibrated against the geological timescale through the fossil evidence. So far as their position on the phyletic tree is concerned, however, fossils must be evaluated in precisely the same way as contemporary animals; the characters of each specimen must be critically examined and interpretations tested against the current model, or models, of the phyletic tree. Should they prove incompatible, the specimen is reassessed and, if necessary, the model revised.

An additional factor enters into the phyletic evaluation of organisms removed from one another in time. This arises from the assumption (accepted, presumably, by the majority of contemporary biologists) that the fauna of each age is in a general sense the descendant of that of the preceding age and that gaps in the fossil record should not be assumed without good evidence to correspond to episodes of catastrophic extinction. This being so, the known fossils from any one period are more or less likely to include the ancestors of forms from the succeeding period, the likelihood varying in proportion to the completeness of the record.

The plausibility of accounting for resemblances between forms in successive faunas by parallelism or convergence rather than relationship by descent therefore depends not only upon the factors to be considered for contemporary faunas, but also upon the likelihood of the earlier leaving no recognizable descendant, and the latter having no recognizable ancestor in the fossil record.

Two points emerge from this consideration of what makes a character useful in phyletic reconstruction. First, the assessment of a character is largely dependent upon judgement of what is 'reasonable', so that the model phyletic trees constructed as working hypotheses are based not upon proof but upon plausibility. This is not to say, of course, that they are untestable; they are tested by their ability to accommodate data other than that from which they were originally constructed; new fossil finds, for instance, or the results of comparative studies upon hitherto unexamined anatomical systems in living forms. There is plenty of room for debate between protagonists of different phyletic schemes and such debate and the careful reinterpretation of characters are equally as important as measuring and describing differences. Such arguments are, of course, distinct from those over the relative merits of alternative classification as expressions of an agreed phyletic model.

Second, the understanding of the functional importance or lack of it of a character that unites or distinguishes two taxa is absolutely fundamental to understanding its phyletic relevance at that level. Without a knowledge of the one, the other cannot be estimated.

The hypothesis to be developed in this paper is that most of the major discordances between the large African monkeys can best be understood as aspects of a single functional complex adaptive to their present (or recent) ways of life.

THE THREE LARGE MONKEYS

A. NOMENCLATURE.

Throughout this paper, separate generic status is maintained for each of the three large monkeys. The generic name for the gelada is unequivocally *Theropithecus*. Following a recent revision (Jolly, 1970) the extinct species hitherto generally referred to *Simopithecus* are put in the same genus. The mandrill and drill are here referred to *Mandrillus* Ritgen rather than to the strictly valid *Papio* Müller; *Papio* Erxleben is than available for the long-tailed, common baboons, rather than *Chaeropithecus* Blainville. Retention of this nomenclature in the face of the Rules has been defended by Hill (1967) on the ground of familiarity and long usage. The hamadryas baboon is not accorded separate generic or subgeneric status.

B. DISTRIBUTION AND INTERNAL DIVERSITY

Mandrillus is confined to the west-central area of the equatorial forest. Two forms are distinguishable: Mandrillus sphinx (the mandrill) and M. leucophaeus (the drill). As their ranges seem to overlap broadly, as they are clearly distinct morphologically, and as no intermediates have ever been described, they are unanimously accorded full specific status except by Buettner-Janusch (1966). The drill has the wider range from the R. Cross in south-east Nigeria in the north to Gabon in the south, including the island of Fernando Po. Most museum records come from an area extending inland about 150 miles from the coast and are concentrated towards the northern end of the range. All the mandrill localities lie south of the R. Sanaga in Cameroon, extending south into Rio Muni. At two localities in Kribi District, Cameroon both mandrill and drill were collected in one case by the same collector (Powell-Cotton). Mandrillus has yet to be recognized in the fossil record.

Theropithecus is at present represented by a single species T. gelada, confined to montane grasslands in Ethiopa. From the fossil records, however, it is clear that this restricted range must be regarded as comparatively recent. In deposits dating from 50,000 years ago and more the genus is found from South Africa to Ethiopia and, in fact, probably extended throughout non-forested regions of Africa south of the Sahara, and perhaps into North Africa also. T. gelada is the smallest of the known species, and differs from the fossil, lowland species chiefly in some details of the construction of the nasal region, perhaps as an adaptation to life at high altitudes. The fossil evidence is at present insufficient to demonstrate synchronous speciation within the genus, although as in living baboon species there was considerable variation between local populations in size and size correlated characters. There is some evidence for the evolutionary elaboration of some of the Theropithecus specialities during the course of the Pleistocene, from the earliest known forms at Makapan (T. darti) to T. oswaldi leakeyi from Olduvai IV and T. gelada. Evolved forms of Theropithecus, range in size from T. gelada, slightly smaller than most living forms of Papio, to T. oswaldi leakeyi, in which the males were probably the size of a gorilla.

Papio is currently the most widespread of the three, its total range being many times larger than those of the other two combined. It occupies all areas of trans-Sahara Africa with the exception of waterless deserts, areas of flat, treeless grassland, and parts of the equatorial rain forest belt (Tappen, 1960; Jolly, 1965; Hall, 1966). The internal taxonomy of Papio has yet to be satisfactorily resolved. Five rather distinct geographically replacing forms can be distinguished and are usually called species, for example, by Hill (1967). All but one of these show internal diversity in size or pelage characters. As Tappen (1960) indicated, the status of these will be satisfactorily settled only when the border areas have been investigated in the field as well as by museum studies, a process that can be immensely time consuming (see e.g. Maples and McKern, 1967). The evidence to date will be presented elsewhere (Jolly and Groves in preparation). The indications are that none of the five is completely isolated reproductively from all its neighbors, but that intermediate populations are not found in all areas of apparently contiguous distribution, indicating the operation of some form of isolating mechanism. In at least one case, reproductive isolation appears to exist in some areas between forms which elsewhere show primary intergradation.

C. NATURAL HISTORY

Until the work of Gartlan (this volume) very little seems to have been added to our knowledge of the natural history of Mandrillus since Tappen's summary of the available evidence. Drills and mandrills are confined to rain forests and are believed to be forest floor animals for the most part, feeding upon fruits, tubers and insects.

The natural history of Theropithecus gelada was virtually unknown until the recent studies of Crook (1966, 1967) and Crook and Aldrich-Blake (1968), although earlier reports indicated that it might be very similar to that of Papio. The recent work, however, indicates significant differences between the habitat utilization of the two genera, particularly in the areas where both occur. *T. gelada* is confined to a region of high plateaux dissected by steep gorges and valleys. The sides of the gorges are used as sleeping roosts at night and refuges during the day. Most feeding takes place on the plateaux above the gorges, a treeless grassland, and at all seasons the blades, rhizomes and seeds of grasses predominate in the diet together with small bulbs of herbaceous plants, leaves of shrubs picked by reaching from the ground and arthropods from under stones. Food objects are mainly small and scattered more or less evenly on the ground. Gelada foraging therefore includes long intervals of sitting, while the area of ground within reach is systematically worked over for food objects which are picked up using a thumb-index grip and collected in the hand until a mouthful has been accumulated. The sitting position frees both hands for simultaneous food- collection, and hence is a most important adaptation for a diet in which each individual food-object is relatively small. Most of each day is spent picking up and masticating tough and soil-covered food, relatively little in travelling or seeking feeding-areas. The daily range is consequently small. Geladas sometimes enter the woodland at the base of the cliffs but were never observed to climb trees, to exploit the foods to be found there or to take shelter from rain. They do, however, feed upon the stones (not the fruits) of wild olives, gathered from the ground.

The data available on the habitat of the lowland, Pleistocene *Theropithecus* are consistent with a similar diet (Jolly, 1967, 1970). All known specimens are from areas which today have a seasonal rainfall and are outside the rain forest. The fauna in each case indicates that Pleistocene conditions were also non-rain forest, ranging from the bushveld of Basal Pleistocene Makapan (Brain, 1958) to the 'lush vegetation' of Olduvai during Bed II times. The most consistent feature in the ecology of Pleistocene *Theropithecus* is its association with indicators of waterside, especially lakeside, conditions. (Jolly, 1970). Under seasonal rainfall conditions shallow lakes like Olduvai and Olorgesailie would fluctuate widely in extent producing flats where the seasonal flooding prevents the growth of trees while promoting that of herbaceous vegetation, especially grasses; areas like the *dambos* of Zambia (Darling,

1960), the edaphic grasslands described by Richards (1952) and the surroundings of Lake Amboseli today. Such areas may indeed have been the only pure grasslands within the present lowland savanna zone before the modification of the indigenous woodlands and deciduous forest by overburning, overgrazing and cultivation by man (Richards, 1952; Hopkins, 1965), mainly after the introduction of agriculture. According to this view, it is a mistake to see Theropithecus, as a genus, as a 'montane baboon' (cf. Buettner-Janusch). It is, in fact, a grassland baboon, one species of which has survived in high altitude grasslands after the virtual extermination of the genus. Possibly it owes its survival to the gorges which afford refuge from human predation. The large, slow moving Theropithecus of the Pleistocene lowlands was a major prey of the early humans who also frequented lake edge sites and it seems very probable that this was a major factor in its extinction.

Papio is today the only large monkey over the greater part of its range which includes a variety of habitats from tropical rain forest to semi-desert scrub, from vegetational zones where trees produce most of the potential food to those dominated by grasses and other herbs. It is therefore to be expected that a wide variety of foods would be exploited, and that great differences in diet should exist between populations of different regions. Groups living in forests feed largely on tree-products such as wild figs. The Nairobi National Park baboons reported by Washburn and DeVore are largely dependent upon grasses together with tree-products. The reported range of dietary variation within the genus has been reviewed by Jolly (1965) and Hall (1966). One of the most outstanding features of the genus is its dietary catholicity and adaptability. Small and large food objects, tree-products and ground-products, are all utilized at the present. The current lack of competition for Papio must however be regarded as a fairly new phenomenon, dating from the virtual extinction of Theropithecus in the Upper Pleistocene, since the two genera occur together in sites in both South and East Africa. At the early South African sites at least three genera now extinct also occur. Since the Papio of Makapan differ little from those of the present day, we must assume that the genus evolved to its present form under conditions where its econiche was far more restricted by competition from other large monkeys, especially Theropithecus. To reconstruct the Pleistocene econiche of Papio it therefore becomes very important to examine the habitat utilization practised by the two genera in that small area of Ethiopia where they are still partially sympatric. Fortunately, this has recently been reported in outline by Crook and Aldrich-Blake (1968). In this region, Papio anubis lives in the woodlands below the cliffs occupied by the geladas. It climbs freely in trees and a high proportion of its foods are tree- or shrub-products. Very significantly, grass is not among the plants listed. Most of its foods come as large items requiring preliminary preparation before chewing, especially removal of an outer husk, biting into hunks, gnawing from a twig or branch, or, in the case of the flesh of a succulent leaf or fruit, nibbling from the outer rind. The fact that its food occurs as large but relatively dispersed items means that less time is spent in food collection and mastication, but more in travelling between feeding places. In contrast to the habitual sitting of the gelada, the anubis baboons more often adopt a tripedal stance when feeding on the ground, the large food-object being grasped in one hand while the other supports the weight of the forequarters. Papio hamadryas, a 'semi-desert' form, also ranges occasionally into gelada country. While its habitat probably indicates a greater exploitation of ground resources than with the woodland dwelling P. anubis, its diet includes tree-products, and the sedentary grass feeding habit of the gelada has not been observed.

From these observations, it might be suggested that the original adaptation of Papio was to the exploitation of both tree- and ground-products in about equal proportions, a way of life best fitted to deciduous forests, woodlands and wooded savannas, rather than to treeless grasslands.

Here again we must constantly bear in mind the drastic deforestation that is believed to have affected much of tropical Africa under the relatively recent influence of man. Apart from the edaphic grasslands most of what is today treeless grasslands or tree savanna would have been savanna woodland, deciduous forest or even evergreen rain forest under preagricultural conditions. The impact of man has been especially great upon the seasonal rainfall vegetation types since these can be burned during the dry season and hence modified without tree felling. Much of the present day habitat of Papio is covered with fire climax vegetation including far fewer tress than the climatic climax type. Unfortunately, most of the studies of baboon ecology and social organization have been carried out in just such deforested areas because the open habitat makes it easier to observe the animals. This has led to the myth, now firmly entrenched in the popular anthropological literature, that Papio and even Macaca are 'open country' animals which rarely climb trees. Even under present day conditions with no competition from other large monkeys over most of their range, populations of Papio reach much higher densities in woodland and groups foraging into open country usually include areas of woodland or riparian forest among their 'core areas'. Of the species of Papio, only Papio hamadryas is predominantly an open country, cliff-roosting from, and as we shall see is apparently acquiring some gelada-like adaptive features as a consequence.

Pleistocene Papio has a wide distribution in Africa. (The supposed occurrence of the genus in Asia is based upon Procynocephalus Schlosser, a baboon-sized animal probably more closely related to Macaca than to Papio.) Like Theropithecus it occurs in sites which must have been under seasonal rainfall conditions durring the Pleistocene as today, but there is a distinct indication that Papio avoided the open, treeless flats surrounding playa lakes where Theropithecus was at home, preferring the woodlands beyond the floodline (Jolly, 1970).

The exploitation of seasonal woodlands and deciduous forests differs from that of both open grassland and evergreen rain forest in that it requires versatility of feeding behavior, to exploit both arboreal and terrestrial, large-object and small-object food resources according to the time of year. We might expect to find that species adapted to such an environment would be behaviorally flexible and structurally unspecialized, 'all-purpose' animals, intermediate in many respects between their rain forest and open country relatives. Such adaptability would enable Papio (like the the African Cercopithecus aethiops and the rhesus and bonnet macaques of Asia) to exploit the new conditions brought about by human activity, and hence it would be the best candidate for survival in the new, relatively impoverished ecosystem. At the same time, its ecological center of gravity would shift in favor of ground- rather than tree-products, with the spread of anthropogenic savannas at the expense of woodland and deciduous forest, and the virtual extinction of *Theropithecus*.

The exploitation of the major African habitats by the large monkeys at the present time is shown diagrammatically in Fig. 2 and compared with a reconstructed scheme for the Pleistocene before the emergence of *Homo sapiens*. Other large Cercopithecidae (*Cercopithecoides*, *Gorgopithecus* and *Dinopithecus*) were also present during the early Pleistocene in seasonal habitats, but their econiches cannot be reconstructed on present evidence. Their presence is likely to have limited the breadth of the econiches of both *Theropithecus* and *Papio*.

DISTINGUISHING CHARACTERS AND THEIR INTERPRETATION

The comparative material available effectively limits comparisons to characters of the skull, dentition, extremities and external characters, all of which have in the past been used to distinguish the three genera.

In both *Papio* and *Theropithecus* there is considerable variation in size between adults of the same sex of different populations, as well as distinct sexual dimorphism in size and in the development of such structures as the canine teeth and temporal musculature. Differences in size between populations and sexes are associated with differences in proportion and shape that are themselves allometrically determined. In *Mandrillus* there is little interpopulational difference in size between adults of the same sex (although the mandrill is somewhat larger than the drill) but differences in shape correlated with sexual size dimorphism are very striking. It is therefore more satisfactory where possible to express the differences between genera in terms of developmental and allomorphic trends rather than of simple dimensions or indices. This applies especially to the shape of the skull.

1. THE DENTITION

The dentition may be considered in functional terms as consisting of three major regions. The first two include teeth whose function is primarily digestive; the incisal region which is concerned with preliminary preparation of food-objects, and the cheek-teeth which are concerned with finer mastication. The third region includes the canine and the anterior part of the third lower molar, and is primarily concerned with fighting, not mastication.

The three genera differ in the relative development of the two masticatory dental groups, and in the structure of the teeth in each group. *Mandrillus* has broad and high-crowned incisors, while the cheek-teeth are comparatively small and have low, rounded cusps. In spite of their large unworn size, the incisors wear down very rapidly in life, being reduced to stumps, and often lost in adults. Wear in the cheek-teeth is pronounced, but less rapid than in the incisors.

In *Theropithecus*, on the contrary, the incisors are relatively small, slender teeth, but the cheek-teeth are large and high-crowned, and have a distinctive structure with high, rather columnar cusps, large anterior and posterior cingular ridges, and deep and often complex infoldings of enamel from the buccal and lingual surfaces. Either because wear is unusually rapid, or because of delayed eruption of the third molars, or both, the more anterior molars, especially the first, are usually severely worn before the third is fully erupted.

In *Papio* the structure and proportions of the masticatory teeth are intermediate between those of the other two genera. The cheek-teeth tend to be more high-crowned and angular than those of *Mandrillus*, but less so than those of *Theropithecus*, while the relative size of the incisors is intermediate. Because of the rapid wear of the incisors in *Mandrillus* and *Papio*, it is difficult to illustrate the incisal dominance in their dentition from the dimensions of the teeth themselves. In Fig. 3, the length of the upper molar row, measured mesio-distally between the mesial face of M^1 and the distal face of M^3, is plotted against incisal breadth of the maxilla, measured straight across at the alveolar level, in males of the three genera. Almost all the mandrills and drills (which are not differentiated in the diagram) have an incisal breadth greater than molar row length. Three samples of *Papio* males are shown on the plot; the filled circles represent the smallest (*Papio cynocephalus kindae*) of the savanna baboons, and one of the largest, the gray-footed chacma, currently known as *Papio ursinus griseipes*. The open circles represent *Papio hamadryas* males. As will be seen, the majority of *Papio* individuals fall between the 100 and the 75 percent lines, and there is little or no indication of a size-related trend in proportions. It is, however, noticeable that the hamadryas tend to cluster towards the low-index side of the *Papio* distribution, all but one lying below the dashed line connecting the mean-points of the two savanna baboon samples. All the *Theropithecus* males, represented by crosses, have relatively narrower incisors than any of the *Papio* individuals. The isolated individual in which the incisal breadth is less than 50 percent of the molar row length is a male, from the Pleistocene site of Kanjera, assigned to *Theropithecus oswaldi* (Jolly, 1970).

It is not clear, on present evidence, whether the broad incisal region of *Mandrillus* is simply an accommodation for large incisor teeth. While this would seem to be a common-sense explanation, the maximum breadth is not developed until full adulthood, by which time the incisors have been greatly reduced in size by wear, and the great incisal breadth is filled mainly by the lateral incisors, which lean very strongly medially. This medial inclination is less marked in *Papio* and not seen in *Theropithecus*. Thus the appearance is rather of the teeth being displaced to fill an expanding incisal region, rather than vice versa, and perhaps the explanation for the great incisal breadth is rather to be found in a high degree of alveolar stressing during ontogeny. Whichever explanation is correct, it emphasizes the importance of incisal function in mandrills and drills, and its progressively lesser importance beside cheek-tooth function in *Papio* and, more especially, in *Theropithecus*.

In both *Mandrillus* and *Theropithecus* the larger forms show their distinctive dental proportions in more exaggerated form. In *Mandrillus* the males have greater incisal breadth than the females, but their molars are scarcely larger. In *Theropithecus* the opposite trend is even more striking, with the largest Pleistocene forms (*Theropithecus oswaldi mariae* and *T. o. leakeyi*) having no greater incisal breadth than *T. gelada*, although they were animals about the size of a female gorilla, and had, of course, very much larger cheek-teeth than the gelada (Jolly, 1970). In *Papio*, the condition is again intermediate, incisors and cheek-teeth increasing in proportion with an increase in general size.

The characters of the masticatory dentition which distinguish the three genera are comprehensible on the grounds of diet; that of *Theropithecus* consists mainly of small objects that require little or no incisal preparation, but do need prolonged chewing between the cheek-teeth, thus favoring distribution of dental material towards the posterior rather than the anterior dental region. The high crowns of the cheek-teeth obviously prolong the period of functional effectiveness of the dentition, while the deep infoldings of enamel and rather columnar shape of the cusps maintains a functional chewing surface composed of alternating bands of enamel and dentine until an advanced stage of wear. Such a surface provides a self-sharpening mincer adapted to chopping up grass blades, and is seen in a variety of grass-eating animals as diverse as elephants and voles. In the forest, on the other hand, the great bulk of feed resources (apart from leaves) consists of forest fruits and tubers, in both of which the edible portion is enclosed in a rind or husk which must be removed by incisal action. In the case of tubers, the outer rind is also soil-covered, hastening the process of incisal attrition. Once freed of the outer rind, however, such foods require relatively little further processing by the cheek-teeth beyond a simple crushing which can best be provided by blunt, rounded molar cusps. Such a diet would therefore favor the development of the incisor-dominant dentition seen in *Mandrillus*. *Papio* in both structure and diet is intermediate.

The situation with the canines and P_3 is complicated by the occurrence of strong sexual dimorphism in the development of of these 'weapon' teeth. Within each sex, however, there is evidence that the relative size of the canine parallels that of the incisors. Relatively the largest canines are found in males of *Mandrillus*, and the smallest in the big, Pleistocene *Theropithecus*. There is also some evidence that the relative size of the canines in *Theropithecus* males decreased in the course of the Pleistocene (Jolly, 1970). The difference in the size of the male canines between the three genera is difficult in terms of the function of the canines themselves; for instance it does not support the hypothesis that the canines will be larger in open-country forms where there is greater need for protection of the group against predators. Possibly the reduction of the canines in *Theropithecus* is merely a secondary effect of incisal reduction, either by genetic pleiotropism (that is, the genes responsible for incisal reduction have a secondary effect upon canine size), or as an adaption to the reduced size of the anterior muzzle, itself a consequence of incisal reduction. Of the two mechanisms, the second seems the more likely since in the early *Theropithecus* from Makapan, incisal reduction was already fully developed while the canines were still large; the canines became gradually reduced during the course of the Pleistocene.

2. MASTICATORY MUSCULATURE AND CRANIAL SHAPE

The chief determinant of the striking differences between the three genera in the shape of the skull seems to be the orientation of the major mass of the temporal muscles. In *Theropithecus* the temporals are set well forward. Their line of action is more or less perpendicular to the occlusal plane of the cheek-tooth, and almost parallel (in the sagittal plane) to that of the pterygoids and masseters.

Thus, the vertical component of their action is large, and the horizontal small. This is an economical arrangement for bringing the power of the temporals to bear with the masseters and pterygoids upon the task of side-to-side chewing of small, tough objects between the cheek-teeth. The forward situation of the temporals is directly responsible for various features of the cranial bones, and can therefore be recognized in fossil as well as extant *Theropithecus*.

One of the most striking indices of the large size and anterior position of the temporal musculature is the narrow postorbital construction. Fig. 5 shows the minimum intertemporal dimension plotted against mastoid breadth in a series of Papio and Theropithecus skulls. The stippled area represents the distribution of 323 adult male savanna baboons of all major subgroups. The open circles represent adult males of P. hamadryas which again show a tendency towards the Theropithecus condition. Theropithecus males are represented by crosses, and their distribution falls well apart from that of the Papio series. The single larger specimen is again the Kanjera male of T. oswaldi. Mandrillus skulls have not been included on the plot; their distribution is about the same as that of savanna baboons, but because of the development of temporal flanges in adult males the measurement of minimum temporal breadth is not really comparable in this group.

Other indications of the anterior situation of the temporals in Theropithecus are the strong ridge or crest in the temporal fossa which marks its anterior margin, the fact that the temporal lines in adult males tend to meet well forward on the cranial vault, and the wide angle between the zygomatic arch and the lateral pillar of the orbit.

In males, which develop a sagittal crest when adult, this crest often decreases in height as it approaches the inion, indicating that the bulk of the muscle lies further forward. In the larger Theropithecus, the forward position of the temporals is maintained in spite of an allometric increase in facial (nasion-prosthion) length. Consequently, the ascending ramus of the mandible becomes tall (though still vertical), and the posterior maxilla deep.

The condition in Mandrillus contrasts to that in Theropithecus in that the mass of the temporal muscle is set obliquely, with a strong part of the muscle originating in the region of the inion. The line of action of the muscle is consequently almost perpendicular to that of the 'chewing' muscles, and the horizontal component in its pull is large. The muscle fibres are mostly long, enabling large objects to be grasped in the wide gape. More importantly, the action of the muscle in bilateral adduction is to retract the mandible, an action used in incisal 'nibbling', and also to resist forces tending to displace the mandibular head forwards, as when objects are held in the hand and stripped through the front teeth. In the large Mandrillus males, the ascending ramus is low and oriented obliquely to the corpus, and elongation of the face is accomodated by lengthening rather than deepening the maxilla, so that the tooth-row is also lengthened. Thus the action of the temporal muscles, more or less in line with the occlusal plane, is maintained. Again, Papio is intermediate in the orientation of its temporal muscles, although perhaps rather closer to Mandrillus than Theropithecus. Probably in relation to the major forces generated in occlusion, the shape of the supraorbital region differs in the three. In Theropithecus the brow-ridges tend to be prominent, perhaps to absorb the predominantly vertical forces generated in chewing with the cheek-teeth. In Mandrillus the anterior muzzle is heavily-built and buttressed, but the supraorbital region is relatively lightly-built, possibly because the forces generated by the largely horizontal action of the temporals are absorbed by the cranial vault. Papio is intermediate between the two extremes in facial shape and the orientation of the masticatory muscles.

3. POSTCRANIAL SKELETON

All three large monkeys show the basic characters of quadrupedal Cercopithecines. The chief differences between them lie in the body-proportions and the structure of the extremities. Theropithecus differs from the others in being rather shorter in the hindlimbs. This criterion applies especially to the large Pleistocene forms.

This character is unusual in the Cercopithecoidea and might suggest "brachiating" adaptations were it not for the indications of extreme terrestrialism in the rest of the skeletal features of the genus. It may be suggested that the short hindlimb is related not so much to locomotion as to sitting-while-feeding.

The structure of the extremities differs in the three forms in much the same way as the dentition and cranial anatomy. In Theropithecus the phalanges of both hand and foot are very short and stout, while in Mandrillus they are much longer and more slender, and in Papio they are intermediate. (Fig. 6). In Theropithecus, the foot is organized about the stout middle digits (III and IV), and the hallux is scarcely stouter than digit V, is short and probably almost immobile. The ligamentous connections between the metatarsals are strong and the whole foot is built for stability. In Mandrillus, the hallucial is the stoutest metatarsal, and the foot is organized about an axis running between the first and second digits. The hallux itself is very long, stout and abductable. The joint surfaces of the tarsus and metatarsus are smaller and often raised, presumably to allow greater freedom of movement. Again, Papio is intermediate in these respects. The same relationship extends to the musculature of the foot and leg, especially the proportions of the peroneal group and the digigal flexors (Jolly, 1965). These features of the foot and its musculature are clearly related to the variation between genera in the degree to which tree-climbing is practised. The mandrill foot, with its long digits and large, powerfully-muscled hallux is obviously much better adapted to grasping branches either in climbing or in branch-walking. Its structure strongly suggests that Mandrillus will be found to be more of a tree-feeder, and less of a forest-floor animal, than is generally thought to be the case.

The hand of Theropithecus shows a peculiarity of structure that is probably related to feeding rather than locomotion. This feature is the extremely abbreviated, but not atrophied, index finger, which combined with the rather long pollex, gives the animal the highest "Opposability Index" among the monkeys (J. R. Napier, pers. comm.). This feature was first noted by Pocock (1925), but its significance only became obvious with the work of Crook (especially his filmed close-ups of the gelada hand in feeding). The high Opposability Index can be related to the gelada's habit of using an index-pollex grip in gathering the small objects on which it feeds (Fig. 7).

4. INTEGUMENTAL CHARACTERS

As Crook and Gartlan (1966) have pointed out, the pectoral rather than the perineal position of the female sexual skin in the gelada may well be adaptive to the sedentary feeding habits of the species, in which case it probably occurred also in the fossil species. The unique, fat-filled cushions (first noted by Pocock, 1925) probably act as accessory sitting-pads and are thus attributable to the same habit.

In an earlier publication (Jolly, 1963) it was suggested that the one-male group type of social organization of the hamadryas baboon was a consequence of small-group foraging which itself was an adaption to sparse resources in an 'open' habitat, and that the male mane was a consequence of sexual selection acting within this social organization. At that time, the social organization of the gelada had not been described, but the same argument applies with equal force to this species, in which the male is again maned and holds a harem. If the reports (Gartlan, this volume) which suggest that the drill is also a one-male group animal are confirmed, it will be most instructive to examine the nature of the social bond between overlord and females in this species. If the hypothesis developed from the hamadryas is correct (Jolly, 1963), and the hypertrophy of the male mane in this species and the gelada is due to its attention-binding quality for potential mates, then one would expect to find that grooming of the male by the females is a less prominent activity in the drill than in these species. If, on the other hand, the mane is rather to be interpreted as an organ of threat to other males, as Washburn (1964) has suggested, we should not expect to find any significant difference in grooming behavior.

The functional importance of the facial and posterior ornamentation of the two Mandrillus species is still obscure, as is the use of the glandular sternal tuft. In a general way, the mandrills agree with the rule that colorful pelage and striking species-specific patterning are most strongly developed in those Anthropoidea which live in forest where visibility is generally poor and allied species are sympatric. It is noteworthy that the muzzle-ridges of the male drill and mandrill, like the mane of the hamadryas, have an unelaborated homologue in the male savanna baboons. In the latter, marked longitudinal protuberances develop in adolescence, accommodating the unerupted germs of the massive maxillary canines. The swellings become less noticeable as the canines erupt, reducing to a variable rugosity on the maxillary angles. It seems that it is this feature, initially an incidental effect of canine dimorphism, which has been elaborated as an epigamic feature in the male *Mandrillus*.

E. *THEROPITHECUS*-LIKE FEATURES IN *PAPIO HAMADRYAS*

The hamadryas baboon is, of all species of *Papio*, the closest to the gelada in habitat and habits. If our adaptational hypothesis is correct, we should expect to find *Theropithecus*-like characters developed by convergence. This in fact is the case. The hamadryas tends consistently to fall to the *Theropithecus* side of the *Papio* range in characters related to the small-object feeding complex: relative size of incisors and molars (Fig. 3), forward position of the temporal crests, small postorbital constriction (Fig. 5), and concave facial profile. (The material is insufficient to examine postcranial characters). In none of these features, however, does the hamadryas fall clearly apart from other *Papio*, which perhaps suggests that its excursion into the open country habitat is more recent, or less complete, or both, than that of *Theropithecus*.

This conclusion might be supported by aspects of the social organization of the two forms. Both have a one-male group organization in which males round up their females and try to prevent sexual access by other males. In the hamadryas, the male maintains a physically compact harem by rounding-up and chastising straying females. In the gelada, harems seem to mingle within the herd, the females following the male when he leaves the group. It appears, from published accounts, that gelada society is less tense and more flexible in the spatial disposition of its members, and that the male expends less energy to achieve the same degree of fidelity than in the hamadryas. This situation might be regarded as a higher stage of adaptation to one-male group life, and hence to open-country habitat.

PHYLETIC AND TAXONOMIC DEDUCTIONS

It would therefore seem that all the major differences between the three groups can be plausibly explained in terms of their current (or very recent) ecology. In particular, the many discordances between *Papio* and *Theropithecus* are now explicable in functional terms, and therefore are of less phyletic significance than was previously believed (Jolly, 1966), when the ecology of the two genera was thought to be practically identical. Thus, there is no need to postulate a very ancient separation between them to account for their differences. On the other hand, there is little evidence for common ancestry for the three genera apart from other living monkeys. The 'long faces' which all three have are of no phyletic significance, since facial elongation is a feature accompanying increased body-size in most if not all Cercopithecidae (Freedman,1962; Jolly, 1965), and the shape of the elongated face is quite different, at least in *Theropithecus*. So far as cranial shape is concerned, it would be easier to derive each genus independently from a relatively small, generalized monkey in which the growth patterns either of a *Theropithecus* or a *Mandrillus* had yet to be established. *Cercocebus*, or a small, primitive species of *Papio* or *Parapapio* would be equally plausible as ancestors.

There is no fossil evidence for the time of differentiation of *Mandrillus* from the common ancestor.

Until recently, the upper limit of the time of divergence of *Theropithecus* and *Papio* was set by the occurrence of both genera in association at Makapan, a site of indeterminate, but probably basal Pleistocene, date. The *Theropithecus* from this site is somewhat primitive in its large male canines, but has the cheek-tooth structure and incisal reduction of the genus. More recently, Howell (1968) has reported both *Papio* and *Simopithecus* (=*Theropithecus*) among the provisional fauna of the Brown Sands locality of the Omo beds. The stratum from which the *Theropithecus* were derived can be dated at about 2.5 m.y. from a potassium-argon date of a superimposed tuff (Howell, 1969). Providing that these dates and the association of the fossils with the datable materials are reliable, we can therefore confidently push back the time of divergence of Papio and Theropithecus into the Pliocene.

It is suggested that a taxonomic scheme compatible with the degree of morphological divergence between the large monkeys, and with the phyletic model which postulates divergence from a common ancestor sometime in the middle Pliocene, would be to maintain full, separate generic status for each of the groups within a tribal grouping that includes all 42-chromosome, five-cusped-M_3 Cercopithecinae. The name proposed for such a group previously was Cercocebini (Jolly, 1966). Cercocebus was selected as type genus because of its rather primitive structure, its morphologically central position in the group, its sound generic status, and the indisputable validity of its name, as well as for reasons of euphony. It now appears that under the new Rules, the tribal name must be taken from the oldest valid generic name, which would make it Papionini.

In view of the proposed sinking of Simopithecus in Theropithecus, and the reinterpretation of the characters distinctive of the genus as adaptive rather than heritage, there seems little reason to retain the separate tribe proposed in an earlier publication (Jolly, 1966) for Theropithecus and its allies. It might, however, become necessary to revive it should the Omo material prove generically distinct from Theropithecus.

ACKNOWLEDGEMENTS

I am indebted to Dr C.P. Groves for much of the data used in compiling Fig.5; to Mrs Lita Osmundsen and the Wenner-Gren Foundation and to Dr J.R. Napier for providing the opportunity to attend the Conference at which this paper was presented.

REFERENCES

Brain, C.K. (1958). The Transvaal Ape-man-bearing Cave deposits. Transvaal Mus. Mem. 11: 1-125.

Buettner-Janusch, J. (1966). A problem in evolutionary systematics: nomenclature and classification of baboons, genus *Papio*. Folia primat. 4: 288-308.

Chiarelli, B. (1968). Taxonomy and phylogeny of Old World Primates with references to the origin of Man. (Ed. Chiarelli, B.). Turin.

Crook, J.H. (1966). Gelada baboon herd structure and movement: a comparative report. Symp. zool. Soc. London 18: 237-258.

Crook, J.H. (1967). Evolutionary change in primate societies. Sci. J. 3(6): 66-72

Crook, J.H. & P. Aldrich-Blake. (1968). Ecological and behavioral contrasts between sympatric ground dwelling primates in Ethiopia. Folia primat. 8: 192-227.

Crook, J.H. & J.S. Gartlan. (1966). The evolution of primate societies. Nature Lond. 210: 1200-1203.

Dandelot, P. (1968). Smithsonian Institution preliminary identification manual for African Mammals. 24: Primates: Anthropoidea. Washington. 1-80.

Darling, F. (1960). *Wildlife in an African territory*. London.

Fiedler, W. (1956). Übersicht über das System der Primates. *Primatologia*. Vol I. (Eds. Hofer, H., A.H. Schultz and D. Starck). Basle: S. Karger.

Freedman, L. (1962). Growth of muzzle length relative to calvaria length in *Papio*. Growth 26: 117-128.

Hall, K.R.L. (1966). Distribution and adaptations of baboons. Symp. zool. Soc. London 17: 49-73.

Hill, W.C. Osman. (1967). Taxonomy of the baboon. In The Baboon in medical research. Vol. 2. (Ed. Vagtborg, H.) Austin: University of Texas.

Hopkins, B. (1965). Forest and savanna. Ibadan and London: Heinemann.

Howell, F. Clark. (1968). Omo Research Expedition. Nature Lond. 219: 567-572.

Howell, F. Clark. (1969). Remains of Pliocene/Pleistocene Hominidae from the Lower Omo Basin, Ethiopia. (Mimeo).

Jolly, C.J. (1963). A suggested case of evolution by sexual selection in primates. Man 222: 178-179.

Jolly, C.J. (1965). Origins and specialisations of the long-faced Cercopithecoidea. University of London Ph.D. thesis.

Jolly, C.J. (1966). Introduction to the Cercopithecoidea, with notes on their use as laboratory animals. Symp. zool. Soc. London 17: 427-457.

Jolly, C.J. (1966). Evolution of the baboons. In The Baboon in Medical Research. Vol. 2. (Ed. Vagtborg, H.) Austin: University of Texas.

Jolly, C.J. (1970). The classification and natural history of Simopithecus Andrews (= Theropithecus Geoffroy). B.M. (N.H.) (In press).

Lydekker, R. (1893). The Royal Natural History. Vol. I. London

Maples, W.R. and T.W. McKern. (1967). A preliminary report on classification of the Kenya baboon. In The Baboon in Medical Research. Vol. 2 (Ed. Vagtborg, H.) Austin:

University of Texas.

Pocock, R.I. (1925). External characters of the catarrhine monkeys and apes. Proc. zool. Soc. Lond. 1479-1579.

Richards, P.W. (1952). The Tropical Rain Forest: An Ecological Study. Cambridge: University Press.

Simpson, G.G. (1961). Principles of Animal Taxonomy. New York and London.

Simpson, G.G. (1963). The meaning of taxonomic statements. In Classification and Human Evolution. (Ed. Washburn, S.L.). Chicago: Aldine.

Tappen, N.C. (1960). Problems of distribution and adaptation of the African monkeys. Current Anthropology 1: 91-118.

Washburn, S.L. (1964). A suggested case of evolution by sexual selection in primates (letter). Man 104: 91.

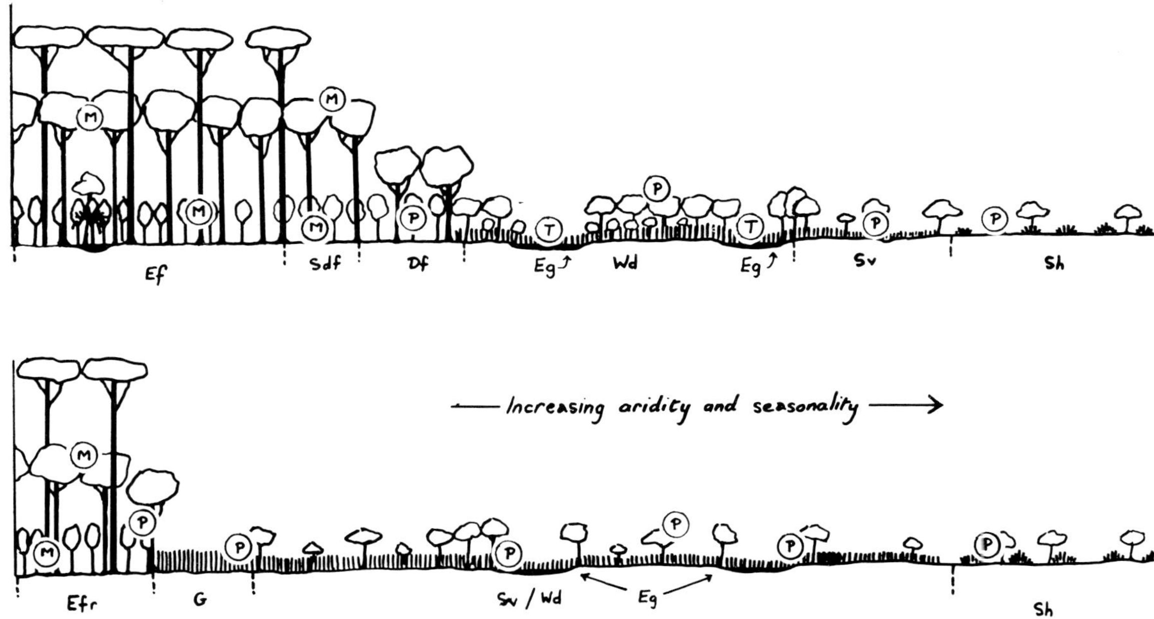

Fig.2 Climatic climax (upper) and fire-climax (lower) vegetations of tropical Africa. Profiles redrawn from Hopkins (1965). Df: deciduous forest. Ef: evergreen forests. Efr: evergreen rainforest remnants, not burned. Eg: edaphic grasslands. G: fire-climax grassland. Sdf: semi-deciduous forest. Sh: sahel. Sv: mixed savanna types. Sv/Wd: fire-climax savannas and woodland remnants. Riverain forests and altitude-related variations not shown. The supposed or actual habitat of each genus is shown by the initial letter of its generic name.

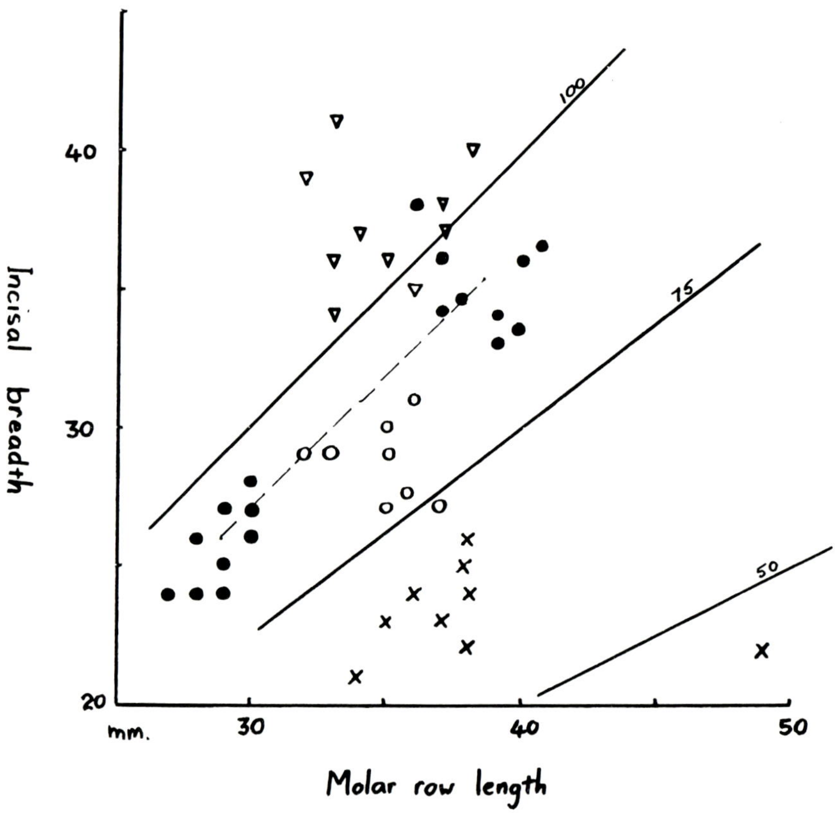

Fig. 3 Tooth-proportions in large monkeys, males only. Triangles, Mandrillus. Solid circles: Papio cynocephalus kindae and P. ursinus griseipes. Open circles: P. hamadryas. Crosses: Theropithecus. The solid lines represent 100, 75 and 50 percent values of Incisal Breadth/Molar Length. The broken line connects the mean points of the two savanna baboon series.

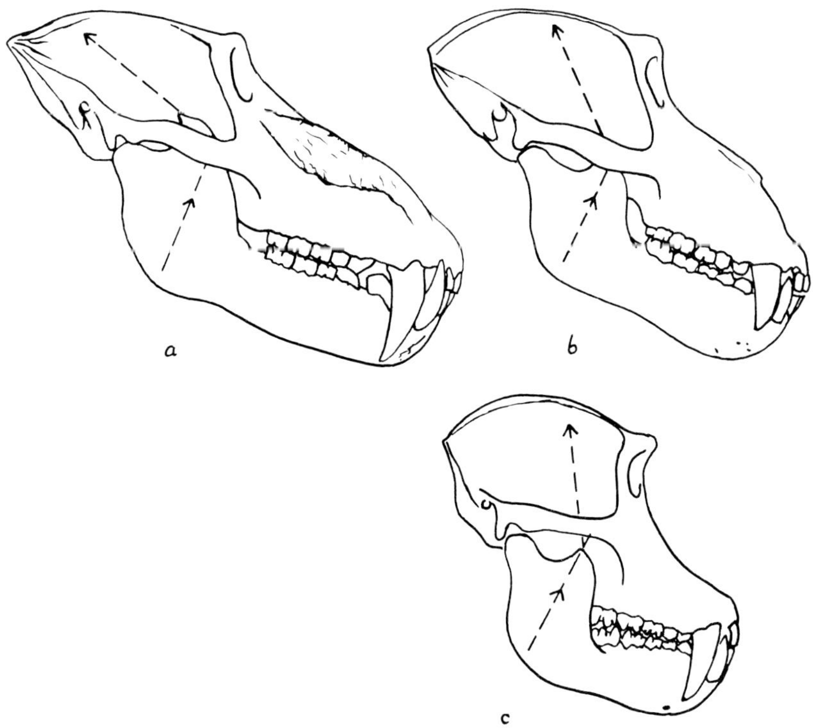

Fig. 4 Lateral aspects of adult male skulls. a: Mandrillus sphinx (B.M. 5.5.23.19). b: Papio anubis (B.M. 7.7.8.12). c: Theropithecus gelada (Field Mus. 8174). Dashed lines represent the approximate lines of action of M. temporalis and M. massetericus in each case. Note also that the length of molar row is approximately equal in each, in spite of the marked differences in the total length of the tooth-row and muzzle. The 'sectorial' form of the P^3 and P_4 in Mandrillus is not unusual.

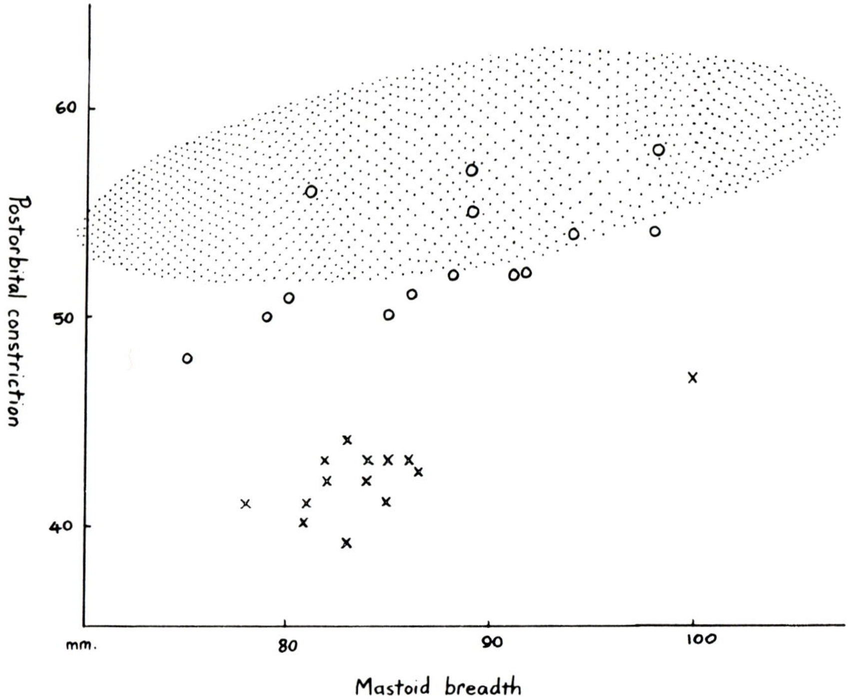

Fig. 5 Relative size of the postorbital constriction in males. Stippled area: distribution of 223 adult male 'savanna' baboons, all subgroups. Open circles: P. hamadryas. Crosses: Theropithecus.

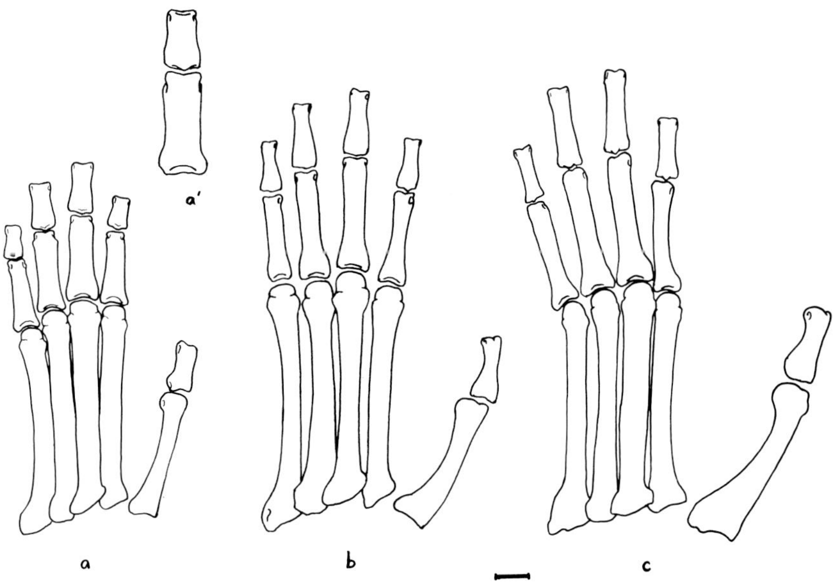

Fig. 6 Left metatarsals, proximal phalanges and middle phalanges of large monkeys.

a Theropithecus gelada, male A.M. 60568.

a' Proximal and middle phalanges, probably of pedal digit III, Theropithecus oswaldi mariae (Pleistocene, Olorgesailie)

b Papio anubis male. AM 82096

c Mandrillus sphinx male. AM 89364

(Scale is one centimeter)

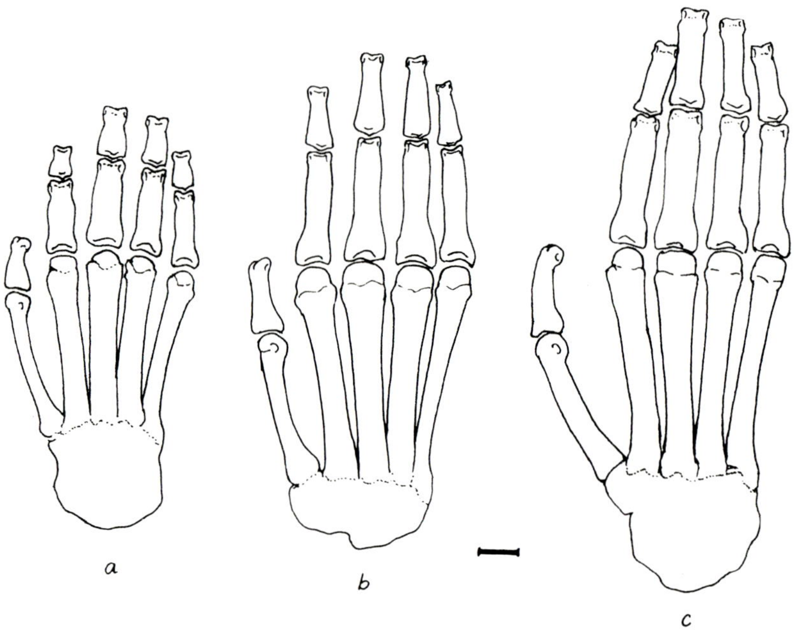

Fig. 7 Right metacarpals, proximal phalanges and middle phalanges of large monkeys.

a *Theropithecus gelada* male AM 60568
b *Papio anubis* male AM 82096
c *Mandrillus sphinx* male AM 89364

(Scale is one centimeter)

Note differences in relative length of index and pollex.

PRIMATE SYSTEMATICS WITH SPECIAL REFERENCE TO OLD WORLD MONKEYS

A Protein Perspective

Vincent M. Sarich

INTRODUCTION

In recent years it has become apparent that a great deal of evolutionary information is available in the form of the amino acid sequences of modern protein species.* Proteins evolve over time just as do the species lineages of which they are a part. Their particular utility as a source of evolutionary information is two-fold: (1) Protein evolution would appear to be almost completely a divergent process with parallelism or convergence occurring at very near to statistical expectation, and (2) more importantly, the unit of protein evolutionary change is a single amino acid substitution resulting from a single nucleotide substitution in the DNA of the genome. Thus the difference between two protein species is readily quantified in terms of the number of nucleotide substitutions required to produce it, and a common measure of differentiation among morphologically quite disparate taxa is therefore available. The contrast with comparative morphological studies is evident.

Given an adequate range of sequences it is possible to reconstruct the specific sequence of mutational events which produced the sequences we see today. The process involved in such a reconstruction has recently been reviewed among others by Fitch and Margoliash (1967, 1969) and Dayhoff (1969a, b). One interesting and potentially valuable conclusion arising out of such studies is the possibility of developing the amino acid sequence of, for example, the cytochrome c or hemoglobin α and β chains of the common ancestor of all placental mammals - even though we may have very little idea of where that animal lived and what it may have looked like. The phylogenies resulting from such a process of reconstruction can then serve as a potent investigative tool - a framework upon which comparative anatomical, behavioral, physiological evidence can be analyzed and structured into a meaning-

---

* Protein species is a phrase in common use amongst workers in the field of chemotaxonomy and refers to the different products of the same locus in different organisms. Eds.

ful evolutionary whole.

In theory this is elegant; in practice it is something else again. Sequence data are, we all concede, the ultimate source of protein evolutionary information, but their generation is a lengthy, arduous task which is still something of an art and not usually of particular intrinsic interest. In other words sequencing is a routine task which requires something more than a routine chemical talent for success. The result of such a situation is readily predictable, and thus we find that the sequence data available to us as mammalian systematists lack something in resolving power. The largest body of such data is centered on the fibrinopeptides and hemoglobins and a brief analysis makes the problems evident. Though the fibrinopeptides (approx. 40 amino acids) and hemoglobins (287 amino acids) differ in size by a factor of about seven, their relative evolutionary rates are such that the net mean rate is about one substitution per 5 million years per lineage for the mammalian fibrinopeptides and about one per 3.5 million years for the hemoglobins (α and β chains combined). Thus it would be entirely possible for several millions of years to elapse with no substitutions occurring in a particular lineage containing these molecules. In such a case that lineage would have no existence for the protein chemist. He cannot see a lineage along which no change has taken place. Among mammals even 5 million years is a significant interval of time and 10 million years is 10 to 15 percent of all mammalian history. The obvious solution is to obtain more sequences so that a lack of resolving power in any single protein might be balanced by another, or, alternatively, to sequence proteins containing more evolutionary information; that is to say, the larger ones. Neither approach is likely to be undertaken in significant amounts in the foreseeable future. What is necessary is that the formidable technical problems involved in obtaining the amino acid sequences of large, rapidly evolving (and therefore more informational) proteins be at least partially overcome. The only readily apparent and feasible approach would appear to be the immunological one, but here a complex, indirect, and often suspect procedure is being substituted for the elegance of real sequences. The basic question is to ask how well do immunological data reflect the evolutionary information present in sequences, and what is the resolving power of the immunological method?

IMMUNOLOGICAL DATA

The immunological data presented in this paper were obtained using the microcomplement fixation (MC'F) technique (Sarich and Wilson, 1966). Antiserum pools were made by mixing the individual antisera (at least three per albumin species) in reciprocal proportion to their MC'F titers. Generally the degrees of cross-reaction obtained have been given as indices of dissimilarity (I. D.). The I.D. is defined as the factor by which the antiserum concentration must be raised in order to produce a heterologous curve of complement fixation, equal in height to that produced by the homologous antigen. It has already been suggested that log I. D. is proportional to the number of amino acid sequence differences between two protein species (Sarich and Wilson, 1967b; Sarich, 1968). To facilitate comparisons on a linear scale, the term _immunological distance_ (defined as 100 x log I. D.) will be used in this paper. These immunological distances range from zero for reactions of immunological identity, to approximately 180 units for the maximum differences measurable between mammalian albumins using the MC'F technique.

VALIDITY OF THE IMMUNOLOGICAL APPROACH

At present I can see three tests of the immunological approach and we have available a certain amount of data in all three areas - sensitivity, reciprocity, and scale. The ultimate test of sensitivity asks whether single amino acid substitutions can be detected immunologically. This matter has been looked into directly for at least three proteins: human hemoglobin (Reichlin, et al, 1966; Reichlin, 1970), _E.coli_ alkaline phosphatase (Cocks and Wilson, 1969) and _E.coli_ tryptophan synthetase (Murphy and Mills, 1969); in each case, using the microcomplement fixation (MC'F) technique a substantial proportion, if not all, of the single amino acid substitutions were detected. In the case of the hemoglobins, where some substitutions could not be detected, it can be shown that they generally occurred in areas where the human and rabbit sequences were identical and which would then not be expected to represent potential antigenic sites. In general, then, it is probably fair to conclude that if a substantial portion of the protein being tested differs in sequence from that of the rabbit (or other antiserum producer) even single amino acid substitutions will be detected - at least using the MC'F technique.

Next the internal consistency of the immune system can be looked into. In this area a test is available whose power is not sufficiently appreciated. Given the same protein from two different species, A and B, wd can ask to what degree the extent of the anti-A with B reaction compared with the anti-B with A reaction. This is usually termed reciprocity. Particularly for two distantly related species the two reactions are really quite different measurements of the same set of differences between the two protein species. I have collected data, both published (Sarich and Wilson, 1967a, b; Sarich, 1968; Sarich, 1969a, b) and unpublished involving 21 antiserum pools (using usually three rabbits per albumin species) to various primate, carnivore, and artiodactyl albumins. Fifty-two pairs of anti-A with B and anti-B with A tests of reciprocity were performed. The total of the anti-A and anti-B means was 3396 immunological-distance units, the sum of the differences between the anti-A and anti-B results was 221 units, and the mean difference was thus 6.5 percent of the mean observed cross-reaction. These results certainly suggest that (at least for the albumin anti-albumin system) the antisera to the two different albumin species must contain antibody populations directed against very much the same proportions of the two albumins. The further fact that the observed degree of reciprocity is not a function of strength of the cross-reaction, even for very distantly related albumins in different mammalian orders, indicates that large portions of the albumin molecule are being sampled.

Finally one can ask whether a proportionality exists between the amount of sequence difference separating two proteins and the extent of the cross-reaction between them and, if so, what is the proportionality constant? There are both direct and indirect data available on these points. First one can look at those relatively rare situations where both sequence and immunological data are available for the same protein. Unfortunately this means the various human hemoglobins (discussed in Sarich, 1968) and incomplete studies on avian lysozymes (Arnheim, Prager and Wilson, 1969). There is a consistent correlation between the number of sequence differences and the degree of cross-reaction and an indication from the hemoglogin data and some unpublished lysozyme work that the form of the relationship is S (measured in amino acid differences) ID (measured in immunological distance units).

A further test of scale makes use of the fact that two New World monkey albumins, those of *Aotus* and *Cebus*, have undergone quite different amounts of evolutionary change since they last shared a common ancestor. As *Aotus* and *Cebus* stand in monophyletic origin relative to any non-platyrrhine species, the amount by which *Aotus* albumin has changed less than that of *Cebus* (measured in terms of amino acid sequence) and therefore the amount by which it will be more similar to the albumin of any non-platyrrhine remains the same, no matter what the reference species. The pivotal test for the sequence difference (S) - immunological distance (D) relationship stems from this observation, for if an f(D) =kS relationship exists, then the *Aotus* - *Cebus* difference expressed in f(D) units should remain the same no matter what the reference species from which this difference is measured. Table 1 records the results. It is apparent that only when the *Aotus* - *Cebus* difference is expressed in immunological distance units does it remain rather constant over a broad range, approaching the maximum distances measurable between mammalian albumins using the MC'F technique - where some immunological insensitivity might have been expected. In summary then the available tests of sensitivity, reciprocity and scale indicate that immunological comparisons can provide a sensitive and reliable approximation to the structural differences which exist between mammalian albumins.

THE IMMUNOLOGICAL APPROACH

The development of a primate phylogeny requires one assumption - that the primates stand in monophyletic relationship to any non-primate species (in our case, certain carnivores will represent these outside reference species). Given this assumption, it will be demonstrated that an internally testable and internally consistent immunological phylogeny for the living primates, with specific reference to the Old World monkeys, can be developed.

The protein studied was serum albumin. It was chosen because it possesses certain advantages which make it particularly suitable for an evolutionary study. It is a single polypeptide chain coded for at a single locus with no evidence of multiple forms or ontogenetic variation. It is well defined and readily purified and characterized. It is perhaps the best studied large nonenzyme protein known and has been used in a large variety of research procedures. It is quite antigenic, and is large enough, and evolves rapidly enough so that a statistically reliable sample of differences is accumulated even between relatively closely related taxa. With almost 600 residues it is among the longest polypeptide chains known. Finally, and of particular importance, is the fact that albumin contains no other covalently bound, potentially antigenic components, such as carbohydrates and lipids -- a property which places it in a unique position among the various serum constituents. This insures that the comparisons are actually made on structures whose mode of evolution is known, and which are a direct reflection of the genetic material itself.

In general, the albumins used for immunization purposes were purified from the respective sera or plasmas by the heat-caprylate procedure of Hoch and Chanutin (1954). The only albumin for which this proved inadequate was that of _Lemur_, which was precipitated. It was then redissolved in 0.03M ph 5.5 NH_4Ac buffer, dialyzed against this buffer, and chromatographed over a Whatman CM52 column equilibrated with the same buffer to elute the pure albumin. We have also found that purer albumins are obtained in the heat-caprylate procedure when the sera are diluted only 50 percent instead of the 1:1 dilution originally used.

Antisera to the following albumins prepared in rabbits (at least four per species) were used in this investigation:

HOMINOIDEA: _Homo_, _Hylobates lar_

CERCOPITHECOIDEA: Macaca mulatta, Papio papio, Cercopithecus aethiops, Presbytis entellus

CEBOIDEA: Aotus, Pithecia monachus, Ateles geoffroyi, Saimiri, Cebus capucinus, Saguinus oedipus, Callithrix jacchus

PROSIMII: Nycticebus coucang, Lemur fulvus, Tupaia glis chinensis

CARNIVORA: Ursus americanus, Genetta genetta, Arctogalida trivirgata, Hyena hyena

The immunization schedule was as follows: On the first day, 2-5 mg per rabbit in Freund's complete adjuvant distributed among three or four intradermal back sites; at five weeks, 2 mg per rabbit in C'F buffer (0.14 M NaCl, 0.01 M Tris, 0.5 m M $MgSO_4$, 0.15 m M $CaCl_2$, ph 7.45) by way of the ear vein; at 7 weeks, 3 mg per rabbit in Freund's incomplete adjuvant intradermally in three back sites; at 12 and 13 weeks, 2 mg per rabbit in C'F buffer by way of the ear vein. The rabbits were bled 5 - 7 days after the last injection. The antisera were heated at 60°C. for 30 minutes and stored at -10°C. In addition I have found that centrifuging the stock dilutions of antisera at 37,000 g for 30 minutes before use tends to minimize anticomplementary effects and allows quite distant crossreactions to be consistently measureable.

RATES OF PROTEIN EVOLUTION

The sequence data which have accumulated over the past several years suggest very strongly that there is appreciable regularity in the rate of evolution of any specific protein - even though different proteins certainly do evolve at different rates. We have already pointed out that a similar regularity is evident in immunological studies of primate (Sarich and Wilson, 1967a; Sarich, 1968), carnivore (Sarich, 1969a, b) and artiodactyl (Sarich, Wilson, and Bennett, 1968) albumins. In 1968 I concluded (p.114-5) that this "regularity" meant that the probability of an amino acid substitution occurring in a given length of time has been the same for each primate albumin lineage and I would now extend this to cover all mammalian albumin lineages. The mathematical demonstrations by Kimura (1968) and King and Jukes (1969), which are more simply discussed in Crow (1969), of a mechanism through which such a situation could develop - the random incorporation of selectively neutral mutations into the ongoing gene pools of diverse lineages - make the empirical observations theoretically respectable. King (personal communication) has now analyzed all the available sequence data in a fashion analogous to that discussed in our 1967 papers and has reached the same probabilistic conclusion. We extended this demonstration of regularity to the development of an immunological time scale for hominid evolution (Sarich and Wilson, 1967b), placing the time of divergence between man and the African apes at 5 million years. Recently we have had occasion to point out that subsequent protein evidence is fully supportive of our original proposal (Wilson and Sarich, 1969). This concept of molecular evolutionary clocks has been severely criticized by Buettner-Janusch (1969), Kirsch (1969), and Seal (1970) - all of whom accuse us of "assuming" this regularity - an accusation made seemingly tenable by the convenient device of ignoring the paper by Sarich and Wilson (1967a) where the specific data, involved in developing the regularity hypothesis, are presented. Those data - and other accumulated since that time - indicate that the process of protein evolution has a major stochastic, time-dependent component with a barely perceptible selective (irregular) overlay. The data presented in this paper would appear to be entirely with these conclusions and, thus, are considered in this light. That is they are seen as end results of a stochastic process with the few irregularities possibly indicating selection and/or immunological imperfections.

PRIMATE CLADISTICS, PATRISTICS, AND TIME

In 1965, Mayr cogently discussed the limitations implicit in the structure of phylogenetic trees, listing three types: Cladogram, phenogram, and the orthodox phyletic diagram (true dendrogram). He also listed the sum of evolutionary information that would be present in the ideal phyletic diagram (1965: 80-81).

"(A) Cladogenesis - (1) What are the major furcations of the ancestral lineages that are responsible for the formation of the existing (or fossil) taxa? (2) At what time in the geological time scale did each branching occur?

(B) Anagenesis - (3) How much evolutionary change has occurred in each line since the last furcation?

We are thus trying to convey in a two dimensional diagram three sets of information: (a) the connections (at furcations) of phyletic lines, (b) the absolute geological time at which the furcations occurred, and (c) individual evolutionary rates in each daughter line after a given furcation."

It is in fact possible to present these three pieces of information on a single diagram by the simple expedient of indicating, on each of the various limbs (lineages) of our phylogeny, the amount of evolutionary change (in our case expressed in immunological-distance units) which has taken place during its extence. The derivation of the required information can be illustrated (in a case deliberately chosen where the amounts of change are not very similar) by choosing four primate albumins (*Homo sapiens*, *Nycticebus coucang*, *Lemur fulvus*, *Propithecus verreauxi*) and four carnivore albumin reference species (*Ursus americanus*, *Genetta genetta*, *Hyena hyena*, *Arctogalida trivirgata*). The relevant data are presented in Tables 2 and 3.

It has already been assumed that *Homo*, *Nycticebus*, *Lemur* and *Propithecus* share, as primates, a common ancestral lineage subsequent to the divergence of the lineage leading to the four modern carnivores. We next must apportion the various experimental immunological distances along the various lineages which themselves must be delineated. The *Homo-Nycticebus* albumin immunological distance is 112 units and this difference must have accumulated along the two lineages leading to the modern forms since they last shared a common ancestor. Referring to Figure 1(a), then:

a + b = 112 — but also: a + c + d = 37 + b + c + d (since <u>Homo</u> albumin is 37 units more different from the carnivore pool than is that of <u>Nycticebus</u>)

thus: $a + b = 112$

$a - b = 37$

then: $2a = 149$, $a = 75$, $b = 37$

Now we ask where does the <u>Lemur</u> lineage fit? Refer to Figure 1(b) and repeat the process outlined above for <u>Homo</u> and <u>Lemur</u>.

now: $e + f = 125$ — and $e + g + h = 27 + f + g + h$ or $e - f = 27$

$e - f = 27$

$2e = 152$, $e = 76$, $f = 49$

Therefore lineage <u>a</u> is for all practical purposes coincident with <u>e</u>; that is, the time at which the <u>Homo</u> albumin lineage diverged from that leading to <u>Nycticebus</u> must very closely approximate the time at which it diverged from that leading to <u>Lemur</u>. Of course if <u>Lemur</u> and <u>Nycticebus</u> were somewhat more closely related than either is to <u>Homo</u> (as has been suggested both on anatomical (Le Gros Clark, 1960) and immunological (Goodman, 1967) grounds this result might be expected. However note that <u>b</u> + <u>f</u> = 86 which is essentially equal to the <u>Nycticebus</u>-<u>Lemur</u> albumin immunological distance. Thus it is seen that the relative similarity of the slow loris and lemur albumins compared to that of man is due not to any special evolutionary relationship but only to the fact that they are both relatively less evolved than is that of man. The simplest picture consistent with these data is given in Figure 1(c), which also includes <u>Propithecus</u> and has the <u>Homo</u>, <u>Nycticebus</u>, and <u>Lemur</u> plus <u>Propithecus</u> lineages diverging from one another at approximately the same time. This conclusion is substantiated when antisera to the albumins of five different simians (Homo, Cercocebus, Presbytis, Cebus, and Saguinus - involving a total of 27 rabbits) are used to measure the relative distances to Nycticebus distance is 115 units; the simian-Lemur distance is, as might be expected, greater at 127 units. The five simian albumins are a mean of 26 units more evolved than that of Lemur and 36 more than Nycticebus. In this case adding these further tests only confirms an already clear picture. In other cases the immunological data will show somewhat less internal consistency and the use of different immunological perspectives will give somewhat varying results. However only a single phylogeny for albumin (or any other polypeptide) can describe evolutionary reality and the use of sufficient immunological perspectives should result in a closer and closer approximation to this reality.

Unfortunately I do not have a sufficient quantity of Tarsius albumin for antiserum production and thus all Tarsius measurements are unidirectional. On balance, however, we see Tarsius albumin as only slightly more distant (2 units) than that of Lemur from those of the four carnivore reference species, and about as distant from Lemur and Nycticebus and would be expected if Tarsius had diverged from those lineages at about the same time as did the simians (Fig. 2).

The tupaiid data, just to bedevil us further on the question of tupaiid origins, are perversely the one case among all the primate albumins tested where the immunological picture is ambiguous. The problem is that Tupaia albumin is unreasonably more similar to that of Homo than can be logically defended on the basis of any phylogeny. That such a situation might develop is not entirely unexpected, especially if it noted from Table 2 that the albumin of Tupaia glis chinensis is as evolved as that of the average simian. This means that both albumin groups then have changed more than those of the average primate and the immunological distances between them are beginning to get into a suspect range; i.e. beginning to approximate average inter-ordinal distances which we are well aware to be often difficult if not impossible to measure using the MC'F technique. To sum up - of all prosimian albumins, that of Tupaia is by far the most evolved and yet, on the average it still is no more distant from the albumins of either the various prosimians or simians than would be expected if the ancestral tupaiid lineage originated as part of the original radiation leading to the modern primates (Fig. 2). I would state my conclusion this way (although perhaps a trifle more uneasily than in almost any other conclusions in this paper): that phyletically the tupaiids (at least those genera tested here: Tupaia and Urogale) are primates in the sense that it cannot be demonstrated that the non-tupaiid primates share any period of common ancestry not also shared by Tupaia. If the Lemuriformes, Lorisiformes, and Tarsius are primates, then phyletically Tupaia and Urogale are primates as well.

The requisite data for continuing the reconstructive process are given in Tables 2 - 8, their analysis provides the cladistic set given in Figure 2. The reliability of these conclusions is, as pointed out above, not rate-dependent, being solely a function of the quality of the comparative data. In the absence of adequate non-molecular cladistic evidence, such quality is perhaps best estimated by means of reciprocity tests and by the degree of internal consistency of the final phyletic scheme. Reciprocity has already been discussed; internal consistency refers to the question of how well the calculated immunological distances (obtained by adding them up along the phylogeny) agree with the experimental o ones used to derive that phylogeny. It is readily verified by inspection that the agreement is excellent - another indication of the efficacy of the immunological approach. Even if markedly discordant rates of evolution occur the patristics and cladistics of a particular set of evolutionary relationships are readily developed, but such events would complicate matters as far as the third of Mayr's requirements - the development of relative and absolute times of divergence - is concerned.

PLACENTALIZATION HYPOTHESIS

It would perhaps be appropriate here to comment on the "placentalization" hypothesis of Goodman (1961 et seq.) who has suggested that proteins whose synthesis begins in fetal life should evolve more slowly than those whose synthesis begins later (e.g.

albumins as compared to the immune globulins). As an extension, he also suggested that those species with more intimate transplacental contacts should contain less evolved proteins (again, albumin was the major example used). In this fashion it was possible to accommodate an ancient (30 million-year) divergence between man and gibbon with the marked similarity of their albumins. It is interesting to note that this elegant and logically appealing argument was never tested by Goodman in terms of his own data - even though it could readily have been - without reference to paleontological and anatomical opinion. Thus the hypothesis still appears viable (e.g. Kirsch, 1969; Goodman, 1969) even though we showed (Sarich and Wilson, 1967a) that the albumins of apes, man, and New and Old World monkeys showed quite similar amounts of evolutionary change since they last shared a common ancestor. It could still be argued that at that time the contrast existed between the prosimians and higher primates. The data presented in this paper make it clear that, if anything, the albumin of the average simian is more evolved that is that of the average prosimian. Thus the placentalization hypothesis previous untested and presently disproven - at least as far as primate albumins are concerned. There is also no indication that primate albumin evolution proceeds under different probability considerations than those of other mammalian orders.

The moral to be drawn is that one's data should, insofar as possible, be tested in its own terms. Otherwise one runs grave risks of circularity and reinforcement of current opinion through misinterpretation of new (and old) lines of evidence. A current example is the argument that short generation time lineages should contain more evolved proteins (Laird, McConaughy, and McCarthy, 1969). An analysis of rodent protein evolution makes it obvious that the assumptions of specific divergence times among the murid and cricetid rodents (5 million years for rat-mouse and 10 for rat-hamster) has no real basis in the available paleontological data and is readily disproved when the protein data are considered (Weigle, 1969; Sarich and Wilson, 1970). Again, the points raised could have been tested internally, but again, they were not.

RATES OF EVOLUTION AMONG PRIMATE ALBUMINS

Amounts of evolutionary change within a monophyletic taxon

are measured with reference to a species known to be outside that taxon. Thus for each major group studied one reference point is required. As discussed above, given that reference point, the internal phylogeny of the group can be determined and so internal reference points become available - the various prosimians for the simians, the catarrhines for the platyrrhines and vice versa and the Old World monkeys for the apes and man.

To obtain a broad perspective on the primates as a group, a set of antiserum pools (prepared in 16 rabbits) to the albumins of four carnivore species was used. The four species have albumins that are 50-60 immunological distance units separated from one another; and this fact, in combination with the use of a number of rabbits, helped to insure that the perspective was both broad and reliable. The data obtained are given in Table 2 and it is quite clear that the primate lineages sampled show quite an appreciable range for the amounts of albumin evolution they have undergone. What then, one might ask, of the hypothesis of "regularity"? As we have pointed out at an earlier stage of our research (Sarich, 1968), this "regularity" would appear to have a large random component whose evolutionary outcome is best described in terms of a Poisson distribution. An appropriate homology is the process of radioactive decay. Let us, for example, posit a C^{14} sample which over a period of 30 days provided 1500 decay events. This total of 1500 would not, however, be made up of a set of thirty 50-count days, but of a set of daily counts whose distribution would approximate that given in Fig. 3 where the Poisson mean is 50. The Poisson distribution describes the expected results of a process where the probability of a given individual event remains constant over time and is thus appropriate to the radioactive decay situation - where each C^{14} atom has an equal probability of decaying in a given interval of time. It is also appropriately applied to the process of protein evolution which would appear to be, in large part, one where the probability of an amino acid substitution occurring in a given span of time is the same for each lineage in which that protein is found. Such substitutions presumably reflect those mutations "unseen" by the selective process - unseen because they are functionally and therefore selectively equivalent to the original amino acid. As King and Jukes have so elegantly put it (1969:788):

> "Natural selection is the editor, rather than the composer, of the genetic message. One thing the editor does _not_ do is to remove changes which it is unable to perceive."

Beyond this quota of surviving neutral mutations, whose evolutionary accretion will follow a statistically determined pattern, will be those substitutions conferring a change in fitness. One might have amino acid changes that increase the fitness in a particular environment; conversely the relative proportion of maladaptive to neutral mutations might be increased in some evolutionarily conservative context. Such positive or negative selection will then introduce a nonstochastic element which should then be seen as "irregularity". On the basis of the presently available data, both immunological and direct sequence, one suspects that this nonstochastic element must be rather a minor one.

Looking now at the overall picture of primate albumin evolution from our carnivore perspective, we see the amounts of change ranging from a low in the _Nycticebus_ lineage to intermediate values from _Lemur_, _Galago_, _Tarsius_, and _Aotus_, and appreciably higher ones for the other simians and also _Tupaia_. If one for the moment ignores _Tupaia_, it could be argued that this is precisely what one might expect with the more primitive prosimians showing, as well, a more primitive molecular state and the progressive simians as more advanced one. However, it should be noted, again apart from the _Tupaia_ situation which has the most "primitive" prosimian showing the most evolved prosimian albumin, that the various simian albumins do not necessarily represent a large number of lineages having independently undergone accelerated albumin evolution. As is evident from Fig. 3, all simian albumins share a long period of evolutionary history over which some 45 immunological units of change occurred prior to the origins of the catarrhine and platyrrhine lineages. It is then more than just possible that these simian "advanced albumins" represent nothing more than a period of common ancestry shared by all of them (the 45 units already referred to) over which somewhat more change than the average occurred - and the stochastic model just discussed would in fact require that such statistical fluctuations exist. Now the point might be made that just about any moderate amount of variation about the mean could be viewed in such terms and that this somehow makes for a certain lack of rigor in the logic involved. The answer is that a stochastic process just cannot be squeezed into a deterministic mold. The power of the stochastic, presumably neutral mutation mediated, model of protein evolution is precisely in the vast majority of data for which it provides an essentially complete explanation and that explanation requires that we see ranges of variation about our means - thus the earlier use of the word "regularity" where we could see the pattern but not the model.

The various New World monkey albumin lineages provide the most extensive illustration of the workings of this stochastic process, representing as they do taxa long distant from one another. The detailed cladistic analysis of New World monkey relationships is not yet complete and not really part of this paper, but it does indicate that the various New World monkey genera, with three sets of exceptions (Ateles-Lagothrix, Pithecia-Cacajao, and to a lesser extent Callimico-(Cebuella-Callithrix)) represent lineages that diverged from one another at about the same time. Using catarrhine and prosimian albumins as reference points, we can then measure the relative amounts of change shown by the various platyrrhine lineages (Table 4).

Again a relatively restricted range about the mean (which in this case is about 20 units of albumin evolution for the average lineage since the platyrrhine adaptive radiation began) is seen with Aotus just possibly representing a real case of selection against change; i.e. reducing the relative proportion of albumin locus DNA mutations which were in fact neutral. On the other hand it is but a single case which can just as easily be seen as being on a tail of the Poisson distribution. Finally we can look at the catarrhine picture from a platyrrhine perspective (Table 6).

To summarize, then, the primate (discussed above), carnivore (Sarich, 1969a, b) and artiodactyl albumin immunological data plus the large body of available sequence data on a variety of proteins (summarized in Dayhoff, 1969) are best viewed in a stochastic context. Given this conclusion, the obvious next step is the construction and calibration of an albumin evolutionary clock for primate evolution.

PROTEIN EVOLUTIONARY CLOCKS

It should be noted here that the reality of the stochastic process just discussed has no real bearing on the use of the observed regularity of evolutionary change at the mammalian albumin locus as an evolutionary clock. The facts are that the stochastic model has an operational reality, that we have a measurable parameter of change equally appropriate to all species, and that we know how to convert our immunological measurements into accurate assessments of the amounts of evolutionary change the various albumin lineages have undergone. What we want now is a measure of the constant $\underline{k}$ in the following equation:

Immunological distance (between two mammalian albumins)	= k x time	(10^6 years since the species in which the albumins are found last shared a common ancestor.)

Here the carnivore and primate data are relevant as the marked regularity in the long-term evolution of both the primate and carnivore albumin loci suggests that the rates of albumin evolution in the two orders must be comparable. It seems highly improbable that such a result could reflect two different patterns of regularity which are somehow maintained among so many diverse lineages. Thus $\underline{k}$ should have a value appropriate equally to a derivation of both primate and carnivore times of divergence. I suggested in 1969 that an albumin immunological distance of 100 units refers to a most probable (in the Poissonian sense) time of separation of 60 million years. Though the data involved in that derivation were not as extensive as those now available (there were, for example, no interordinal measurements possible) the indicated $\underline{k}$ value of 1.67 units/10^6 years of separation or 0.83 units per 10^6 years per lineage still seems as reasonable an estimate as can be made. Thus the Poissonian mean of approximately 55 units of change per average primate lineage would suggest the placement of the earliest primate adaptive radiation (Figure 2) at some 65 million years. In view of the necessary inclusion of the tupaiids in the primates this data could hardly be much more recent; in view of the fossil evidence concerning the earliest primates (e.g. Szalay, 1968) it is equally difficult to extend it much further back in time. To this argument must be added the observation that the mean interordinal distances among mammalian albumins are in the area of 50 percent greater than the mean maximum intraordinal distances (data in this paper and also Sarich, 1969a, b; Weigle, 1961). In view of the linearity argument developed from the Aotus-Cebus data (Table 1), this would place the beginnings of the placental radiation in the area of 90-100 million years. Not too long ago this possibility might have been readily discounted, but recent paleontological studies (Szalay, 1968; Lillegraven, 1969) have indicated that many placental orders may well have had a lengthy history prior to the beginnings of the adaptive radiations leading to the modern forms. On the other hand, the time of the placental-marsupial divergence is now indicated as early mid-Cretaceous, 115 million years ago (Clemens, 1968), and thus a number of the calibration points necessary to the derivation of absolute protein clocks are being more and more approximated. As I have already discussed (Sarich, 1969b), an application of our derived k value to the carnivore data give a time of divergence of 52 million years (early Eocene) between the feloid and nonfeloid carnivores. Again the fossil record is not sufficiently precise to provide a direct test, but as in the primate case a significantly lower figure is ruled out because then other divergence points would be definitely too recent and any significantly higher figure is dubious because of the lack of appropriate miacid ancestors (Romer, 1966). Thus a precisely defined calibration may not yet be possible, but it is hard to see the figure chosen ($k = 1.67$) as being seriously compromised in the future.

The application of this figure (with appropriate corrections where statistical deviations from the Poissonian mean along the various lineages are indicated) to the data in Tables 2 - 8 gives us the time scales of Figures 2 and 4. There would appear to be no necessity for an independent consideration of each divergence time, and thus only some of the more important or controversial ones will be individually discussed. Again, the accuracy of the cladistic conclusions is a function of the information content of the comparative protein data, the temporal conclusions are, to a first approximation, Poissonian means - subject to adjustment when clear evidence of discordant amounts of change is available.

CATARRHINE-PLATYRRHINE DIVERGENCES

There are no data indicating that any significant corrections need be made among the simians; thus the mean platyrrhine-catarrhine albumin immunological distance of some 59 units suggests a time of divergence between the higher primates of the Old and New Worlds amounting to roughly half that which separates the Anthropoidea and the major prosimian groups from one another, i.e., no more than 35-40 million years. Thus the living Anthropoidea are allowed some 30 million years of common ancestry subsequent to the divergence of any line leading to a modern prosimian. The commonly held derivation of the New and Old World monkeys from independent prosimian stocks now appears less tenable; on the contrary it would appear that the Anthropoidea of the late Eocene or early Oligocene had probably reached a monkey or near-monkey grade prior to the catarrhine-platyrrhine divergence.

The relative recency of this data renders it quite suspect, particularly if the ancestor(s) of the New World monkeys are held to have reached South American from the north (e.g. Napier, this volume). It would be next to impossible to imagine the maintenance of any sort of genetic continuity between the primates of the Old and New Worlds through the north some 35-40 million years ago. However this overlooks the possibility that the common ancestor of the New World monkeys arrived in South America from Africa - an event which has been appreciably more palatable by recent continental drift studies on the antiquity of the opening of the South Atlantic. As a recent article in Nature points out (Smith and Hallam, 1970):

> "Non-marine ostracod faunas are virtually identical at species level in late Jurassic and early Cretaceous deposits of Brazil and West Africa. This implies a continental connexion across at least part of the South Atlantic until middle Cretaceous times. The geometry of the two continents and their separation paths may have been such that West Africa remained close to Brazil during much of the opening of the South Atlantic."

The point is that unless we are to assume that the New World monkeys have been a distinct group since the Paleocene or earlier, a possibility the available albumin, transferrin, and hemoglobin data appear to make utterly untenable, then their reaching South America was a very unlikely event. The only other modern orders of placental mammals to have duplicated such a feat were the bats (who had the advantage of flight) and the cavimorph rodents (whose ancestry is more obscure but who might also be of African derivation). No matter whether the New World monkeys came from the North or the East they came by water, and if one plots the course of minimum Africa-South America distance through the Tertiary one finds that in late Eocene times there may well not have been all that much difference in the extent of the two water gaps - one of which they must have crossed.

The Old World monkey, ape, and human albumins share only some 11 immunological distance units of common ancestry subsequent to the catarrhine-platyrrhine divergence. This close relationship among the modern Anthropoidea is reinforced by our unpublished transferrin studies using the antisera prepared by Wang et al (1968) which give a mean human-Old World monkey difference of about 30 units and a mean human-Old World monkey difference of 50-55 units where the prosimian distances are greater than 100 units. Also Boyer et al (1969 have recently reported that the β chains of several New World monkey hemoglobins (Ateles, Saimiri, Saguinus) are hardly more different from that of man than is that of the rhesus monkey (Matsuda et al, 1969), while it is well known that the various prosimian β chains are separated by many more mutational events and generally look hardly more like those of the higher primates than do those of the non-primate mammals tested (summary in Dayhoff, 1969). Thus, all the available molecular data indicates a very long period of common ancestry for the living Anthropoidea subsequent to their divergence from the ancestors of the modern prosimians and a rather short period shared by the modern catarrhines subsequent to the catarrhine-platyrrhine divergence. I would estimate the ratio between these spans of time as 3 or 4:1.

HOMINID-PONGID DIVERGENCE

The cercopithecoid-hominoid divergence then must have occurred in latest Oligocene or earliest Miocene times about 22 million years ago. This is obviously a rather low figure as it has recently been proposed that a number of specifically hominoid lineages (those leading to Homo, Pan, Gorilla, and Hylobates) were already distinct by this time (Simons, 1967; Pilbeam, 1968). However, a large body of molecular data concerning these latter taxa is in existence and it would perhaps be wise to consider it before passing judgement on the overall validity of the albumin approach to primate systematics. We have already pointed out that the albumin immunological data indicate a date of approximately 4-5 million years for the hominid-pongid divergence - a figure in appreciable disagreement with the paleontological estimates referred to above (Sarich and Wilson, 1967b). We have recently had occasion (Wilson and Sarich, 1969) to point out other molecular data (in particular those of Matsuda et al (1968) on the rhesus monkey hemoglobin sequence) are in complete accord with the conclusions drawn on the basis of the albumin data. In addition Doolittle and Mross (1970) have demonstrated the identity of the human and chimpanzee fibrinopeptides A and B. As they point out:

> "Previously the fibronopeptides from four old world species of monkeys had been characterized (rhesus, baboon, green monkey and drill). These have been found to differ from the human fibrinopeptides in five to eight of the thirty amino-acid positions compared. The identity of the chimpanzee and human fibrinopeptides re-emphasizes the extraordinarily close relationship of these two species on the molecular level. Recently Wilson and Sarich (1969) on the basis of immunochemical and other molecular criteria, have argued that the time of divergence for these two species may have been as recent as 5 million years ago, in contrast to the 20-30 million year figures postulated by palaeontologists and others. Our data would support their contention, the chimpanzee and human - as judged by their fibrinopeptide structures - being more closely related than such pairs as horse-donkey, water buffalo-Cape buffalo, cat-lion, or dog-fox, all of which have

at least one aminoacid difference in their fibrinopeptides."

The available molecular data are summarized in Table 9. One notes the interesting fact that the albumin difference is both absolutely and relatively the largest found among these four proteins. Thus it does not appear that the albumin result is somehow unique - a result we should have expected from our earlier conclusion that there is but a single phylogeny for any group of species and that the phylogeny of a protein locus must be coincident with the phylogeny of the species themselves. Nevertheless, in view of the relative novelty of the protein approach to evolutionary history, the relative concordance among these various protein comparisons is comforting - particularly as it suggests that the albumin date might be, if anything, an overestimate rather than the gross underestimate that it has generally been thought to be.

It might be instructive to reflect upon what the molecular data and the paleontological times of divergence would lead us to conclude. In the albumin case, using a human-chimp divergence time of 20 million years (Simons and Ettel, 1970), we should expect about 33 ID units of difference. We find 7. In the case of the fibrinopeptides, where the average rate of change is about one substitution per lineage per six million years, we might expect 6 differences between man and chimpanzee. We find none. For the hemoglobins (α and β chains combined) the average is about one substitution per vertebrate lineage per 3.5 million years (Kimura, 1969). Between man and chimp we might then expect 10 differences. We find none. We have already shown that this lack of evolutionary change in the albumins (Sarich and Wilson, 1967a, b; Sarich, 1968) and hemoglobins (Wilson and Sarich, 1969) cannot be due to retarded evolution in the chimp and/or human lineages. In view of the generally stochastic nature of protein evolution one might consider that much of the argument here is, in large part, an accident of history. It is entirely possible to imagine a situation where the molecular evidence might have come first and the paleontological and comparative anatomical data at some later time. Then suggestions of any appreciable antiquity for the African ape-hominid divergence might be greeted with as much skepticism as have been our protein conclusions (e.g. Pilbeam, 1968; Buettner-Janusch, 1969).

The point is that one has to be very careful in distinguishing between data and interpretation. There are molecular and paleontological and anatomical data, but these data often mean different things to different people - thus there are also molecular, paleontological, and anatomical interpretations of those data. There is, however, but a single evolutionary history or phylogeny for a group of species - therefore all interpretations of what happened and why and how it happened in the history of that group must use this single phyletic framework. One cannot legitimately use one phylogeny to explain the molecular data, another for the paleontological, and still a third for the anatomical. Conversely, one cannot legitimately derive three or more phylogenies from three such sets of data. We require that the various lines of evidence be used to place more and more marked constraints upon interpretations derived from other areas and in this cybernetic fashion to more closely approach evolutionary reality. No single way leads to perfection; at present the protein picture lacks something in resolution, the fossil record is necessarily rather incomplete, and the anatomical data often difficult to interpret. I have already discussed the particular advantages of the protein approach in developing the phyletic picture. As might be expected, then, it is not particularly difficult to interpret the available paleontological and anatomical evidence in terms of the short time scale protein phylogeny (Washburn, 1968; Lewis, 1969) but I have yet to see any suggestion as to how a 20-million-year date for the origin of the hominid line can possibly be used to explain the molecular evidence. To put it as bluntly as possible, I now feel that the body of protein evidence on the *Homo-Pan* relationship is sufficiently extensive so that one no longer has the option of considering a fossil specimen older than about 8 million years as a hominid *no matter what it looks like*.

This short time scale makes the understanding of the course of recent (post-Miocene) pongid and hominid evolution in the context of the available comparative anatomical and paleontological data a relatively straightforward affair. On the basis of the molecular data, then, one concludes that the albumin hominid phylogeny provides an evolutionary picture that is elegant in its simplicity (Figure 4). The numerous branches of the widespread group of

Miocene dryopithecines must have left but a single survivor. A small form of this surviving lineage became uniquely successful through the development of the locomotor-feeding adaptation usually termed brachiation. The adaptive success of this development and the subsequent radiation of the group possessing it, made this group the only survivors of the many apes present throughout the tropical and sub-tropical Miocene forests of the Old World. The period of common ancestry shared by the living apes and man after the initial success of this development seems, on the basis of both the immunological and paleontological data, to have been relatively limited. This is as expected, for the rapid development of a new structural-behavioral complex or grade and its subsequent adaptive radiation is the common evolutionary pattern. The products of such an adaptive radiation show similarities in basic pattern, reflecting the attainment of a new grade or organization, but differences in detail, representing the various independent lives of development comprising the adaptive radiation. The unity of the modern Hominoidea, then, is based on the relatively short period of time a major new adaptation was being evolved; and their diversity on the relatively long periods of time during which each line has been evolving independently of the others.

The short period lies within that span of time which separates the Miocene apes whose postcranial remains are represented in the fossil record from the beginning of the pongid adaptive radiation set at approximately 10-12 million years ago. These Miocene apes are apes in terms of their dentition alone for the evidence of the limb bones and the apparent presence of a tail in the Pliopithecus-Limnopithecus group suggests forms not unlike the modern Old World monkeys in their locomotor adaptations; (Napier and Davis, 1959; Washburn, 1968). Such data pose major problems for the student of hominid evolution if an early Miocene or late Oligocene time for the hominid-pongid and intra-pongid separations is used, for then the obvious similarities in morphology among the living Hominoidea have to be explained in terms of synchronous parallelism or convergence -- though such phylogenies have rarely considered what sorts of different adaptations could have produced such similar results in different lineages. The albumin time scale resolves such problems by demonstrating that they do not exist.

OLD WORLD MONKEYS

The hominoid-cercopithecoid divergence in the early Miocene now fits quite comfortably into such a picture of recent catarrhine evolution. It must be considered that it is the cercopithecoid dentition which is the specialized adaptation and whose symmetrical bilophodont molar cusp pattern can be quite readily derived from the more generalized and asymmetrical patterns seen in those Oligocene catarrhines represented in the fossil record. I see no reason to even attempt an apportioning of the Oligocene forms among lineages leading to modern species on the basis of either the molecular or available paleontological evidence. I have already discussed the molecular evidence concerning the origin of the various hominoid lineages; I would now suggest that the molecular and paleontological data show the greatest concordance in an integrated picture if these Oligocene forms are considered as generalized catarrhines ancestral to both the apes and Old World monkeys.

The colobine-cercopithecine divergence is placed at about 16 million years which is more or less reasonable in terms of the other available data and requires little further discussion. It is when we reach the questions of intra-cercopithecine and intra-colobine relationships that the albumin picture again begins to make a significant contribution to the general picture of catarrhine evolution. Again at this point the hominoid evidence becomes relevant because the maximum intra-hominoid, intra-cercopithecine, and intra-colobine albumin differences are both comparable and very small. The data in Tables, 7, 10, and 11 indicate these differences and the relative amounts of change shown by the various albumin lineages in these three taxa since they last shared common ancestors.

RECENT CATARRHINE EVOLUTION

The available data, when developed into a phylogeny, indicate that all the living catarrhines derive from only three ancestral species which lived no more than about 10 million years ago. The adaptive radiations leading to the modern forms must be considered, within the limitations of the method, as roughly synchronous in nature - although it is probable that the hominoid radiation began slightly earlier than the other two. This synchrony of major evolutionary developments among a set of taxa whose members even today are in some ecological niche conflict with one another is difficult to dismiss as coincidental. It would appear much more likely that a major event which reshaped the whole pattern of catarrhine relationships was in progress toward the end of the Miocene and into the early Pliocene (5-10 million years ago). The molecular data suggest that most of the Miocene taxa represented in the fossil record left no lineal descendants and that, in fact, only three of these lineages (one leading to the modern Colobinae, a second to the modern Cercopithecinae, and the third to the modern Hominoidea) showed continued evolutionary success. What this major event might have consisted of is at present obscure; it must at present be reconstructed from data available on modern species -- a risky procedure in the absence of a fossil record. Logically, the oldest adaptive radiation should represent the primary primate component of this event (interactions with non-primate flora and fauna are beyond the scope of this approach). It is therefore unfortunate that the albumin data do not clearly discriminate as to the relative ages of the three radiations, but it is probable that at least the gibbon divergence (and perhaps also that leading to the orang) antedates the colobine and cercopithecine radiations. Further detailed studies, and here the use of a second protein would be most helpful, are probably needed to resolve this issue.

Subsequently the relatively recent adaptive radiation of the cercopithecines provided the impetus for a profound alteration of the nature and extent of the hominoid and colobine ecological niches somewhere in the latter part of the Pliocene. As I have previously pointed out it seems most likely that the African hominoid radiation of the late Pliocene which led to the chimpanzee, gorilla, and human lines was the result of the pre-emption of the small ape niche in Africa by the arboreal lineages of these later cercopithecine radiations. Thus only larger pongids were able to survive and larger size can then be viewed as a preadaptation to a terrestrial way of life -- a way of life to which all three of the living African hominoids are well-adapted.

OLD WORLD MONKEYS

Detailed phyletic conclusions among the Old World monkeys are difficult; difficult among the colobines because of the lack of material, and among the cercopithecines because of the over-abundance of very closely related taxa, a situation which the techniques discussed in this paper has not yet been able to explain. Basically the problem would appear to lie in the fact that antisera to only four cercopithecine albumins (M. mulatta, P. papio, C. galeritus, and C. aethiops) have been prepared; and these do not begin to deal with the cercopithecine complexity in an adequate manner. Certain definite conclusions can be drawn, it is true, but the picture is not as satisfyingly complete as for the other taxa we have looked at. Until the work is done the extent to which further antiserum production would resolve some of these issues is of course a moot point, but a few thoughts on the potential of the protein approach (either sequence or immunological) for dealing with very close mammalian relationships is in order.

If the cercopithecine data (Table 11) are taken at face value, then a number of taxa studied have albumins that are separated from one another by about 5 immunological distance units. It is probable that one ID unit is approximately equal to one amino acid substitution (Sarich, 1967), thus these albumins differ from one another by 4-5 amino acid substitutions. The reliability of the immunological approach in this area, particularly when only a relatively few antisera are used, is certainly suspect and I would be very careful in pushing the data too far. As I pointed out in the beginning of this article, the molecular systematist cannot see a lineage if no change has occurred along it, and if a lineage is only 2 or 3 amino acid substitutions in length anyhow (i.e., 2-3 million years for the albumins) this is certainly approaching the resolving power of the method. In addition, only slight and statistically quite expectable variations in the amounts of evolution experienced might well make a lineage show no substitution where one, two, or three might be expected. For example, in the roughly 4 million years since man and chimpanzee last shared a common ancestor the two lineages have undergone no hemoglobin evolution (in the α and β chains) so that today most human and chimpanzee adult hemoglobins are identical in sequence. Given a mean rate of change of one substitution per 3.5 million years per mammalian hemoglobin lineage (Kimura, 1969), the expected most probable difference between chimp and human hemoglobins would be 2-3; statistically zero is not very different from two and actually to find no changes is hardly surprising. Such variations could, in an immunological context, place the number of differences between two albumins in a not generally detectable range, and probably the retardation would be lost in the area of uncertainty present because of the individuality of the antisera.

In addition in these cases, the problem of intraspecific polymorphism becomes, a significant one as any rapidly evolving, large protein will necessarily show appreciable polymorphism within a given gene pool. This may well be electrophoretically undetectable as it is doubtful that more than 10 percent of all _viable_ amino acid substitutions will involve charge changes but this does not mean that it doesn't exist. In fact we have substantial as yet unpublished evidence that electrophoretically undetectable albumin alleles exist in _Macaca_ and _Aotus_. Since at this time it is not technically feasible to determine the genetic homogeneity of the immunizing albumin by other than electrophoretic methods (the _Macaca_ and _Aotus_ cases just being matters of luck) it is quite likely that some small degree of fuzziness is implicit in any protein approach to the question of extremely close relationships. This of course in no sense compromises the approach; it just means that we must accept limitations in it at both ends of the time scale - at the short end because of the effects just discussed and at the long end because eventually the albumins become so different from one another that immunological recognition of homology is no longer possible. Given these cautions we can gingerly approach the actual data.

The Colobinae: Among the colobines, serum samples are available from only four genera: _Presbytis_, _Pygathrix_, _Nasalis_, and _Colobus_. The data in Table 7 indicate that _Presbytis_ and _Nasalis_ shared a common ancestor subsequent to the divergence of the _Colobus_ and _Pygathrix_ lineage(s) and that the colobine radiation is contained within the last 10 million years; and that the dichotomy between _Presbytis_ and _Nasalis_ occurred approximately five million years ago. It is not possible, on the basis of the available data, to say that Pygathrix shares any period of common ancestry with the other Asian colobines subsequent to the Colobus divergence (but see Groves, this volume). In fact, the relatively less changed nature of Colobus and Pygathrix albumins might very tentatively be taken as a phenetic indication of their phyletically closer relationship. Within the cercopithecines I will deal with the major groups individually and then try to provide a tentative synthesis.

Macaca (including Cynopithecus). With antisera to M. mulatta albumin only the Celebes forms are seen as sharply divergent. Otherwise the small amount of intraspecific variation cannot really be distinguished from the comparable amount of interspecific variation (M. fuscata, M. irus, M. silenus, M. radiata, M. nemestrina, M. sylvanus have been investigated). The distance of the albumins of C. niger and M. maura are both about 5 units and from tests with the antisera to other cercopithecine albumins it is evident that the Celebes forms have albumins which are more changed than those of the macaques proper. Small, but experimentally real, differences in the niger and maura reactions demonstrate that their albumins are different from one another, but no antisera have been prepared to that of either species so that no measure of this difference is available. Almost certainly, then, a very recent common ancestry for all the macaques is indicated (probably within the Pleistocene) with the Celebes forms having a recent common derivation from some Macaca lineage.

Cercopithecus (including Miopithecus and E. patas). Only a set of six antisera to the albumin of a single C. aethiops is presently available. These antisera clearly distinguish, at about the same level (4 to 6 ID units), every other Cercopithecus tested (cephus, diana, mona) and Miopithecus, but not Erythrocebus. There would appear to be some very small difference between the albumins of the patas monkey and the aethiops monkey when antisera to other cercopithecine albumins are used but there is no doubt that the patas monkey belongs in the genus Cercopithecus and, of those species tested, on the aethiops lineage. The conclusion is that the aethiops and patas monkeys share a common ancestral lineage subsequent to the divergence of the other species tested.

Cercocebus. Only two species, galeritus and torquatus, have been looked at and their albumins seem to be immunologically identical. The placement of the mangabeys within the Cercopithecinae is complicated by the relatively less changed albumins of the group which makes them show a greater similarity to those of Cercopithecus than do the albumins of the macaques and baboons. However, when this "conservatism" is taken into account the mangabeys are seen to belong to the same line as the macaques and baboons. In other words, the 42-chromosome monkeys would appear to be of monophyletic origin relative to any given Cercopithecus lineage though the available data do not allow the converse conclusion to be drawn; i.e. that the various species in the genus Cercopithecus are of monophyletic origin relative to the macaques, baboons and mangabeys. Neither, of course, do they disallow it. The matter is just not yet clear and requires the preparation of antisera to many more Cercopithecus species for clarification. Neither can it be concluded that the mangabeys have a special relationship to either the macaques and baboons, or that the macaques and baboons are more closely related to each other than either group is to the mangabeys.

"Baboons". The various "baboon" albumins fall quite sharply into two groups: Drill and mandrill in one and all the others tested (P. papio, P. anubis, P. hamadryas and T. gelada) into the other. No antiserum yet prepared, including those to the albumins of Macaca, Cercocebus, Cercopithecus and Presbytis, as well as Papio, shows any difference between the albumins of Papio and Theropithecus. On the other hand, they all sharply distinguish between the Theropithecus - Papio albumin as compared to those of the drill and mandrill. In view of the proposed antiquity of the Theropithecus - Papio divergence (Jolly, Simons, this volume), it is interesting to note that the identity of their albumins cannot be accounted for by a lack of evolutionary change. The albumins of Theropithecus and Papio are, in fact, the most sharply divergent of all cercopithecine albumins, reacting appreciably less well (on the average, about 4 units) than the drill and mandrill albumins when tested with antisera to various non-baboon albumins. Thus one is forced to conclude that Theropithecus and Papio share a common ancestral lineage along which a substantial amount of albumin evolution occurred *subsequent* to the divergence of all other lineages leading to the modern cercopithecines. I would then be seriously inclined to question the attribution of various African Pleistocene baboons to the *Theropithecus* lineage (Jolly, this volume) and think it much more logical to see *Theropithecus* as of quite recent derivation from a specifically *Papio* lineage. This disagreement is, in spite of its smaller dimensions, quite similar to the *Homo-Pan-Gorilla* situation. However where the albumin picture for the hominoids has been quite consistently confirmed by more recent molecular studies, the *Theropithecus-Papio* data are based only on albumin data. Further protein studies should make this another pivotal test case for the immunological approach.

The simplest picture consistent with these data would then have the modern cercopithecines divided into two groups, one including the macaques, baboons and mangabeys and the other the various members of the *Cercopithecus* complex. As discussed above, the first group would appear to be a good monophyletic taxon; the second at present is a more tenuous association. Within the former group we would then have at least four lines diverging from one another at the same time: *Macaca* and *Cynopithecus*, *Cercocebus*, *Papio* and *Theropithecus*, *Mandrillus* with a subsequent divergence between the Celebes forms and one of the various *Macaca* lineages. I would also reiterate that a number of these conclusions, as specifically discussed earlier, are drawn from the limits of the immunological approach. To more clearly fix the detail, several times as much data as are presently available will have to be collected.

The important findings, however, are those which indicate the recency of the various evolutionary events which have given rise to the modern catarrhine lineages. This recency places marked restraints on the range of interpretations which might be made of the comparative anatomical and behavioral evidence, but at the same time it allows the logical reconstruction of recent catarrhine history to proceed. The large variety of modern species arising from so recent a set of common ancestors should make it possible to draw some significant conclusions as to the anatomical and behavioral attributes of these common ancestors even in the absence of an adequate fossil record.

This article thus illustrates what can be done to quantify phylogeny by means of the comparative molecular approach. The molecular data aid us in developing the history of the modern organisms under study. This history is then a framework upon which we can properly assess the meaning of the evidence we glean from other areas of research. The molecules at this early stage tell us nothing about their own adaptations, but they do allow us to make more meaningful statements about the adaptations of the organisms containing them. This history is essential to any evolutionary study. For the student of catarrhine primate evolution it allows a coherent picture to be developed that both accounts for the available facts and points toward rewarding new areas of research.

ACKNOWLEDGEMENTS

The research summarized in this contribution had its inception in 1965. Many individuals and institutions have contributed support and materials. Dr. Allan C. Wilson of the Biochemistry Department, University of California, Berkeley has made his laboratory facilities available, introduced me to a number of the approaches used, and acted as a collaborator throughout. His support has come through NSF grants GB-3839 and GB-6420. In addition I received a USPH5 predoctoral fellowship 1-F1-GM-30, 454-01 during 1966, and a University of California Summer Faculty Fellowship for 1968. Technical assistance has been provided, mostly through the OEO work-study program, by Linda Ferguson, Jule Griebrok, Karen Reinheimer, Richard Rutherford, Emma Rivera, and most recently and most ably by Peter Hornbeck.

No study of this type could prosper without the generous co-operation of investigators willing to provide serum samples. In particular I would like to thank Dr. Morris Goodman of Wayne State University who, even though we may have our disagreements as to the interpretation of results, has always been generous with material which would have been very difficult to obtain otherwise. Others who have responded generously to our many requests are Drs. Robert Cooper and Lester Nelson of the San Diego Zoological Society; Dr. Robert Hummer, Southwest Foundation for Research and Education; Dr. Theodore Hayashida, U.C. Medical Center, San Francisco; Dr. Michael Crawford during his stay at the University of Washington; Drs. Charles Sedgwick, Joseph Bernstein, and Nathan Gale of the Los Angeles Zoo; Dr. John Buettner-Janusch of Duke University; Dr. Curtis Williams of Rockefeller University; Dr. Clifford Jolly of NYU; Dr. Margaret La Salle of the Oregon Regional Primate Center; the Delta Regional Primate Center; and the Davis National Primate Center.

REFERENCES

Arnheim, N., E.M. Prager and A.C. Wilson (1969). Immunol logical prediction of sequence differences among proteins: chemical comparison of chicken, quail, and pheasant lysozymes. J.Biol.Chem. 244: 2085-2094.

Boyer, S.H., E.F.Crosby, T.F. Thurman, A.N. Noyes, G.F. Fuller, S.E. Leslie, M.K. Shepard and C.N. Henderson (1969). Hemoglobins A and A_2 in New World primates. Science 166: 1428-1431.

Buettner-Janusch, J. (1969). The nature and future of physical anthropology. Trans. N.Y. Acad. Sci. 32: 128-138.

Clemens, W.A. (1968). Origin and early evolution of marsupials. Evolution 22: 1-18.

Cocks, G.T., and A.C. Wilson (1969). Immunological detection of single amino acid substitutions in alkaline phosphatase. Science 164: 188-189.

Crow, J.F. (1969). Molecular genetics and population genetics. Proc. XII Int. Cong. Genetics 3: 105-113.

Dayhoff, M.O. (1969). Atlas of protein sequence and structure. Silver Spring, Maryland: National Biomedical Research Foundation.

Dayhoff, M.O. (1969). Computer analysis of protein evolution. Sci. Am. 21(1): 86-91.

Doolittle, R.F., and G.A. Mross (1970). Identity of chimpanzee with fibrinopeptides. Nature Lond. 225: 643-644.

Fitch, W.M., and E. Margoliash (1967). The construction of phylogenetic trees. Science 155: 276-284.

Fitch, W.M., E. Margoliash, and K.S. Gould (1969). The construction of phylogenetic trees. II. How well do they reflect past history? Brookhaven Symposia in Biology 21: 217-242.

Goodman, M. (1961). The role of immunological differences inthe phyletic development of human behavior. Human Biology 33: 131-162.

Goodman, M. (1962). Evolution of the immunologic species specificity of human serum proteins. Human Biology 34: 104-150.

Goodman, M. (1963). Serological analysis of the systematics of recent hominoids. Human Biology 35: 377-436.

Goodman, M. (1967). Deciphering primate phylogeny from macromolecular specificities. Am. J. Phys. Anthrop. 26: 255-275.

Goodman, M., Sorenson, M.W., Farris, W., and E. Poulik (1969). Immunodiffision systematics of Tarsius syrichta. Paper presented to American Association of Physical Anthropologists. Mexico City. 11 April 1969.

Hoch, H., and A. Chanutin (1954). Albumin from heated human plasma. Preparation and electrophoretic properties. Arch.Biochem.Biophys. 51: 271-276.

Kimura, M. (1968). Evolutionary rate at the molecular level. Nature Lond. 217: 624-626.

Kimura, M. (1969). The rate of molecular evolution considered from the standpoint of population genetics. Proc.Nat.Acad. Sci. 63: 1181-1188.

King, J.L. and T.H. Jukes (1969). Non-Darwinian evolution. Science 164: 788-798.

Kirsch, J.A.W. (1969). Serological data and phylogenetic inference: the problem of rates of change. Syst.Zool. 18: 296-311.

Laird, C.D. McConaughy, B.L., and B.J. McCarthy (1969). Rate of fixation of nucleotide substitutions in evolution. Nature Lond. 224: 149-154.

Le Gros Clark, W.E. (1960). Antecedents of man. Chicago: University Press.

Lewis, O.J. (1969). The hominoid wrist joint. Am.J. Phys. Anthrop. 30: 251-268.

Lillegraven, J.A. (1969). Latest Cretaceous mammals of upper part of Edmonton formation of Alberta, Canada, and review of marsupial-placental dichotomy in mammalian evolution. Article 50. University of Kansas Paleontological Contributions.

Matsuda, G., M. Tetsuo, H. Takei, H. Ota, M. Yamaguchi, T. Miyauchi, and M. Migita (1968). The primary structure of adult hemoglobin from Macaca mulatta monkey. J. Biochem. 64: 279-282.

Mayr, E. (1965). Numerical phenetics and taxonomic theory. Syst. Zool. 14: 73-97.

Murphy, T.M. and S.E. Mills (1969). Different specificities in antisera prepared against Escherichia coli tryptophan synthetase α subunit. J. Immunol. 103: 1377-1384.

Napier, J.R. amd P.R. Davis (1959). The forelimb skeleton and associated remains of Proconsul africanus. Fossil Mammals of Africa, No. 16. London: British Museum (Nat. Hist.).

Pilbeam, D.R. (1968). The earliest hominids. Nature Lond. 219: 1335-1338.

Reichlin, M. (1970). The distribution of specificity in rabbit antisera directed toward human hemoglobin. Immunochemistry 7: 15-28.

Sarich, V.M. (1967). A Quantitative immunochemical study of the evolution of primate albumins. Unpublished Ph.D. thesis. U.C. Berkeley.

Sarich, V.M. (1968). Hominid origins: An immunological view. In Perspectives on Human Evolution (Eds. S.L. Washburn and P.C. Jay). New York: Holt,Rinehart & Winston.

Sarich, V.M. (1969a). Pinniped origins and the rate of evolution of carnivore albumins. Syst. Zool. **18**: 286-295.

Sarich, V.M. (1969b). Pinniped phylogeny. Syst. Zool. **18**: 416-422.

Sarich, V.M., and A.C. Wilson (1966). Quantitative immunochemistry and the evolution of primate albumins. Science **154**: 1563-1566.

Sarich, V.M., and A.C. Wilson (1967a). Rates of albumin evolution in primates. Proc. Nat. Acad. Sci. **58**: 142-148.

Sarich, V.M., and A.C. Wilson (1967b). Immunological time scale for hominid evolution. Science **158**: 1200-1203.

Sarich, V.M., and A.C. Wilson (in prep.). Generation time and rates of molecular evolution with reference to the rodents.

Sarich, V.M., Wilson, A.C., and A. Bennett (1968). Paper presented to the American Society of Mammalogists. June 18, 1968.

Seal, U.S., and N.I. Phillips, and A.W. Erickson (1970). Carnivora systematics: immunological relationships of bear serum albumins. Comp. Biochem. Physiol. **32**: 33-48.

Smith, A.G., and A. Hallam (1970). The fit of the southern continents. Nature Lond. **225**: 139-144.

Szalay, F.S. (1968). The beginnings of primates. Evolution **22**: 19-36.

Simons, E.L. (1967). The earliest apes. Sci.Am. 217(b): 28-35.

Simons, E.L., and P.C. Ettel (1970). Gigantopithecus. Sci.Am. 222(1); 76-87.

Wang, A.C., J. Shuster, A. Epstein, and H.H. Fudenberg (1968). Evolution of antigenic determinants of transferrin and other serum proteins in primates. Biochem.Genet. 1: 347-358.

Washburn, S.L. (1968). The Study of Human Evolution. University of Oregon Press.

Weigle, W.O. (1961). Immunochemical properties of the cross-reactions between anti-BSA and heterologous albumins. J.Immunol. 87: 599-607.

Table 1

Antiserum pool to the albumin of: Aotus-Cebus difference

Cross-Reactions

	Index of dissimilarity			
	Aotus (ID_A)	Cebus (ID_C)	$ID_C - ID_A$	100 ($\log ID_c - \log ID_a$)
Homo sapiens	2.75	4.9	2.15	25
Presbytis entellus	2.70	4.6	1.9	23
Nycticebus coucang	10.0	15.2	5.2	18
Lemur fulvus	16.1	30.2	14.1	25
Ursus americanus	21.8	37.9	16.1	24
Genetta genetta	29.4	47.5	18.1	21

Table 2

Amounts of albumin evolution shown by the various primate albumin lineages relative to that of *Aotus trivirgatus* as seen from a carnivore perspective.*

Albumin species	Relative change
Nycticebus coucang	-15
Lemur fulvus	- 5
Propithecus verrauxi	- 5
Galago senegalensis	- 3
Tarsius spectrum	- 3
Aotus trivirgatus	0
Ateles geoffroyi	+ 9
Hylobates lar	+15
Saguinus oedipus	+16
Tupaia glis chinensis	+16
Pithecia monacnus	+18
Homo sapiens	+22
Cercocebus galeritus	+24
Cebus capucinus	+24
Macaca mulatta	+26
Presbytis entellus	+28

*The antisera used (16 rabbits were directed to the albumins of *Hyena hyena*, *Genetta genetta*, *Ursus americanus*, and *Arctogalida trivirgata*. The figures given indicate the mean amounts (in immunological distance units by which the albumins of the various species reacted more (-) or less (+) strongly than that of *Aotus*.

Table 3

Albumin immunological distances among the Prosimii

	Nyc	Le.f.	Tup	HP*
Nycticebus coucang	0	86	130	115
Galago senegalensis	45	95	NT	127
Lemur fulvus	86	0	130	127
Lemur catta	87	30	NT	129
Propithecus verrauxi	89	65	NT	131
Tarsius syrichta	105	126	103	123
Tupaia glis chinensis	130	130	0	123
Urogale everetti	NT	NT	31	125

*Data in first three columns obtained with antisera to the albumins of *Nycticebus coucang*, *Lemur fulvus*, and *Tupaia chinensis*. All figures for comparisons where both antisera are available are means. The last column (HP) gives mean distances of the albumins of *Homo*, *Cercocebus galeritus*, *Presbytis entellus*, *Saguinus oedipus*, *Cebus capucinus* and those of the prosimians listed. Data obtained with antisera to the albumins of the five simians (27 rabbits). NT indicates that the experiment has not been performed as yet.

Table 4

Amounts of albumin evolution shown by the various platyrrhine lineages relative to that of *Ateles geoffroyi*.*

Albumin species	Relative change
Aotus trivirgatus	-11
Callicebus cupreus	- 5
Pithecia monachus	- 4
Cacajao rubicundus	- 2
Callimico goeldii	0
Ateles geoffroyi	0
Leontideus rosalia	+ 2
Saguinus oedipus	+ 2
Callithrix jacchus	+ 3
Aloutta caraya	+ 4
Saimiri sciurea	+ 6
Cebus capucinus	+10

*The antisera used (18 rabbits) were directed to the albumins of *Homo*, *Presbytis entellus*, and *Nycticebus coucang*.

Table 5

Amounts of albumin evolution shown by various simian albumin lineages relative to that of *Aotus trivirgatus*.*

Albumin species	Relative change
Aotus trivirgatus	0
Macaca mulatta	+ 9
Homo sapiens	+13
Cebus capucinus	+22

*The antisera used (7 rabbits) were directed to the albumins of *Nycticebus coucang* and *Lemur fulvus*.

Table 6

Amounts of albumin evolution shown by various catarrhine albumin lineages relative to that of *Hylobates lar*.*

Albumin species	Relative change
Homo sapiens	+ 2
Hylobates lar	0
Presbytis entellus	- 3
Macaca mulatta	- 2

*The antisera used (17 rabbits) were directed to the albumins of *Aotus trivirgatus*, *Pithecia monachus*, *Ateles geoffroyi*, *Cebus capucinus*, and *Saguinus oedipus*.

Table 7

Albumin immunological distances among the Cercopithecoidea

	Anti-*Presbytis* albumin	Relative amounts of change*
Presbytis entellus	0	0
Pygathrix nemaeus	7	-5
Nasalis larvatus	9	-2
Colobus polykomos	9	-5

*Amounts of change shown by the four colobine albumin lineages since they last shared a common ancestor relative to the amount shown by the *Presbytis* lineage. Data obtained with antisera to the albumins of *Macaca mulatta*, *Cercopithecus aethiops*, *Papio papio*, *Cercocebus galeritus*, and *Homo Sapiens* (total of 24 rabbits). The mean cercopithecoid-hominoid distance is 35 units (Sarich and Wilson, 1967a).

	*Presbytis entellus***
Macaca mulatta	27
Papio papio	26
Cercocebus galeritus	22
Cercopithecus aethiops	25

**Data obtained with the five antiserum pools mentioned above.

Table 8

Albumin immunological distances among some platyrrhines*

	Ao	At	Ce	Sa
Aotus trivirgatus	0			
Ateles geoffroyi	35	0		
Cebus capucinus	37	40	0	
Saguinus oedipus	39	38	40	0

*Data obtained with antisera to the albumins of all four species (14 rabbits). The mean platyrrhine-catarrhine distance (obtained with antisera to the albumins of Homo, Pan, Macaca, Cercopithecus, and Presbytis (17 rabbits) and albumins of the platyrrhine genera Aotus, Pithecia, Callicebus, Ateles, Callimico, Saguinus, Lagothrix, Alouatta, Cacajao, Saimiri, and Cebus) is 59 units (Sarich and Wilson, 1967a).

Table 9

Protein Differences between *Homo*, *Pan*, and the Old World Monkeys

Protein	*Homo*-*Pan* difference	*Homo*-Old World Monkey difference (mean)	Scale used
Albumin	7	35	ID units
Hemoglobin	0	15	minimum mutational distance
Transferrin	3	30	ID units
Fibrinopeptides	0	7	minimum mutational units

Table 10

Albumin immunological distances among the Hominoidea*

	H	PT	G	P	Hy
Homo sapiens	0				
Pan troglodytes	7	0			
Pan paniscus	7	0			
Gorilla gorilla	9	7	0		
Pongo pygmueus	12	10	NA	0	
Hylobates lar	15	14	12	12	0
Symphalangus syndactylus	15	14	NA	NA	3

*Data obtained with antisera to the albumins of *Homo*, *Pan troglodytes*, and *Hylobates*. Antisera to *Gorilla* and *Pongo* albumins are in preparation and these data will become available soon. (NA) data not available.

Table 11

Albumin immunological distances among the Cercopithecinae*

	P	M.m.	C.ae.	C.gal.
Papio papio	0	6	12	7
Mandrillus sphinx	5	3	6	4
Macaca mulatta	6	0	9	3
Macaca maura	8	3	13	8
Cercopithecus aethiops	12	9	0	6
Cercocebus galeritus	7	3	6	0

*Data obtained with antisera (20 rabbits) to the albumins of *P. papio*, *M. mulatta*, *C. aethiops*, and *C. galeritus*.

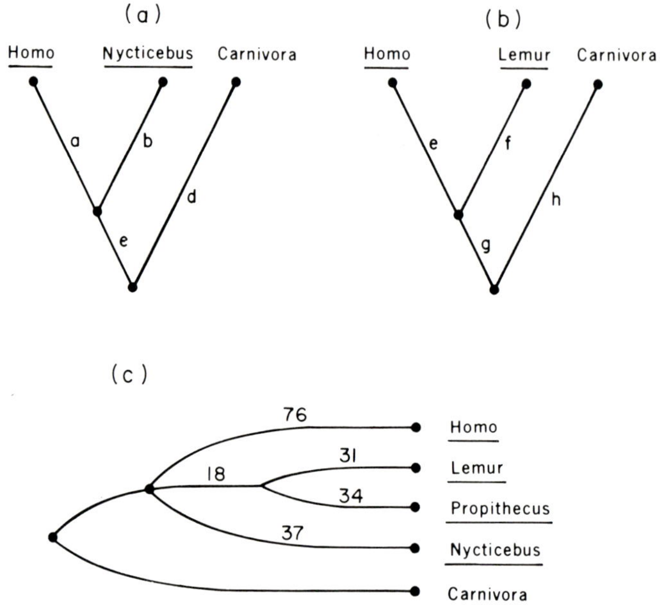

Figure 1 (a) A phylogeny for Homo, Nycticebus, and the various Carnivora. The letters along the lineages indicate the amounts (measured in ID units) of albumin evolution which took place along them.

(b) The same as (a) substituting Lemur for Nycticebus.

(c) An albumin phylogeny for the four genera indicated. The numbers refer to the amount of albumin evolution which took place along those lineages.

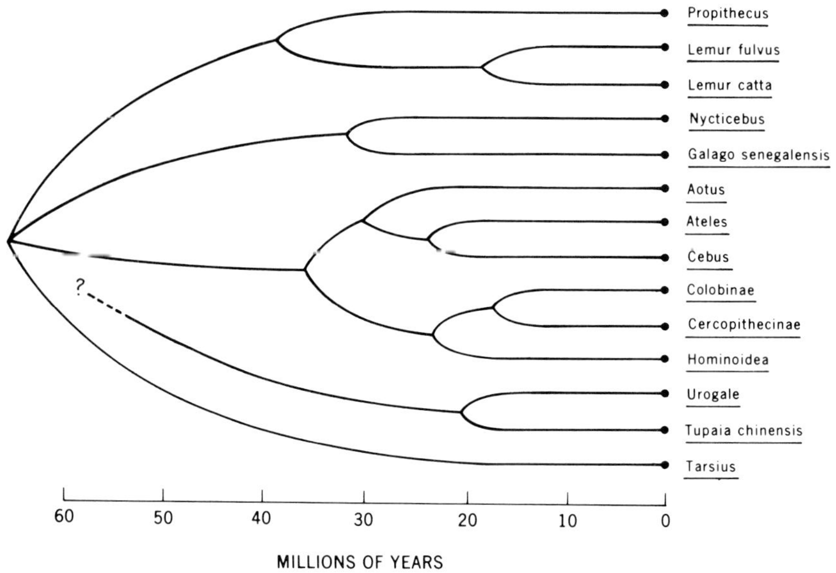

Figure 2 An albumin phylogeny for the primate taxa listed where the time scale is assigned assuming their most recent common ancestor lived around 65 million years ago. More detailed discussions on the Hominoidea, Cercopithecinae, and Colobinae are provided in Fig. 4 and the text.

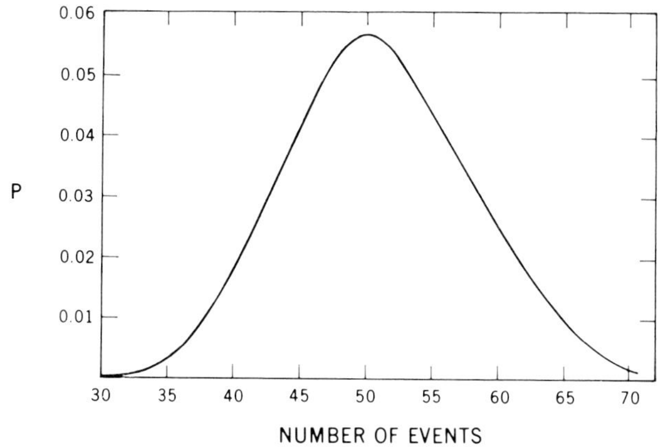

Figure 3 A Poisson distribution where the Poisson mean is 50 events.

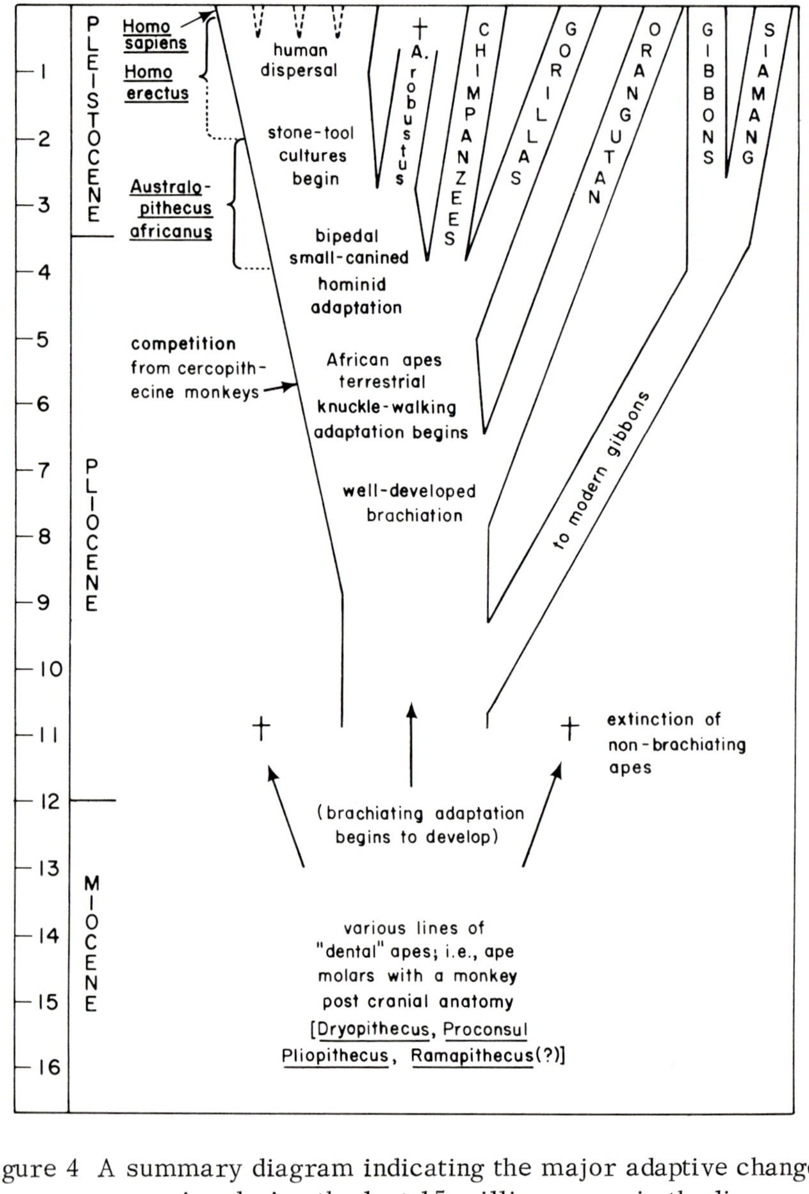

Figure 4 A summary diagram indicating the major adaptive changes occurring during the last 15 million years in the lineages leading to man and his pongid relatives.

PROTEIN STRUCTURE AND THE SYSTEMATICS OF OLD WORLD MONKEYS

N. A. Barnicot and Patricia T. Wade

PROTEINS AND PHYLOGENY

The relevance of genetics to problems of phylogeny probably needs no argument and this is also true for taxonomy in so far as it aims at classifications that are consistent with phylogenetic relationships. The characteristics of the species tend to be conserved by the remarkable fidelity of gene replication while rare deviations of gene structure supply the material for evolutionary divergence. In essence evolutionary change can be regarded as a process of genetical diversification in reproductively isolated populations and it can be argued that the most fundamental evidence of phylogenetic relationships is therefore to be found in the structure of the genome.

Of course in most evolutionary and taxonomic work it is no more than a tacit assumption that we are dealing with characters that depend largely on the genotype rather than with transient phenotypic modifications by local environment. It is seldom feasible or even necessary to verify this assumption by experiments. Even if we are satisfied on this score it is difficult or impossible to go further and specify the genetic differences involved. The anatomical features that are the currency of most taxonomic work are far from ideal genetical material and even in the limited cases where breeding experiments are possible they are unlikely to yield a very detailed picture of genetical differences between taxa.

Until quite recently a comparison of gene structure in distantly related organisms would have seemed a hopelessly ambitious project. The chemistry of genes was not fully known and although it was recognised that they are intimately concerned with the production of enzymes, their precise role in protein synthesis was not clear.

This situation was radically changed by the discovery that genes are made of DNA and the linear sequence of four organic bases in this long, thin molecule acts as a code specifying the sequence of amino acids (primary structure) of protein polypeptide chains. Protein structure thus emerges as an aspect of the phenotype that is separated from the genes by very few developmental steps. Due to the coding mechanism the primary structure of proteins yields information about the structure of genes concerned in their synthesis. An approach to the comparative anatomy of genes which does not depend on breeding experiments thus becomes possible.

The primary structure of one of the two short chains composing the insulin molecule was reported by Sanger in 1951 two years before a definitive structure for DNA was published. The details of the genetic code were worked out in micro-organisms and appear to be valid for higher forms. It is therefore possible to specify the sequences of three bases (codons) that correspond to each of the 20 amino acids. In the last 15 years many other proteins have been sequenced, some of them in a variety of species, though the growth rate of this new branch of systematics is still retarded by rather laborious laboratory methods.

HOMOLOGIES

The work on the respiratory enzyme cytochrome c ranks as a classic example of the application of protein chemistry to phylogenetic problems (Margoliash and Smith, 1965; Smith, 1968; Margoliash, Fitch and Dickerson, 1968). The molecule is quite small consisting of a folded chain of about 100 residues and it occurs in a very wide range of organisms. About thirty species including several mammals, birds, reptiles, insects and plants have now been examined.

In order to compare the array of sequences one must first align them so that corresponding sites are adjacent. Since the chains vary in length (vertebrates, 104; insects, 107; wheat, 112 residues) deletions or additions of genetic material or both must have occurred in evolution. However there is a sequence of eleven residues that is the same in all species examined. If this is aligned certain other scattered sites are also seen to be invariant. In aggregate identical sites comprise about one third of the total. When the aligned sequences are compared it is clear that the degree of similarity between them depends on the closeness of relationship between the organisms as judged on other evidence. Man differs from rhesus monkey at one site only, from horse at 12 sites, from the dogfish at 24, and from yeast at 45. By far the most acceptable explanation of these facts is that all cytochromes c are derived from a common ancestral protein which has undergone divergence in different lineages. In view of the constancy of the physico-chemical and physiological properties of this enzyme it seemed very likely at the outset that it was homologous throughout the array of forms examined. This is confirmed and greatly strengthened by the sequence data. However functional similarities may sometimes be misleading; the bacterial subtilisins resemble pancreatic proteases at this level but are quite different in primary structure (Smith, 1968).

Homologies become more doubtful as the proportion of identical sites decreases and the question arises whether slight resemblances are more likely to be vestiges of a common ancestral structure or whether they could have arisen independently in two or more lineages. It is a reasonable assumption that the most persistent sites and sequences will be those that play some vital part in the functioning of the molecule and can seldom be changed without detriment. Active sites of enzymes tend to be invariant and in hemoglobin some relatively invariant sites are known to play a part in bonding the heme group, attaching subunits to one another or forming interior bonds that maintain the three dimensional form. Presumably such sequence patterns did not arise at one stroke by a happy conjunction of mutations but attained their present state more gradually under the pressure of selection. If the structural alternatives compatible with efficient function are limited, selection would tend to favour the same or similar solutions in different lineages. While selection would not increase the chance of a functionally advantageous mutation turning up it would increase the chance of it being preserved. On the other hand identity of a few sites in otherwise remotely related organisms is best attributed to chance unless some functional significance can be found.

PHYLOGENETIC TREES

Protein sequences lend themselves readily to a quantitative treatment of resemblances. The number of sites identical in composition in two different species can be taken as a measure of similarity. The relationships between a set of sequences can then be displayed as a branching diagram. Such trees (see Fitch and Margoliash, 1967; Dayhoff, 1969) are usually found to resemble quite closely the familiar phylogenetic trees on anatomical and geological data, despite the fact that they are constructed from the evidence of a single polypeptide chain and hence of one gene out of thousands in the genome. Minor discrepancies are to be expected due to limited data and idiosyncrasies in the evolution of a particular protein in certain lineages, but it is to be anticipated that the picture will become more consistent as more species and more proteins are examined.

A more fundamental measure of evolutionary divergence would be the number of mutational events that have led to the observed differences of sequence. For several reasons this "mutational distance" cannot be determined with certainty. Most mutations seem to involve a change in a single base of the set of three (the codon) coding for a particular amino acid. But most of the amino acids have more than one codon and therefore some mutations (synonymous mutations) amounting to about 25 percent of all possible ones, do not result in a change of amino acid and cannot be detected by protein analysis. If a succession of substitutions has occurred at a site we can only detect the last of the series and if this happens to be a change to an amino acid that is present in related species we may be unaware that anything has happened. A count of amino acid differences therefore tends to underestimate the amount of mutational change and the discrepancy will increase as evolutionary divergence becomes greater.

In spite of these difficulties it is still possible to estimate a "minimal mutational distance" by considering the changes in relation to the genetic code. But if a particular amino acid has several alternative codons, and we do not know which is in use, we may underestimate; for example serine to alanine requires either one or two base changes depending on the serine codon chosen.

Several procedures have been described for constructing trees of relationship from measurements of resemblance and difference. A very large number of alternative trees could be constructed from a set of twenty or more items so that it is eminently a job for a computer. We need a criterion for choosing the most acceptable tree and in the interests of standardization we choose the one that has the minimum branch lengths. This assumption of minimal path is arbitrary and may not correspond to the true course of evolution in some instances. Cavalli Sforza Barrai, and Edwards (1964) have applied such methods to anthropometric and blood group frequency data on human populations, and Fitch and Margoliash (1967) and Dayhoff (1969) have published trees based on protein sequences. Fitch and Margoliash start from a matrix of mutational distances and proceed by a method of pairwise comparisons. Dayhoff prefers to reconstruct the most probable ancestral sequences by considering the amino acid most often present at each site in the putative descendants. The topology is then chosen which involves the least number of mutational changes from ancestral to descendant sequences. These procedures have the advantage of objectivity but they necessarily involve arbitrary, though not necessarily implausible, rules and they may sometimes lead to wrong conclusions if evolution has not obeyed these rules.

We have already mentioned the distortion of mutational distances arising from multiple changes at a site. Attempts have been made to correct for this (Zuckerkandl and Pauling, 1965; Margoliash and Smith, 1965) by assuming that changes are equally probable at all sites and then calculating the expectation of double, triple etc., changes. Dayhoff (1969) prefers an arbitrary, but probably more realistic, division of sites into three categories of mutability and gives empirically derived estimates of the relative mutabilities of the various amino acids.

THE RATE OF PROTEIN EVOLUTION

The important question arises of the relation of substitutional or mutational distances to time. Do all proteins evolve at the same speed and for any one protein is the rate of change constant over long periods and in different lineages? If so, protein differences might provide an evolutionary clock which can tell us about dates of divergence when reliable paleontological evidence is not available.

If proteins behave like anatomical characters (Simpson, 1944) there is certainly no reason to expect that their evolutionary rates will be constant. If they evolve mainly under the influence of selection, the rate of change would be essentially unpredictable since the complex circumstances that influence selection pressures are indeterminate. But if random processes (genetic drift) is an important factor the rates might be constant provided mutation rates do not change systematically. Proteins are intricate structures and, where we have sufficiently detailed information, it is clear that parts at least of their conformation are intimately adapted to particular functions. It is perhaps difficult to conceive that their primary structure can change very much without affecting function and, by implication, selection pressures. Nevertheless in laboratory tests cytochromes c from a wide range of species are very similar (Margoliash et al, 1968) and many human hemoglobin variants due to single substitutions show no detectable change of properties; some of these changes are in fact ones which occur in the normal sequence of other species. The molecule of myoglobin though it differs from hemoglobin chains at many sites (80 percent) has a very similar three-dimensional structure. One may either conclude that our tests are too crude to detect functional differences that natural selection might discriminate, or that many of the changes in proteins have little or no effect on function and are neutral from the point of view of selection. The side chains of certain amino acids are similar in size, polarity, hydrophobic properties etc., so that we might expect little change when one of them is substituted for another. However predictions on physico-chemical grounds may not always be reliable especially if the three-dimensional structure of the molecule is not known in detail so that the possible interactions of side chains cannot be appreciated. Zuckerkandl and Pauling (1965), in a thorough review of this problem, took the view that the most frequently observed substitutions were likely to be ones that produce minimal changes of properties but after examining the nature of such "conservative" changes they remarked that "apparently chemists and protein molecules do not share the same opinion regarding the definition of the most prominent properties of residues".

It is clear that all proteins do not evolve at the same rate as the following extracted from Dayhoff's (1969) larger list, shows:

Protein	Accepted point mutations per 100 million years.
Fibrinopeptides	90
Ribonuclease	30
Hemoglobin α chain	11
Hemoglobin β chain	13
Insulin	4
Cytochrome c	3

The accuracy of the figures is subject to various limitations but the differences are too large to be ignored. We might explain the rapid evolutionary rate of fibrinopeptides by saying that they are highly responsive to selection and that their varied structure in mammals is an adaptive phenomenon; or we might take the view that their structure is not critical and that selection has been relaxed. Fibrinopeptides are small fragments lopped off the ends of long fibrinogen chains in clotting. Although their configuration in this molecule is not known, it seems plausible that it is not very important so that the second hypothesis is the more attractive. There is evidence that the connecting piece joining the A and B chains of insulin in the precursor molecule (proinsulin) is much more variable than these chains themselves (Chance et al, 1968; Nolan et al, quoted in Margoliash et al, 1968). This is probably another example of the same phenomenon since the junctional region is eliminated when the active conformation is achieved. Presumably all genes coding for proteins are equally susceptible to mutation. Differences of evolutionary rate may therefore depend in part on variations in the stringency of selection in eliminating variants. We can conclude that changes that have become incorporated in a species are not disadvantageous but whether they are advantageous or merely indifferent is more difficult to determine.

The next question is whether the rate of change in a particular protein is constant over long periods of time and in different lineages. We can only answer it by appealing to paleontological evidence of divergence; for some groups of organisms and for remote periods of geological time this evidence may not be very good. Most workers on protein sequences have adopted a cautious attitude to this problem while admitting that on average rates may be fairly constant. Fitch and Margoliash (1967) point out that in their phylogenetic tree of cytochrome c, the mutational distances between primates and other mammals imply a more rapid rate of change of this enzyme in the former; but the difference is relatively small and might fade away if more species were examined. Margoliash et al (1968) draw attention to the fact that the distances between lemurs and other primates in hemoglobin chains are as great as those between the latter and other mammals, indicating either that the prosimians diverged roughly at the time of the main radiation of mammalian orders or that hemoglobin has evolved faster in the prosimian line. An interesting example which seems to imply inconstancy in the rate of change of the hemoglobin β chain has recently been provided by Boyer et al (1969a) and will be discussed below. Both more paleontological and more sequence data are needed to reach definite conclusions about this problem.

Kimura (1968a, b) approached the question of the role of selection in protein evolution from a different angle. From data on three proteins he estimated the average rate of change to be one substitution in 28 million years for a chain of 100 residues. Assuming that all the DNA in the genome codes for proteins, an estimate of one base-pair replacement per 1.8 years for the genome as a whole is reached. Haldane (1957) calculated the cost, in terms of mortality, of selectively replacing one allele by another and concluded that a replacement rate of more than about one allele in 300 generations could not be tolerated by the species, so that the rate of replacement calculated by Kimura would certainly not be tolerable if it were due to selection. Kimura was therefore led to conclude that most of the mutations involved in protein evolution must be neutral from the point of view of selection and that replacement depend on random processes. In these circumstances, as he showed, replacement rates would depend on and be approximately equal to mutation rates. The probability of a mutation being fixed in a species (that is to say of completely replacing its allele) is an inverse function of population size (strictly, effective size) once it has occurred; but the chance of it occurring is a direct function of this parameter, so population size sould not influence replacement rates.

King and Jukes (1969) have marshalled various lines of evidence in support of Kimura's opinion though they criticize his argument on two scores and have revised his estimate of the observed average rates of protein evolution. Firstly, there is evidence from DNA hybridization studies and other sources that by no means all the genome is concerned in coding for proteins and if this is so Kimura's estimate of overall mutation and replacement rates may be as much as one hundred times too high. Secondly they point out that Haldane's conclusion about the cost of selection may not be valid since, as Maynard Smith has shown (1968), it depends on the implausible premise that selection at different loci is independent and very much higher rates would be tolerable on other assumptions. This would mean that there is no need to assume that a high proportion of the mutations of proteins are neutral. However there are other grounds for thinking that they may be, and among those advanced by King and Jukes (loc. cit.) the following may be mentioned. Taking the data on the sequence differences between various species in cytochrome c and globin chains and excluding invariant sites, distributions of the number of sites that have experienced one, two or up to nine changes, can be obtained. These distributions fit Poisson distributions quite closely which suggests that all these sites are equally changeable and that replacement is at random. Synonymous mutations, according to King and Jukes, are likely to be neutral because no change of amino acid is produced; but the possibility that all codons are not equally efficient and that some mutations of this type may affect rates of synthesis should not be ignored. If the incorporation of particular substitutions is due to selection, then favorable primary sequences would tend to be achieved in the course of evolution irrespective of the genetic code which dictates the number of mutational steps needed to pass from one amino acid to another. On the other hand if most replacements are due to selectively neutral mutations, amino acids which have more alternative codons would tend to occur more often. If an average composition derived from analysis of many different proteins is compared with the composition expected on the basis of random changes in the nucleic acid bases, the two agree quite well with one or two exceptions. But, as Richmond (1970) emphasises in a critique of the concept of non-Darwinian evolution, the genetic code itself has presumably evolved so as to minimize the chance of deleterious substitutions. Mutational changes in structural genes are thus biased towards neutrality.

As King and Jukes themselves point out the rate of protein evolution is such that natural selection is not excluded on this count. The observed rates could be achieved with quite low selective pressures. The rate of evolution by natural selection would not however be independent of population size and in small populations larger selection pressures would be necessary to prevail against genetic drift.

An important consequence of the "non-Darwinian" mechanism is that evolutionary rates would depend largely on mutation rates and would be constant if these were constant; whether they are we do not know. Rapidly changing proteins, such as fibrinopeptides, presumably give us some idea of the rates of change that can occur when selection is partly relaxed. It follows that other, more conservative proteins must experience the same rate of occurrence of mutant forms but that most of these are deleterious and are prevented from spreading by selection.

If protein divergence is mainly due to this non-Darwinian process we should not expect to find proteins that are "living fossils", preserving the structure of an ancestral protein more or less unchanged. In a given time each of a set of descendant proteins should have diverged to about the same extent from their common ancestor. Of course all proteins (or indeed all organisms) may be regarded as "living fossils" in so far as all retain traces of ancestral structure; the point is that we should not expect to find a disproportionate lag in a particular lineage. It is unlikely that we shall be able to test this proposition by direct examination of ancestral proteins from fossils, though it is conceivable in the case of bone proteins. We can however infer the most probable structure of an ancestral protein (Fitch and Margoliash, 1967; Dayhoff, 1969) from the structure observed in a set of descendants, but this procedure involves various assumptions and requires data on as wide a spectrum of related forms as possible. How does the concept of continuous change in proteins fit with the occurrence of "living fossils" at the anatomical level? We know too little about the relations between protein structure and gross morphology to give a definite answer but is is not unlikely that the latter could be stabilized by selection and yet allow a fair amount of selectively neutral change in component proteins.

HEMOGLOBINS OF OLD WORLD MONKEYS

Over the course of several years we have examined material from a fair range of species and have compared it with that from certain New World genera and with hominoids. Ideally one would wish to have abundant material from as many species of Anthropoidea as possible. In order to detect polymorphisms one needs fairly large samples from different parts of the distribution range of each species since the frequency of variants may differ between regions. In practice one is usually restricted to the quite narrow range of species that are used for laboratory work and the exact provenance of the material is not always easy to discover. Thus although we have examined quite large numbers of baboons, certain species of macaques and patas monkeys, we have had few specimens of colobines and few of the genus *Cercopithecus* other than the vervet. The trapping of natural populations, which would provide data of the greatest interest for populations genetics, is not easy and would be a major project.

We hope that our biochemical work will provide data that are useful to primate taxonomists. Reciprocally we need the expertise of taxonomists to get reliable identification of the animals. This is not easy to arrange since our material is apt to come from many different laboratories and primate taxonomists are few and far between.

HEMOGLOBIN STRUCTURE

One of the simplest methods of comparing hemoglobins is by

electrophoretic migration in starch gels, but with rare exceptions, this only detects differences due to net charge on the molecule and gives minimal information about structural differences. Two hemoglobins with different mobilities almost certainly differ at at least one site, but identical mobilities do not exclude major differences of structure.

It is not without interest that the normal adult hemoglobins of most of the Old World species we have examined are very similar to human Hb-A (the normal adult hemoglobin) in mobility at alkaline pH, and for that matter they are like those of most New World forms and hominoids. Why this should be so is not clear but it suggests that the net charge tends to be stabilized by natural selection. There are some exceptions. The hemoglobin of baboons, mangabeys, probably mandrills, and conspicuously the gelada baboon, is slower than Hb-A. In fact both the mandrills we examined had two hemoglobins, and we have since found this to be so in additional samples of mandrills and drills studied in a collaborative project with Dr. Morris Goodman and Dr. Andrew Wilson. The possibility that there is a gene duplication in these species is worth considering.

A little more information can be got by dissociating the hemoglobin into α and β chains by treatment with the mercurial PCMB (Rosemeyer and Huehns, 1967) so that their individual mobilities can be compared in different species. Fig. I shows the result of such an experiment on four species. Both the α- and β-chains of man have the same mobilities as those of the macaque whereas both are different from those in the baboons. The slow mobility of gelada hemoglobin relative to <u>Papio</u> is evidently due to an α-chain difference.

Nearly all humans have a minor component (Hb-A)which is slower in electrophoretic mobility than Hb-A and is present in low concentration (about 2.5 percent). It is composed of α-chains and δ-chains ($\alpha_2^{A}\delta_2^{A2}$) in contrast to Hb-A ($\alpha_2^{A}\beta_2^{A}$). The δ-chain differs from the β at only ten sites and it was therefore suggested by Ingram (1961) that it arose by gene duplication of the β-locus at a relatively recent data as compared with the origin of the β-chain itself by duplication of the α.

Hoffman et al (1967) found a minor component similar electrophoretically to Hb-A_2 in all of 109 chimpanzees and we have found it in all the chimpanzees (44) gorillas (2), orangs (10) and gibbons (15) we have examined. In the last named species there is polymorphic variation of the minor component (Barnicot and Jolly, 1966; Barnicot, Jolly and Wade, 1967). We have made a preliminary study of ape minor components using the fingerprinting method. The globin is enzymatically digested into smaller peptides which are then separated on paper by a combination of electrophoresis and chromatography so that the positions of the peptides can be compared. This work suggests that the ape minor components resemble the human quite closely and it may well be that they have a homologue of the δ-chain.

We have not detected a minor component in any Old World monkey species we have examined, other than Macaca fascicularis (= irus). In this species a minor component similar in mobility and concentration to Hb-A is present in only about 20 percent of individuals. It proves to be a derivative of the α-chain from which it differs by at least four substitutions (Barnicot, Jolly and Huehns, 1966; Wade, Barnicot and Huehns, 1967) and it is therefore clear that it has an independent evolutionary origin from the hominoid δ-chain. If it arose by duplication of the α-chain subsequent to the separation of M. fascicularis (= irus) this would imply that four differences have accumulated in perhaps one or two million years - quite a rapid rate. If it arose earlier we might expect to find its homologue in some other macaque species or even in other genera. So far we have not, but a continued search is needed.

We have also examined fairly small samples of several New World genera (Ateles, Saimiri, Cebus, Aotus, Lagothrix, Saguinus, Leontideus) and we found a minor component in all individuals. Boyer et al (1969a) have found the same in a study on larger samples. Our chain-splitting and hydridization experiments showed that it differed from the major hemoglobin in the non-α chain and to that extent is similar to human Hb-A_2. However our preliminary fingerprinting studies on the minor component of Cebus and Aotus suggest that there may be considerable differences from the human δ-chain. Boyer et al (1969a, 1969b) have now taken the analysis of this problem much further and have published data for three New World genera which we shall discuss below.

The fingerprinting method has serious limitations for species comparisons since differences that cause no change in the electrophoretic or chromatographic mobilities of peptides will remain undetected. A further step is to analyse the amino acid composition of each peptide. Even this may fail to reveal all the differences since although a given peptide in two species may have the same composition the amino acids may not be in the same linear order. It is possible to determine the actual sequence for each peptide but this is very time-consuming and has only been done for two primates, namely man and the rhesus monkey. Most of our information on the primary structure of primate hemoglobin chains is therefore based on peptide compositions, the sequence being inferred by comparison with these two thoroughly analysed species.

Rifkin and Konigsberg (1965) detected no difference between man and the chimpanzee in the α and β-chains. The human and chimpanzee cytochrome c also appear to be identical but this slowly evolving protein is less useful for primate comparisons; man and rhesus monkey differ at two sites only. Recently, however, Doolittle and Mross (1970) have shown that the fibrinopeptides of the chimpanzee and man are identical in sequence in contrast to four genera of Old World monkeys which differ from man at 5 - 8 of the 30 sites compared. The data from chemical analysis are therefore in accord with serological studies (Goodman, 1968; Sarich and Wilson, 1967) in showing close similarities between human and chimpanzee proteins. The most reasonable explanation of these facts is that these two lineages diverged quite recently but this is in marked conflict with current paleontological opinion (Simons and Pilbeam, 1965; Simons, 1965, 1967) which would place the separation of hominids at least as early as the Miocene and perhaps in the Oligocene.

The α and β chains of the gorilla and man are also remarkably similar (Zuckerkandl and Schroeder, 1961; Zuckerkandl in Dayhoff, 1969). On the other hand the thorough sequence study of Macaca mulatta by Matsuda et al (1968) showed four differences in the α chain and eight in the β . Our own work on M. fascicularis (Wade, Barnicot and Huehns, unpublished) shows a similar degree of overall difference between this species and man and at least three differences between the two macaques; this illustrates how deceptive fingerprinting results alone can be, since the peptide patterns of man and macaque hemoglobin are very similar.

Hill et al (1963) remarked on the very different fingerprint patterns of baboon (Papio) and human hemoglobins and we confirmed this (Barnicot et al, 1965). Since then we have compared Papio and Cercocebus by this method and find considerable similarities as shown in Fig. 2 There must be considerable differences of sequence between macaque hemoglobin and those of the baboon and mangabey.

Boyer's (loc. cit.) work on the α and δ chains of ceboids may not be strictly relevant to this symposium but is too interesting to ignore. He felt that the minor components could hardly be much use in oxygen carriage and, since no other function for them has been discovered, they may be good examples of proteins that have evolved largely by random change. The fact that they seem to be present in all hominoids and ceboids suggests, but does not prove, that they may actually have some significant function. The persistence of these hemoglobins in such low concentrations is puzzling but at present we can do little more than speculate as to how it came about. It could be that this particular duplication event (assuming homology) gave rise to a gene that produced less protein than normal, perhaps due to damage to initiator or terminator regions. If so the duplicate may never have been exposed to significant selection pressure and could have become fixed and have diverged by random processes. But if the duplicate initially produced about as much protein as the original gene we have to account for the subsequent decrease. If this was brought about by selection this would reduce the chance of the duplication being fixed in the species by drift.

The sequence differences as determined by Boyer from peptide compositions are shown in Table I and we have added the rhesus monkey and M. fascicularis data. On the whole the human and ceboid δ-chains are not very much alike but they share a common difference from the β-chains in the substitution of arginine and asparagine for histidine at positions 116 and 117. This is suggestive of homology but the presence of glycine at position 5 in all the ceboid β and "δ" chains, instead of proline as in man, supports the opposite interpretation. It is interesting that while β 116 His is found in the rhesus monkey, mouse and rabbit, arginine occurs at this position in various ungulates as it does in primate chains. If the δ-chains had retained in this an ancestral feature of the β-chain this would place the duplication event early in the radiation of mammalian orders and we might then expect to find δ-chain homologues in some non-primate mammals. As far as we know this is not the case, though no very thorough search seems to have been made.

Although the evidence for homology of the ceboid and human δ-chains is rather slight and equivocal we may follow Boyer in believing that this is the most likely interpretation of the data. We note that all the δ-chains differ from the β-chains to about the same extent, which at least suggests that they may have been diverging for roughly the same length of time in the two lineages. It is noticeable that only at position 5 do the three ceboid β-chains have the same differences from the human β. This suggests that most of the differentiation of the ceboid chains occurred after these genera had become distinct. On the other hand their "δ"-chains are identical, but different from the human δ, at seven sites in addition to position 5, so that, by the same reasoning, much of the differentiation of these chains occurred before generic divergence. During this phase little change seems to have taken place in their β-chains. It appears that the rate of change in one or other of these chains, or perhaps both, has not been constant and this should serve as a warning that the protein clock cannot always be relied on to keep correct time.

It is also interesting that the β-chain of the rhesus monkey shows as many, or more (8), differences from the human β than those of the ceboids (6-7). If we assume constant rates of change in these three lineages we might have expected the ceboid chains to be more different from human than the macaque chain on the grounds that the divergence of cercopithecoids from hominoids was substantially later than the divergence of Old World from New World anthropoids. This early phase in the evolution may prove to be ill-founded (See Sarich, this volume); but if the ceboid β chains evolved slowly at first this would account for the discrepancy.

HEMOGLOBIN POLYMORPHISM

Polymorhpic variation of many different blood proteins has been detected in man but the origins and biological significance of this biochemical variability is a metter for conjecture. A large proportion of the variants are quite rare but some attain high frequencies in one or other part of the world and in these cases it is usually argued that we are dealing with a balanced polymorphism maintained by some local selective factor. Almost the only case where there is convincing evidence of a specific selective factor is that of sickle cell hemoglobin which seems to confer significant immunity to malaria in early life when the gene is present in the heterozygous state. The geographical distribution of β-thalassemia, hemoglobin E and glucose-6-phosphate dehydrogenase deficiency suggest that falciparum malaria may also be involved in these polymorphisms.

A study of protein polymorphisms in other primates provides a background against which the phenomenon in man may be viewed. In addition, as Goodman (1968) has stressed, protein variants could tell us much about the population genetics of lower primates. At the present time when intensive field studies on natural primate populations are proliferating, integration of the findings with genetical data is much to be desired, though it is understandable that studends of behavior are not always receptive to plans for wholesale trapping and bleeding of animals. Nevertheless studies on the blood from natural groups could provide important data on such topics as inbreeding in social groups and rates of gene exchange both within and between species. Furthermore the study of nonhuman primates may lead to the discovery of variants with interesting biochemical properties, the polymerizing hemoglobin of Macaca fascicularis (see below) being a case in point.

It appears that Old World monkeys are very variable in the occurrence of hemoglobin polymorphism. Some species, such as Macaca fascicularis, M. nemestrina, M. arctoides (=speciosa) and M. sinica have high frequencies of variants and this may also be true of the Celebes black ape (Jolly and Barnicot, 1966). In others, such as M. mulatta, M. cyclopis, Cercopithecus aethiops, Erythrocebus patas and Theropithecus, variants have not been observed. In the macaques, as in man, there may well be a correlation between the distribution of certain variants and that of simian malarias. We have investigated the variants in M. fascicularis fairly thoroughly and we found (Barnicot et al, 1966) high frequencies of two α-chain variants in material from Thailand and Vietnam. It seems from earlier reports (Lie Injo et al, 1960) that the situation may be similar in Java. One of these variants (Hb-P^{mi})polymerizes as hemolysates age. The other (Hb-Q^{mi})does not do this, but is found to vary considerably in relative concentration when present with the normal hemoglobin (Hb-A^{mi}). The phenotype frequencies in this material were consistent with these variants being determined by alleles at the α-locus. However we have since examined over 200 samples from Malaya and we find that about 80 percent of them have two hemoglobins resembling Hb-A^{mi} and Hb-Q^{mi} and most of the remainder have Hb-A^{mi} alone. These curious phenotype frequencies can hardly be explained by selective elimination of homozygotes since this would require an implausibly high mortality. It seems more likely that it is due to the presence of an α-locus duplication such that one of the duplicate genes produces Hb-A and the other Hb-Q^{mi}. In man the sickle cell gene is virtually lethal in the homozygous state and, in areas where the gene frequency is high, up to 4 percent of the population may be eliminated prematurely. The cost of maintaining the polymorphism is therefore rather high. If a similar selective mechanism is operative in the macaques a duplication producing a phenotype (AQ) like the heterozygote would be advantageous since it would entail no production of Q homozygotes which may perhaps be at a disadvantage. We have recently (Barnicot, Cohen and Wade, in press) found further evidence that such a duplication occurs, for we have detected a few animals which appear to have all three hemoglobins Hb-A^{mi}, Hb-Q^{mi} and Hb-P^{mi} and therefore carry three αchain genes. Since the last-named appears to comprise about 40 percent of the total pigment this suggests that such animals are heterozygous for an α-allele producing Hb-P^{mi} and a duplication producing both Hb-A^{mi} and Hb-Q^{mi}. It is interesting that Oliver and Kitchen (1968) have reported that every individual in a sample of M. arctoides had two hemoglobins and suggested a gene duplication to account for this. Experimental breeding, which is a slow business in monkeys, would of course be very helpful in analysing this situation.

One point that emerges from these studies is that it is rash to generalize about polymorphism from material derived from a limited part of the species range. It would be interesting to know whether chemically identical variants occur in different but related species. It seems likely that an ancestral polymorphism might be retained in descendant lines, at least for a time, though the possibility of independent mutational origin cannot be eliminated. Although the electrophoretically fast variant in M. arctoides resembles Hb-Q^{mi} in its behavior on starch gels it is evidently not the same since the two hemoglobins differ in fingerprint patterns.

Although the Buettner-Janusch's (1964) have reported the occurrence of two hemoglobins in certain ceboids we have not yet observed this and Boyer et al (1969) found only one animal with a major hemoglobin variant among 343 specimens examined. It would be interesting if hemoglobin polymorphism proves to be rare in ceboids since some authorities (Dunn, 1965) believe that the malaria parasites Plasmodium brasilianum and P. simium of these monkeys may have been derived from the human quartan and benign tertian parasites respectively, which would imply that simian malaria is of relatively recent origin in the New World.

RED CELL ENZYMES

We have also compared various species of Anthropoidea with respect to the electrophoretic patterns of five red cell enzymes (Barnicot and Cohen, 1970). These enzymes are not very stable so the material was stored in liquid nitrogen so that we could compare samples received at different times. The number of individuals examined for most of the species was not large and the chance of detecting polymorphism correspondingly low.

A good account of the human red cell enzyme variants will be found in Giblett (1969). They are present in low concentration in the red cells and special methods must be used to show them up on starch gels. Most of these enzymes regularly show not one, but a series of bands, indicating that there are several forms differing in charge or size. The structural relationships between such *isoenzymes* is not in all cases clear. They may arise because the enzyme is composed of two different chains of differing charge and these combine at random to form tetramers, as in the case of lactic dehydrogenase (LDH), because more than one configuration of the molecule is possible or because secondary losses or additions to the gene product occur. Many of the enzymes which have been studied occur widely in other tissues than the red blood cell. In the mature red cell they are no longer being synthesized. Some, such as glucose-6-phosphate-dehydrogenase (G6PD) and 6-6-phosphogluconate dehydrogenase (6PGD), have a functional importance in the red cell since they are involved in the pentose shunt glycolytic pathway which keeps the heme iron and also glutathione in the reduced state.

Simple starch gel comparisons are obviously only a first primitive step in a comparative study. Nevertheless results of taxonomic interest can emerge, for as Tashian (1965) showed in a study of primate red cell esterases and carbonic anhydrases, there are certain distinctive features of the patterns of bands in the three anthropoid superfamilies.

We found that the mobility of G6PD was quite similar in all species studied and we did not detect polymorphism; the mobilities in the tamarin and in the gibbon were the most deviant. Quantitative assays showed that the activity of this enzyme was considerably higher in most of the ceboid genera than in man but this was also true of the orang.

Perhaps the most interesting point about the acid phosphatase pattern (Fig. 3) was that M. fascicularis, Papio, Theropithecus and and the Celebes black ape all showed bands with relatively high mobilities but the langur showed a pattern more like that in hominoids and most ceboids. It is also interesting that our two specimens from Oedipomidas, which is sometimes regarded as only a subgenus of Saguinus, were quite distinct from it in AP pattern and also in certain other respects.

Polymorphic variation of 6PGD was detected in M. fascicularis, Ateles, Saguinus and Oedipomidas. The gibbon differed from the other hominoids in that this enzyme had a relatively slow mobility as in Papio, the mandrill and the gelada while the single band in the langur was unusually fast.

The patterns of phosphoglucomutase (PGM) are more complex (Fig. 4) and it has been shown (Harris, Hopkinson and Luffman, 1968) that in man three gene loci are involved. It is noteworthy that the patterns in the two subspecies of Hylobates lar (identified by Dr. C.P. Groves) were different. A complex polymorphism exists in M. fascicularis, the pattern types differing mainly in the relative strengths of bands. The langurs have a remarkable pattern consisting of a series of faint very fast bands; it is possible that they represent the products of the third gene locus which has been detected in man (Hopkinson and Harris, 1968). The PGM patterns of the ceboids are to a certain extent distinctive in that they usually have only three bands. There is a non-specific oxidase which shows up as a clear area in the gels because it prevents the local development of colorization when the reagent is applied. It is a good taxonomic discriminant. This series of bands is relatively slow in all the Old World monkeys examined as compared with hominoids, other than the orang and gibbon, and with New World monkeys. All seven genera of ceboids could be distinguished from one another by the mobility of this band.

ACKNOWLEDGEMENTS

We are grateful to the MRC Unit in Haemolytic Anaemias, University College Hospital Medical School for the use of finger-printing equipment.

REFERENCES

Barnicot, N.A., C.J. Jolly, E.R. Huehns, & N. Dance. (1965). Red cell and serum protein variants in baboons. In The Baboon in Medical Research. (Ed. Vagtborg, H.) Austin: Univ. Texas Press.

Barnicot, N.A., C.J. Jolly, & E.R. Huehns. (1966). Biochemical studies on hemoglobin variants of the irus macaque. Proc. Roy. Soc. B 165: 224-244.

Barnicot, N.A., C.J. Jolly, & Patricia T. Wade. (1967). Protein variations and primatology. Am. J. Phys. Anthrop. n.s. 27: 343-356.

Barnicot, N,A. & Philip Cohen. (1970). Red cell enzymes of primates (Anthropoidea). Biochem. Genet. 4: 41-57.

Barnicot, N.A., Philip Cohen & Patricia T. Wade. (1970). Evidence for a second hemoglobin α-locus duplication in the irus macaque. (in press).

Boyer, S.H., E.F. Crosby, T.F. Thurmon, A.N. Noyes, G.F. Fuller, S.E. Leslie, M.K. Shephurd. (1969a). Hemoglobins A and A_2 in New World primates: comparative variation and its evolutionary implications. Science 166: 1428-1431.

Boyer, S.H., E.F. Crosby, G.L. Fuller, A.N. Noyes, & J.G. Adams. (1969b). The structure and biosynthesis of hemoglobins A and A_2 in the New World primate Ateles paniscus: a preliminary account. Ann. NY Acad. Sci. 165: 360.

Buettner-Janusch, J.V. Buettner-Janusch. (1964). In Evolutionary and Genetic Biology of Primates. (Ed. J. Buettner-Janusch.) Vol. II. New York: Academic Press.

Cavalli-Sforza, L.L., I. Barrai & A.W.F. Edwards. (1964). Analysis of human evolution under random genetic drift.

Cold Spring Harbor Symp. 29: 9-20.

Chance, R.E., R.M. Ellis & W.W. Bromer. (1968). Porcine Proinsulin: characterisation and amino acid sequence. Science 161: 165-167.

Dayhoff, Margaret O. (1969). Atlas of protein sequence and structure. Silver Spring, National Biomedical Research Foundation.

Doolittle, R.F. & M.O. Mross. (1970). Identity of Chimpanzee with Human Fibrinopeptides. Nature, Lond. 225: 643-644.

Dunn, F.C. (1965). On the antiquity of malaria in the western hemisphere. Hum. Biol. 37: 385-393.

Fitch, W.M. & E. Margoliash. (1967). Construction of phylogenetic trees. Science 155: 279-284.

Giblett, E.R. (1969). Genetic markers in human blood. Oxford & Edinburgh: Blackwell Scientific Publs.

Goodman, M. (1968). Evolution of catarrhine primates at the macromolecular level. Primates in Medicine 1: 10-26.

Haldane, J.B.S. (1957). The cost of natural selection. J. Genet. 55: 511-524.

Harris, H., D.A. Hopkinson & J.J. Luffman. (1968). Enzyme diversity in human populations. Ann. N.Y. Acad. Sci. 151: 232-242.

Hill, R.L., J. Buettner-Janusch & R. Buettner-Janusch. (1963). The Evolution of hemoglobin in Primates. Proc. Nat. Acad. Sci. 50: 885-893.

Hoffman, H.A., A.J. Gottlieb & W.G. Wisecup. (1967). Hemoglobin polymorphisms in chimpanzees and gibbons. Science 156: 944.

Hopkinson, D.A. & H. Harris. (1968). A third phosphoglucomutase locus in man. Ann.Hum.Genet. 31: 359.

Ingram, V.M. (1963). The hemoglobins in genetics and evolution. New York: Columbia Univ. Press.

Jolly, C.J. & N.A. Barnicot. (1966). Serum and red cell protein variations of the Celebes black ape. Folia Primat. 4: 206 - 220.

Kimura, M. (1968a). Genetic variability maintained in a population due to mutational production of neutral and nearly neutral isoalleles. Genet.Res. 11: 247-269.

Kimura, M. (1968b). Evolutionary rate at the molecular level. Nature, Lond. 217: 624-626.

King, J.L. & T.H. Jukes. (1969). Non-Darwinian evolution. Science 164: 788-798.

Lie-Injo, Mau Eng., M. Mansjoer, & H.W.A. Donhuysen. (1960). Two types of hemoglobin in Macaca irus mordax (Thos. & Wrought) monkeys. Communs.Vet. 4: 59-64.

Margoliash, E. & E.L. Smith. (1965). Structural and functional aspects of cytochrome c in relation to evolution. In Evolving Genes and Proteins (Ed. Bryson, V. & H.J. Vogel). New York & London: Academic Press.

Margoliash, E., W.M. Fitch & R.S. Dickerson. (1968). Molecular expression of evolutionary phenomena in the primary and tertiary structure of cytochrome c. Brookhaven Symp. in Biology. 21: 259-305.

Matsuda, G., T. Maita, H. Takei, H. Ota, M. Yamaguchi, R. Miyauchi, & M. Migita. (1968). The primary structure of adult hemoglobin from Macaca mulatta monkey. J.Biochem. 64: 279-282.

Maynard-Smith, J. (1968) "Haldane's dilemma" and the rate of evolution. Nature, Lond. 219: 1114-1116.

Nolan, C., E. Margoliash, O. Hallund & D.F. Steiner. (unpubl.). Quoted in Margoliash, Fitch & Dickerson (1968) loc.cit.

Oliver, E. & H. Kitchen. (1968). Hemoglobins of adult Macaca speciosa: an amino acid interchange (15 gly-asp). Biochem. Biophys. Res. Comm. 31: 749-754.

Rifkin, D. & W. Konigsberg. (1965). The characterisation of the tryptic peptides from the hemoglobin of the chimpanzee (Pan troglodytes). Biochem. Biophys. Acta. 104: 457-461.

Rosemeyer, M.A. & E.R. Huehns. (1967). On the mechanisms of the dissociation of hemoglobin. J. Mol. Biol. 25: 253-273.

Sarich, V.M. & A.C. Wilson. (1967). Immunological time scale of hominid evolution. Science 158: 1200-1203.

Simons, E.L. (1965). New fossil apes from Egypt and the initial differentiation of Hominoidea. Nature Lond. 205: 135-139.

Simons, E.L. & D.R. Pilbeam. (1965). Preliminary revision of the Dryopithecinae (Pongidae, Anthropoidea). Folia primat. 3: 81-152.

Simons, E.L. (1967). The earliest apes. Scientific American 217: 28-35.

Simpson, G.G. (1944). Tempo and mode in evolution. New York: Columbia Univ. Press.

Smith, E.L. (1968). The evolution of proteins. The Harvey Lectures. New York & London: Academic Press.

Tashian, R.E.I. (1965). Genetic variation and evolution of carboxylic esterases and carbonic anhydrases of primate erythrocytes. Am. J. Hum. Genet. 17: 257-272.

Wade, Patricia T., N.A. Barnicot, & E.R. Huehns. (1967). Possible duplication of hemoglobin α-locus in the irus macaque. Nature, Lond. 215: 1485:1487.

Zuckerlandl, E. & L. Pauling. (1965). Evolutionary divergence and convergence in proteins. In Evolving genes and proteins. (Ed. Bryson, V. and H.J. Vogel). New York & London: Academic Press.

Zuckerlandl, E. & W.A. Schroeder. (1961). Amino acid composition of the polypeptide chains of gorilla hemoglobins. Nature, Lond. 192: 984-985.

Table I

Sequence differences between certain primate hemoglobin chains. Only sites at which there are differences are shown. Data from Dayhoff et al (1969) and Boyer et al (1969a).

A = alanine, D = aspartic, E = Glutamic, F = phenylalanine, G = glycine, H = histidine, K = lysine, L = leucine, M = methionine, N = asparagine, P = proline, Q = glutamine, R = Arginine, S = serine, T = threonine, V = valine, Y = tyrosine. It should be noted that the changes at some sites are ambiguous in the absence of full sequence data; Glu-asp 6 or 7, ala-thr 10 or 13, gly-ser-asn 46 or 56, asp-ala 47 or 52, pro-ala 51 or 58, ala-thr 70 - 76, Gln (NW monkeys) 124 or 125, thr-ala and gly-ala in Saimiri and Saguinus δ chains 12 or 16.

Chain	5	6	9	10	12	13	16	21	22	33	47	50	51	56	70	76	86	87	104	116	117	125	126	130	139
Human β	P	E	S	A	T	A	G	D	E	V	D	T	P	G	A	A	A	T	R	H	H	P	V	Y	N
Rhesus β	P	E	N	A	T	T	G	D	E	L	D	S	P	G	A	N	A	Q	K	H	H	Q	V	Y	N
M. irus β	P	E	N	A	T	T	G	D	E	V	D	S	P	G	A	N	A	Q	K	H	H	Q	V	Y	N
Ateles β	G	E	S	A	A	T	G	D	E	V	D	T	P	S	A	A	A	Q	R	H	H	Q	L	Y	N
Saimiri β	G	D	A	A	T	T	G	D	E	V	D	T	P	N	T	A	A	Q	R	H	H	Q	V	Y	N
Saguinus β	G	E	S	T	T	T	G	E	E	V	D	N	P	S	A	A	A	Q	R	H	H	Q	V	Y	N
Human δ	P	E	T	A	N	A	G	D	A	V	D	S	P	G	A	A	S	Q	R	R	N	P	M	Y	N
Ateles δ	G	E	A	A	A	A	G	D	E	V	A	T	P	G	A	A	A	Q	R	R	N	Q	V	F	T
Saimiri δ	G	D	S	A	A	A	S	D	E	V	A	S	A	G	A	A	A	Q	R	R	N	Q	V	F	T
Saguinus δ	G	E	S	A	A	A	S	D	E	V	A	S	P	G	A	A	A	Q	R	R	N	R	V	F	T

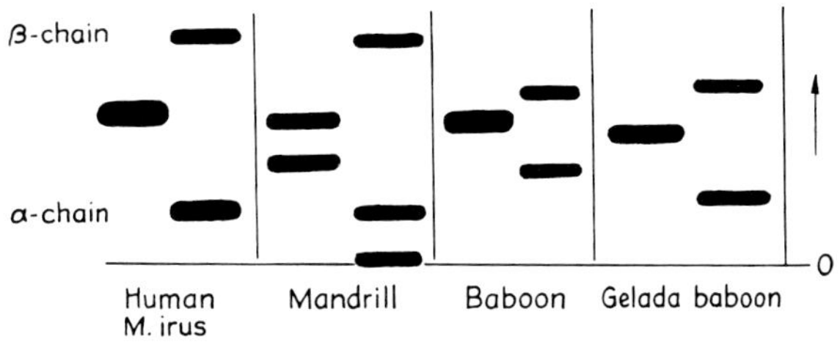

Fig. 1. Relative mobilities of hemoglobin α - and β -chains on starch gel in four primate species as shown by PCMB dissociation. For each species the undissociated hemoglobin is shown on the left and the dissociated chains on the right.

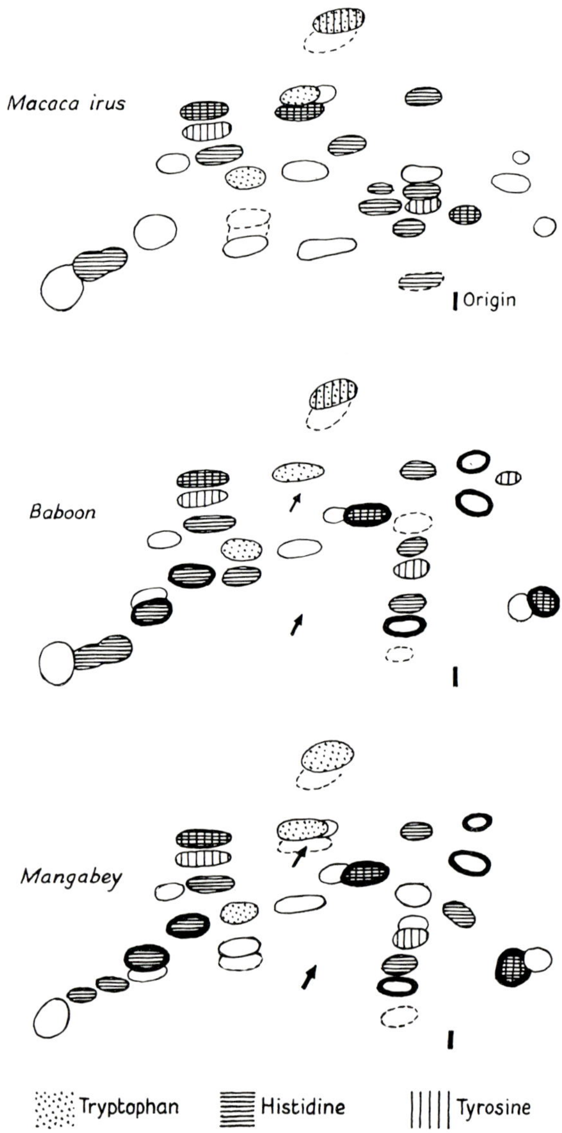

Fig. 2. A comparison of the fingerprint patterns of the soluble tryptic peptides of three Old World monkey species. The electrophoretic direction is from right to left and the direction of chromatography from bottom to top. Peptides reacting positively with specific stains for three amino acids are shown and peptides which have the same positions in the baboon and mangabey but not in the macaque are heavily outlined.

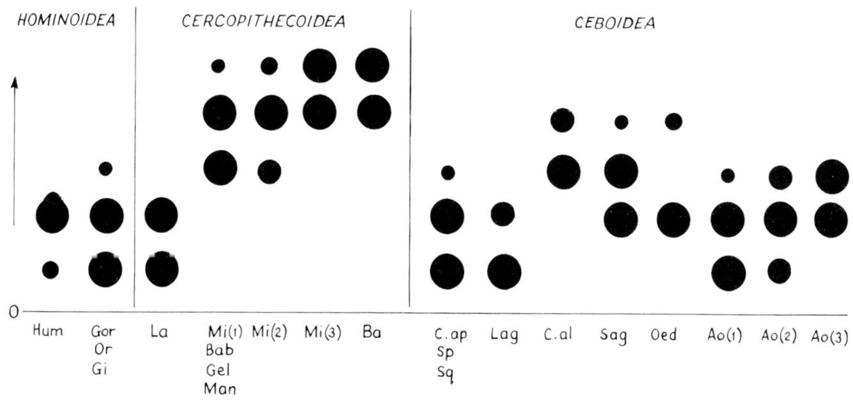

Fig. 3. Acid phosphatase phenotypes of certain Anthropoidea. 0 = origin. Arrow points to anode. Hum = human, Gor = gorilla, Or = orang, Gi = gibbon, La = *Presbytis entellus*, Mi = *Macaca fascicularis*, Bab = *Papio*, C.ap = *Cebus apella*, Sp = *Ateles* Sq = *Saimiri*, Lag = *Lagothrix*, C.al = *Cebus albifrons*, Sag = *Saguinus*, Oed = *Oedipomidas*, Ao = *Aotus*. (Reproduced from Barnicot and Cohen, Biochemical Genetics, 1970, by permission).

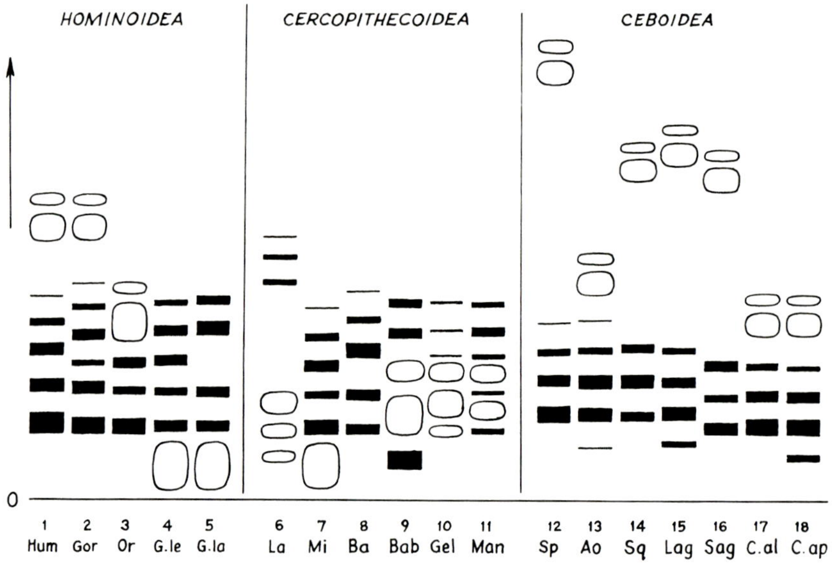

Fig. 4. Phosphoglucomutase phenotypes of certain Anthropoidea.

0 = origin. Arrow points to anode. White areas are achromatic bands.

Species indicated as in Fig. 3 with addition of G. le = Hylobates lar entelloides, G. la = Hylobates lar agilis.

Reproduced from Barnicot and Cohen, Biochemical Genetics, 1970, by permission).

PART III

BEHAVIOR AND ECOLOGY

SOME BEHAVIORAL ELEMENTS OF THE CERCOPITHECOIDEA

Irwin S. Bernstein

INTRODUCTION

Any genetically determined characteristic is potentially useful in establishing the taxonomic position of an organism. To the extent that genes limit the developmental potential rather than determine the exact expression of a trait, and in accordance with the number of available alleles for each locus, individuals will vary. Nonetheless,, systematics is based on the recognition of common features shared by populations of individuals. Numerous morphological characteristics can be selected which find similar expression only among members of a particular taxon. Some traits appear to be conservative, whereas others are more malleable to environmental influences, i.e. some traits find common expression across a broad range of related populations and others find specific expression in a single population. The systematist uses the available information to produce a taxonomy divided into multiple levels, expressing degrees of affinity and separation.

Sometimes the evidence presented by different characteristics is contradictory. This may be due to parallel selective pressures operating on taxonomically distinct forms resulting in the development of analagous structures. In other cases, diverging taxa may be subjected to greatly different ecological pressures such that the rate of change in certain traits in one may be accelerated whereas, in the other, the trait may remain conservative. These problems force the systematist to use all the available evidence in developing any taxonomy. It would be extremely difficult to demonstrate that one characteristic was necessarily more important than another.

Behavioral characteristics can contribute to systematics to the extent that their expression is genetically determined and to the extent that their attributes can be adequately described and measured. An organism evolves to meet the demands of its habitat, and selection processes can genetically encode behavioral as well as morphological adjustments.

On of the behavioral characteristics of Old World monkeys is their social nature. Some type of relatively stable social organization can be found in every species, and the form of the social organization appears responsive to environmental changes whereas the broad limits are determined for any given species (Kummer, 1968; Crook, 1967; Crook and Aldrich-Blake, 1968; Hall, 1963).

Individuals within a society coordinate their activities and maintain the society through a system of social communication. Altmann (1962a) even uses communication patterns to define a society as a group of intercommunicating individuals surrounded by boundaries of less frequent communication. This communication depends on well developed social response systems, and these systems appear to be genetically determined in nonhuman primate societies, (Hines, 1942; Crook, 1967; Marler, 1967; Mason et al, 1968).

Hockett and Ascher (1964) say it was the opening of the call systems which was one of the major events in human evolution. Nonhuman primates may have some limited ability to recombine signal elements, (Van Hooff, 1963; Mason et al, 1968). They may vary the use of signals (Crook, 1967), and they may learn the function, use and significance of the innate signal responses, but they do not seem able to add new elements. Ordinarily, they neither create new signals nor use the signal responses of other species which are not already in their own repertoires. In one long term experiment using captive monkeys living in a mixed taxa assemblage at the Yerkes Primate Center, no evidence was found that any individual had incorporated the signals of another species into its own signal system, although the initial confusion caused by inappropriate responses to the signals of extraspecific individuals did diminish. The animals appeared to learn the significance of extraspecific response patterns and learned to respond appropriately to them, but did not modify their own signal patterns.

The form of message units thus appears to be relatively fixed within an individual and within a given species population. Recombination and multiple use of the available responses requires that some additional signals mediate to indicate the particular intended use of a response pattern in any specific instance. Whereas the context often provides this information, Altmann (1962a) has shown that some signals are indeed meta-communication signals, that is to say they provide information about what other communications signals will, or do, mean.

The societies that the signal patterns maintain and coordinate, however, do seem to be capable of rapid response to ecological changes. An individual primate lives in a group which may be subject to seasonal or longer term environmental variations, and the group may alter its social organization to meet these new requirements (Kummer, 1968; Crook and Aldrich-Blake, 1968; Hall, 1963). The organization changes, but the communication responses do not. If the message units also changed, all members would presumably learn the new responses as well as the new social system. If, however, another member of the species were to come into contact with the group, either briefly, or as a new group member, we could expect communication difficulties with regard to the new signal patterns which had developed. The new member would be much in the same position as an extra-specific animal who also had to learn the signal patterns, and as Marler (1961, 1967) writes, it is the diversity of signal patterns which helps maintain species isolation and integrity, and designates the intended respondents to the signals. Thus, if each primate group in a species developed a modified signal pattern, we might expect rapid speciation, but in the Order Primates it is the generality of adaptation rather than narrow specific adaptations that have been selected. Specializations suited to a particular ecological niche would limit the ability of a species to expand rapidly into a new area available as a result of changing conditions, and would handicap populations living in areas subject to rapid changes. In order to insure wider adaptations and continuity of a widespread species population, means of genetic exchange between local populations have developed, typically an adult, usually male, member of a group leaves his natal group and joins another. A fixed communication repertoire insures a lack of communication barriers between groups so that basic social messages can be exchanged with a newcomer immediately upon his arrival. The lack of a common communication repertoire may be one of the mechanisms blocking hybridization between sympatric primate species which will interbreed under disturbed conditions, (Bernstein,1966).

The particular signals used in a communication repertoire may indeed have evolved through some process unrelated to communication, as Andrew (1963) suggests, and only later functioned as social signals must have been selected to suit the ecological requirements of the natural habitat of a taxon. Despite the broad basic similarities recognized in all major primate divisions, as a population speciates, some signals are no longer adaptive or pertinent and may drop out or be modified to suit new requirements. Thus, related taxa should retain certain "conservative" signal patterns whereas more distantly related taxa within the order may have evolved distinct signals to serve the same purposes. Naturally, the problems of parallel evolution found in morphology will also apply to behavioral attributes.

If an analysis of social responses is to make a useful contribution to a discussion of the systematics of Old World monkeys, we must be able to define and describe consistent response elements. Only a limited amount of data is presently available and it is often difficult to compare the work of any two investigators because of the different levels of description selected in establishing their behavioral taxonomies. Some have been concerned primarily with interpretation and therefore use function as the dominant criterion, whereas others have used structurally defined elements. Struhsaker (1967) recommends the use of both, but even so, behavioral scientists disagree concerning which elements are to be combined or differentiated. Some investigators define response patterns in accordance with who the intended recipient is, what context it is used in, and the body position of the signaling animal. The same facial pattern is thus considered to have a different signal value if the animal is standing, lying, sitting or moving. Problems of intensity and the differential response of the recipients to what appears to be the same signal further confuse matters. Inasmuch as any action may have communication potential, every action may be included in the communication repertoire and even patterns of parturition have been considered as part of behavioral taxonomies.

Despite these difficulties, and the fact that many responses are really steps on a continuum, it is often possible to correlate independently derived behavioural taxonomies. The similarities are truly remarkable considering the widely divergent starting points of the various contributors. Struhsaker (1967) has attempted a preliminary analysis and indicates wide overlap of defined responses. Communication is recognized as one of the most stable behavioral patterns within a species and his comparisons of _Cercopithecus aethiops_ (Amboseli) with _Macaca mulatta_ reveal an overlap of 59%; with savannah baboons (Amboseli), 63%; and with _Erythrocebus patas_, 54%. In contrast, the South African baboon and Kenya baboon populations showed an overlap of 75%.

If the differing signal patterns are to be useful, they must not only be capable of differentiating primate taxa, but they must also recognize affinities among related taxa. Greater differences should be seen between genera than within a genus. Species within a genus should show differences in response repertoires related to the extent of evolutionary divergence. It should be possible to recognize subspecific variation, although some overlap is to be expected, and, of course, individual variation should not be so large as to obscure differences between closely related but distinct populations.

As discussed during this conference, and so ably presented by Hans Kummer (this volume), the study of primate behavior may be conceptualized as having four logical divisions. These divisions range from the relatively concrete descriptions of response units, through frequency counts of the use of such elements, to sequential analysis relating the elements one to another, and finally to the resulting social organization which provides the framework within which these processes operate. In this paper, only the first level of analysis, the descriptions of response units, will be under consideration. It may be assumed that the inclusion of information at the other levels would add to the value of behavior in differentiating systematic groups. Davis, et al (1968) have already demonstrated how even the use of gross behavioral categories can provide distinctive activity profiles for a variety of primates. In their study every individual could be identified as a member of only one of the seven subject populations through a process of elimination using non-overlapping frequency distributions.

In this paper, a selection of responses which might be visually or tactually mediated signals in primate communication systems will be examined. Particular forms of expression will be examined to identify those behavioral elements which might be of use in primate systematics.

The data to be presented have been collected on both captive groups and wild troops, but the work of Bernstein (1967), Eisenberg and Kuehn (1966), Kummer and Kurt (1965), Rowell (1967) and Southwick (1967) and others, suggests that the forms of social expression remain the same in both situations, although frequencies may vary. The data available include information on: *Macaca mulatta*, *M. nemestrina*, *M. sylvanus*, *M. radiata*, *M. fascicularis*, *Cynopithecus niger*, *Cercocebus atys*, *Theropithecus gelada*, *Erythrocebus patas*, *Cercopithecus aethiops* (two forms), *Papio hamadryas*, *P. anubis*, *P. cynocephalus* and *P. ursinus*. Limited information is also included on *Macaca fuscata*, *M. assamensis*, *M. silenus*, *M. arctoides*, *M. maurus*, *Mandrillus leucophaeus* and *M. sphinx*. The data represent a conglomerate of available published data and information drawn from certain studies still in progress at the Yerkes Field Station. The elements discussed are not an exhaustive list of the response repertoires of these taxa.

THE RESPONSE UNITS

Any response which conveys information from one individual to another may be thought of as a communication response. Some patterns may serve solely as signals, but much information is transmitted via responses which also serve other functions. If a response modifies the behavior of other individuals in a recognizeable way, we have no difficulty recognizing it as a signal, but in many cases the effects are subtle and varied and inclusion of any particular pattern may be a matter of investigator interpretation.

For our purposes, we will consider those patterns which appear common to several individuals in a population in a relatively fixed form and which occur in situations where their communication potential seems high. Such a definition would eliminate idiosyncratic responses and would include both structural and functional considerations. The use of the same structural pattern in multiple contexts, and varying responses to the same pattern, however, might suggest a near endless definition of responses. I shall therefore treat responses within a species by structural definition indicating several functions where appropriate. When considering different species I will compare roughly similar patterns serving the same function, and differentiate nearly identical patterns apparently serving different functions. The labels used to designate particular responses are to be considered as mnemonic devices only; those selected are the ones most frequently used or most easily recognized.

The analysis of behavioral response elements will begin with a consideration of broad categories common to almost all the Cercopithecoidea, if not indeed common to all mammals. Such patterns may have discrete details of expression which could be found useful in describing the genera within the Cercopithecoidea, but no attempt will be made to do more than indicate the existence of such patterns. More careful consideration will be given to response elements which have been described for only a limited number of primate taxa.

BROAD CATEGORIES WIDELY FOUND IN THE PRIMATES

Some response elements may be considered individual in the sense that they are often seen outside of a social context. Many of these responses are common to a broad range of primates and variations in such patterns as feeding and drinking are likely to be associated with anatomical specializations. The manipulation of objects is another category common to almost all primates but specific forms have been differentiated (Kaufman and Rosenblum,

1966) as have patterns of locomotion and resting postures. There are also a wide variety of self-directed activities among primates, such as: vigorous shaking of the head and/or body, masturbation, scratching, self grooming, and digit sucking. Certain patterns may be idiosyncratic stereotypies, but others are found in only a few taxa where their function remains obscure. For example, two distinctive patterns seen repeatedly in gelada groups are head stands and lying on the back holding one or both feet in the hands. Such postures may have social significance and Hinde and Rowell (1962) have indicated the importance of sitting and resting postures when considering dominance positions.

Responses such as the erection of the penis, ejaculation and the ingestion of genital secretions are often elements in social exchanges, but these individual units appear to have meaning only in context.

Other responses which occur only in a social context can only be interpreted by the context and are otherwise found in almost all vertebrates. These would include: passing another, leaving, approaching, following, progressing, proximity and such patterns as "guard" (Kaufman and Rosenblum, 1966), return, look, watch, glance, check and ignore. The last named response is particularly difficult to define in that it requires the investigator to decide that a subject is not responding to a stimulus which the investigator feels ought to elicit a response. Related to this, but requiring a more positive action are categories such as: reject, stop, end, deter, prevent and turn away. Most of these patterns, by their very nature, are virtually universal, but once again, particular forms of expression may vary.

Individual display patterns are common among the primates. Most include some form of demonstration such as: branch shaking, bouncing, banging of objects and display leaps. The exact function of such responses may be questioned, but the broad pattern is consistent. Elements vary in their expression and frequency but no analyses have been made comparing cercopithecoid patterns.

Other displays such as yawning, which exposes the awesome canines in various contexts), takes from, mouthing, maintaining distance from another, fleeing, withdrawal or avoiding, piloerection, charge (or attacking run or lunge), chasing, manual attacks including slapping, hitting, pushing, pulling and grabbing), biting and kicking. Now, kicking may be very distinctive, as in mangabey aggression and dominance assertion, and biting may involve slashing or careful use of incisors on selected areas of the body, but some form of all these responses appears to function in much the same way in all primates.

Threat behavior is one of the better studied patterns and a rich variety of responses serving this function have been reported. The open mouth gesture seems to be found at some stage in almost all cercopithecoids. Amond the drills and mandrills however, the mouth is closed and the head is jerked forward and down sharply in situations where other species use the open mouth gesture. Bobbing of the head may be seen as part of rhesus, bonnet and green monkey displays but the closed-mouth bob may be common only to *Papio* and *Mandrillus*. The bobbing pattern also seems more elaborate in green monkeys as compared to the macaques and this may represent the selective development of a common pattern.

OTHER WIDELY FOUND BEHAVIORAL CATEGORIES

Many social responses might also prove to be universal among the Cercopithecoidea when sufficient data have been obtained, although their specific details may vary. Such responses include mouth-to-mouth touching; various forms of initiating and soliciting grooming; grooming; the visual and manual examination of wounds, body orifices and especially the genitalia; postures in response to grooming; non-nutritive sucking, the prevention of sucking; deliberate touching, including the baboon light touch (Hall, 1962); rapid and gentle patting of infants (Lahiri and Southwick, 1966); close visual inspection of an infant after first pulling its head back; aiding another; cradling; struggling to escape restraint; positioning an infant ventrally; and lipsmacking, especially when associated with grooming. A response such as lipsmacking occurs in a wide variety of contexts and its functional aspects may vary considerably from taxon to taxon. If each function it served were examined separately, e.g. male approaching estrous female, subordinate approaching social superior, adult or immature animal approaching infant, grooming and mounting contexts, etc.; then considerable interspecific variation could be described. Several variations in this pattern have been noted and will be discussed further.

Manual and/or oral holding of the nipple may be characteristic of only some species, but intensive observation would be required to demonstrate this. Similarly the infant sitting posture facing away from the mother may or may not be widespread. The absence of tooth grinding or gnashing of teeth in Macaca radiata has been suggested by Simonds (1965), although this pattern is prominent in a wide variety of macaques, baboons and mangabeys; it probably functions to sharpen the rear edge of the canines by grinding them against the lower first premolar and front surface of the upper canines.

Patterns of enlisting aid may be present in all cercopithecoids, although Hall and DeVore (1965) suggest the sidewise jerking glances of baboons do so only incidentally.

A dorsal embrace gesture may be seen in many species where the young ride dorsally with some frequency. In geladas the female merely crouches and looks back at the infant whereas in rhesus (Altmann, 1962b) the female may guide the infant with her hand. The variation in frequency with which this infant carrying pattern is seen in populations of the same species makes comparisons difficult. Hall et al (1965) state that this pattern is absent in patas monkeys. The presence or absence of signals to release embrace, as Altmann (1962b) has described for rhesus, are also difficult to evaluate.

The grimace has been reported in almost all Cercopithecoidea studied so far, although its exact form varies. In patas, Hall et al (1965) differentiate between the crouch face and grimace and say the latter is absent. The crouch or other rigid posture is also widely found but minor variations may be distinctive. Infant and adult convulsive jerking have been seen in a number of macaques and baboons and it is possible this whole body convulsive movement may be a widespread response to pain or frustration.

LESS WIDELY REPORTED RESPONSE ELEMENTS

Sexual response patterns may belookedto as an obvious aid to systematics, and wide variations have been reported in sexual initiation and mounting patterns. Multiple versus single mounts in a copulatory sequence are one source of variation and even differentiate the chacma and Nairobi baboons (Hall and DeVore, 1965). The sexual present of Erythrocebus patas appears distinctive, although sexual responses in the most closely related genus, Cercopithecus, are not well known. The locomotor pattern and cheek inflation and wheeze blowing described for the soliciting Erythrocebus patas female (Hall et al, 1965) is sufficiently distinctive that it should be readily recognized if it appeared in other cercopithecoids. The crouching position of the patas female may be reminiscent of crouching presents described for hamadryas females (Kummer and Kurt, 1965) and also seen in pigtail monkeys who are fearful of a male; but in these cases it does not serve the function of a precopulatory present and is probably the outcome of combining crouching tendencies with a sexual presentation.

The absence of a foot or ankle clasp in any observed mount in patas or geladas may indicate distinct adaptations in both these genera. Variations in functions and circumstances of mounting behavior have been reported and a number of varieties of inadequate mounts have been described. The macaques and baboons appear to have used various reproductive sequence responses in other contexts to serve a variety of functions. Positioning or waist grasping, however, does seem common to all the Cercopithecoidea so far studied although, in some, the male literally positions the female, whereas in others a quick tapping touch is used as a signal. Once mounted, however, the position of the hands is much the same in all species so far described.

The sexual gaze and tail entreaty described for Macaca radiata by Simonds (1965) is difficult to evaluate and may reflect distinct adaptations. Other sexual responses such as the incidence of oral-genital stimulation may be common but have only been described for some of the better known species. Reaching back towards the mounted animal has been described in several macaques and in the hamadryas baboon (Kummer, 1956) but is denied for patas monkeys (Hall et al, 1965). On the other hand, the grimace and squeal of the mounted male has been reported to occur erratically in a number of macaque and baboon populations and may be present or absent in different individuals in the same troop. The female after mount vocalization and characteristic quick stepping, or running away, has also been found in an erratic pattern. The geladas may be said to have a distinct pattern of vocal signals during copulations, in that the female emits a pre-copulatory vocal just prior to the male mounting. The mounted male vocalizes and the female also emits a post-copulatory vocal. The quality of the vocalizations has not been considered but certainly the vocal responses in the various primate taxa will differ both as a function of anatomical differences and behavioral specializations. The above example is cited not to note differences in the quality of the vocal signals, but to indicate the presence or absence of a vocal component accompanying the response sequences associated with a chain of frequently associated response elements.

Some patterns which have been reported in only one or a few species probably reflect inadequate sampling and low frequencies rather than basic differences. Others may reflect anatomical features such as the presence or absence of a long tail. Clearly patterns such as "hang on tail" seen in rhesus and described for bonnets (Kaufman and Rosenblum, 1966) could not occur in short tailed species.

Raised tails, however, can be seen in long tailed as well as short tailed species such as the Japanses and pigtail macaques. This tail posture may indicate subordination, as in baboons (Struhsaker, 1967), or alertness or confidence. In mangabeys it may not be related to either fright or assertion and in pigtails the function may also be questioned. The elevated position of the rhesus tail (Altmann, 1965) is perhaps distinctive and may be readily recognized and distinguished from other tail postures.

In mangabeys, tail postures have been suggested as species differentiators (Jones and Sabater Pi, 1968). Tail arching as seen in sooty mangabeys seems a distinctive pattern which is only approximated by *Cercopithecus aethiops* and *Theropithecus gelada*, neither of which lowers the tail tip past the top of the head. The function in the three taxa also seems to differ in that it is an "alert stance" for the sooty mangabey and an aggressive display in the green monkey, whereas it may only be a grooming invitation stance in the gelada.

Retraction of the eyebrows and/or ears has been described both as an element found in multiple patterns and also as an individual response appearing alone, as in frowning or scowling. These responses are especially pronounced among macaques and baboons, and eyebrow retraction exposing the eyelids is a striking display where it exposes contrasting lighter skin. This occurs, however, in *Cercocebus atys*, *Cercopithecus aethiops*, *Papio cynocephalus* and *Macaca radiata* and *M. fascicularis*, whereas it may be missing in closely related forms. Such a response, although distinctive and dramatic, would therefore appear to have either evolved independently (including the anatomical pigmented difference) or to have been lost in a nonsystematic fashion by a number of taxa.

The distinctive head-bob seen in the genus *Mandrillus* may be an outgrowth of head-bobbing seen in macaque threats where it is found as part of a continuum of intensity from staring to the open mouth gesture. The slapping of the ground may grade into lunges or incipient chases and in other baboons it may occur without the open mouth gesture. In *Cercopithecus aethiops*, Struhsaker (1967) describes it as part of the continuum leading through distinctive jerking to a sitting position to quadrupedal and bipedal alternation, to lateral displays involving alternate turning of the flank and even the Red, White and Blue display seen in only some subspecies, (for example, seen in Amboseli but not at Lolui or among West African forms where, incidentally, the anatomical signals for such a display are missing). Bipedal threats, as described for patas (Hall et al, 1965) and baboons (Hall, 1962) and as seen in geladas, may also be associated with this continuum, but appear to have become quite distinctive and versions even appear in new world forms as the spider monkeys. A number of incipient and intermediate forms have also been described for baboons.

Rotating movements of the hands on the ground, slapping-away reactions with the hand, throwing aside of the arms, shoulder shrugging and muzzle wiping have been dexcribed for savanna baboons by Hall and DeVore (1965). These patterns are not noted in other descriptive lists but it certainly seems these may be common to macaques too. Muzzle wiping may be the exception.

Similarly, pacing threats, backing threats, and crouched approach described by Hinde and Rowell (1962) for rhesus may be more widely found, although as yet unreported. Showing the hind-quarters is also reported and it can be seen in drills, mandrills, pigtail macaques and perhaps others.

Communications avoidance, or averting the eyes, and apprehensive looking are also no doubt found in nearly all macaques and baboons. The sudden lunge-grab reported for patas (Hall et al, 1965) may or may not be widely found and further description and study is required.

Other problem areas involve one sided or mutual clasping and embracing. These patterns may be highly developed in some species and the forms as described for *Macaca arctoides*, *M. sylvanus* and *M. radiata* may indeed be distinctive. *Cynopithecus niger*, however, shows a similar pattern but without the clonic jaw movements seen in the others.

Some patterns are described more by social context and function than by specific structural elements. Departure restraint, protective interference, withdrawal and physical danger retrieval, fall in this classification. Defensive aggression may not, however, if a defensive threat combining elements of open mouth threat and grimace is specified and defined. Nonetheless, this pattern may be widely found among the primates.

Variations in lip, mouth and tongue movements may be seen

in several species. Clonic jaw movements described by Kaufman and Rosenblum (1966) for _Macaca radiata_ can also be seen in _M. arctoides_ and _M. assamensis_. This pattern has also been described in _Macaca sylvanus_ as a combination of lipsmacking and teeth chattering but it has not been seen in such well studied macaques as _M. mulatta_, _M. nemestrina_ or _M. fascicularis_.

Another variation in mouth movements is seen in the male sooty mangabey. The pattern consists primarily of rapid tongue protrusion with the lips drawn back and minimal mouth opening. The head is moved slowly from side to side as the male walks. If the male is approaching a female, he visually orients on her genitalia. A mount or genital inspection may follow. Alternatively, a subordinate seated male may show this response to an approaching superior. This pattern seems distinct as compared to those seen in the macaques.

Another variation of lipsmacking is seen in _Cynopithecus niger_ where ordinary lipsmacking is also seen. The variation occurs in aggressive interactions and consists of more rapid mouth movements, somewhat greater jaw movement and less protrusion of the tongue. Hall (1967) describes a similar response for patas monkeys and states that ordinary lipsmacking in these animals is extremely rare although it may be seen in grooming contexts. Lipsmacking also seems rare in drills and mandrills, but a slow chomping together of the jaws and some lipsmacking can be seen associated with grooming.

Other responses, which may be useful taxonomic aides when better understood, include head rocking described by Kaufman and Rosenblum (1966) for two macaques and the affectional present described for rhesus (Hansen, 1966). The "silly grin" of rhesus macaques, however, has also been described for pigtails, (Bobbitt et al, 1964) and both the "silly grin" and affectional present are structurally similar to other responses and may only differ in that they function in the special infant retrieval context, much as the grimace of an adult rhesus male approaching an estrous female is distinguishable from other grimaces only on the basis of context and presumed function.

Penile display has been described in vervets (Struhsaker, 1967) and baboons where it may be associated with genital stomach nuzzling (Hall and DeVore, 1965); with the information presently available it seems unlikely that this is the same response seen in both taxa. Similarly, neck chewing has been described in vervets (Struhsaker, 1967) and bonnet monkeys (Simonds, 1965). These similar patterns, reported only in some distantly related taxa may reflect sampling difficulties with low frequency responses, or analogous rather than homologous patterns. Hopefully, sampling problems could be overcome by collecting additional data and specifically looking for such responses in all taxa studied. Problems of analogous response patterns might be solved through more detailed descriptions of the patterns.

Another pattern reported in only a few species, and then rarely, may indeed be a homologous response. Nape biting has been dexcribed in *Papio hamadryas* (Kummer, 1968) and is said to occur in another context in South African baboons, but is rarely if ever seen in other baboons (Hall and DeVore, 1965). The function in the two populations seems different; in the first, nape biting is used by males to retrieve females, whereas in the second it is used by males to induce females to present for mounting. To decide now, however, that these structurally similar but functionally different responses were homologous would reflect our knowledge of taxonomy rather than aid in establishing one.

POSSIBLE SPECIES SPECIFIC RESPONSES

Some responses have, as yet, only been described for a single species, but their low frequency and nature would make us hesitate before claiming them as species specific, e.g. two handed, carrying of non-clinging immature animals as seen in pigtails; tweaking in baboons (Hall and DeVore, 1965); the scandalized look of rhesus; and back fur nuzzling in baboons (Hall, 1962). Other patterns such as teasing departure (Kaufman and Rosenblum, 1966), rear embrace and harassment in baboons (Hall and DeVore, 1965), flank to flank postures of *Cercocebus atys* facing in opposite directions with arms and legs draped over one another, tail whipping of *Macaca radiata* (Simonds, 1965) and reciprocated anogenital contacts in *Macaca radiata* (Kaufman and Rosenblum, 1966) may prove to be species specific responses. Presenting with lowered hindquarters in *Papio hamadryas* (Kummer and Kurt, 1965) however may only prove to be a combination of crouching and presenting as seen also in pigtail monkeys.

The gelada, the only living species of its genus, also seems to have a number of distinctive behavioral patterns, some of which are associated with anatomical specialties; for example the male may hold a female by the waist and place his muzzle forward of her rear leg and examine the abdominal beads; a female may back towards another, and place one or both feet on the shoulders of the other; an individual may pull the head of another animal until its muzzle touches the red chest area; excited and presumably submissive geladas evert the upper lip and threatening males chomp the jaws in a chewing movement with mouth partially open and lips drawn over and covering upper and lower teeth.

The pigtail pucker (or LEN, jaw thrust, Flehmen face) has already been noted by several authors as distinctive, and its function in this species seems multiple. *Cynopithecus niger* has a distinctive grin involving retraction of the upper lip exposing the upper teeth alone with the mouth closed. The drills and mandrills share a distinctive, figure-of-eight, smiling in which the lips are parted only at the corners of the mouth and the head is rotated from side to side. This usually signifies a friendly approach and is not to be confused with the snarling seen in an aggressive *Macaca maura*, which is not combined with head shaking.

Pursing of the mouth has been described in patas (Hall et al, 1965) and variations appear in *Cercocebus torquatus* and *C. atys* as well as in New World monkeys. Insufficient descriptive data is presently available to decide if these patterns are all distinctive, as they may well be.

Tail-twining has been seen in different contexts in several primate taxa. Its function and significance are generally not understood and it may only be a contact response in longtailed monkeys with good muscular control of the tail. Variations appear in a number of New World monkeys, and in longtailed infants riding in various carrying positions.

Intensive observation of a group will reveal other responses found in few individuals. For example two facial expressions in pigtails have been found in so few individuals that their use as possible taxonomic aids will require vastly augmented data sampling. Some patterns will undoubtedly be idiosyncratic, and some individuals may develop patterns of elements of questionable function. Inasmuch as monkeys can "create" new response patterns by recombining elements, simple structural descriptions may be insufficient, and until such time as the functional relations of a pattern are understood, the pattern should be withheld from a social response list.

DISCUSSION AND CONCLUSIONS

This paper offers a preliminary overview of potential contributions to systematics which might be expected of studies of behavior. Social responses are among those primate behavior patterns receiving the most attention, but as yet too few species have been examined and too few responses described. The levels of description often differ and some investigators report and describe patterns ignored by others. Ideally, a dictionary of recognized responses ought to be available to every investigator who would use it as the starting point of an investigation of a new species, adding to it where required and noting the non-occurrence of described patterns.

Such a systematic approach to the study of primate behavior may not be realized for some time. Basic agreement on behavioral units and the level of analysis is still in the future. Many observers prefer to use molar categories such as attack or fight, which consist of multiple possible combinations of elements. Other observers define responses by the context or the identity of the animal producing the response. The criteria used in selecting the level of definition depend on the requirements of each study. Nevertheless, certain elements have been described repeatedly by different investigators, although the labels selected have varied.

Some of these elements appear to be species specific, others occur only in closely allied species, but the majority seem to be common to broad classifications of animals.

From the data presented we can only indicate what additional data might suggest. The large number of species specific patterns in *Theropithecus gelada* would emphasize the distinct taxonomic position of this animal. *Erythrocebus patas*, on the other hand, has several distinctive responses, but some affinity to *Cercopithecus* responses is suggested. Presumably distinctive responses such as gape yawning may be seen in the genus *Cercopithecus*, but I have seen it in one male *C. diana*. This only serves to indicate how little is known of most of the members of this genus.

The macaques and baboons are among the best known monkeys but even among the macaques, many species are virtually unknown. Some similarities are suggested for *M. assamensis*, *M. radiata* (data are lacking for *M. sinica*), *M. arctoides* and *M. sylvanus* which are absent in other macaques, but it is still not possible to even suggest that the behavioral patterns of *Cynopithecus niger* would support its inclusion or exclusion from the genus *Macaca*. Distinctive patterns, such as the upper lip retraction, can also be found in certain macaques -- the pucker in *M. nemestrina* and snarl in *M. maurus* for instance.

The potential contribution of primate behavior to systematics can only be suggested. Far more data will be required to properly evaluate the relationship between behavioral and taxonomic studies. Cataloging the response elements for any appreciable number of primate species has yet to be accomplished. The identification and definition of response elements is basic to any study of primate behavior and mandatory before quantitative measures can be undertaken. Frequency and duration measures of defined responses can be used to produce activity profiles for each species showing the relative distribution of responses among the available activities. Preliminary limited data suggests that the activity profiles are distinct for each species, and more similar among cogeneric forms than among more distantly related forms. Thus, even where the quality of response expressions do not differ significantly, the relative frequencies of such expressions can be used to differentiate populations. Further, the degree of differentiation appears to be related to the degree of phylogenetic differentiation. Data to test these hypotheses are presently being collected in a number of independent studies.

ACKNOWLEDGEMENTS

This research was supported by National Institute of Mental Health Grant 1 RO1 MH 13864 and in part by National Institutes of Health Grant FR-00165.

REFERENCES

Altmann, S.A. (1962a). The social behavior of anthropoid primates: an analysis of some recent concepts. In "Roots of Behavior" (Ed. Bliss, E.L.). New York: Hoeber Medical Division of Harper Brothers.

Altmann, S.A. (1962b). A field study of the sociobiology of rhesus monkeys, *Macaca mulatta*. *Ann. N.Y. Acad. Sci.* **102** (2): 338-435.

Altmann, S A. (1965). Sociobiology of rhesus monkeys. II. Stochastics of social communication. *J. theor. Biol.* **8**: 490-522

Andrew, R.J. (1963). Evolution of facial expression. Science 142: 1034-1041.

Bernstein, I.S. (1966). Naturally occurring primate hybrid. Science 154: 1559-1560.

Bernstein, I.S. (1967). A field study of the pigtail monkey (Macaca nemestrina). Primates 8: 217-228.

Bobbiott, R.A., Jensen, G.D and Gordon, B.N. (1964). Behavioral elements (taxonomy) for observing mother-infant-peer interaction in Macaca nemestrina. Primates 5: 71-80.

Crook, J.H. (1967). Evolutionary change in primate societies. Science Journal 3: 66-70.

Crook, J.H. and Aldrich-Blake, P. (1968). Ecological and behavioural contrasts between sympatric ground dwelling primates in Ethiopia. Folia primat. 8: 192-227.

Davis, R.T , Leary, R.W., Casebeer Smith, M.D. and Thompson, R.F (1968). Species differences in the gross behaviour of nonhuman primates. Behaviour 31: 326-338.

Eisenberg, J.F. and Kuehn, R E. (1966). The behavior of Ateles geoffroyi and related species. Smithson. misc. coll. 151: 1-63.

Hall, K.R.L. (1962). The sexual agonistic and derived social behaviour patterns of the wild Chacma baboon Papio ursinus. Proc. zool. Soc. Lond. 139: 283-327.

Hall, K.R.L. (1963). Variations in the ecology of the chacma baboon. In "The Primates" (Ed. Napier, J. and Barnicot, N.A.), London, Symp. zool. Soc. Lond. No. 10: 1-28.

Hall, K.R.L. (1967). Social interactions of the adult male and adult females of a Patas monkey group. In "Social Communication among Primates" (Ed. Altmann, S.A.). Chicago and

London: Univ. of Chicago Press.

Hall, K.R.L., Boelkins, R.C. and Goswell, M.J. (1965). Behaviour of Patas monkeys, Erythrocebus patas, in captivity with notes on the natural habitat. Folia primat. 3: 22-49.

Hall, K.R.L. and DeVore, I. (1965). Baboon social behavior. In "Primate Behavior. Field Studies of Monkeys and Apes," (Ed. DeVore, I.). New York: Holt, Rinehart and Winston.

Hansen, E W. (1966). The development of maternal and infant behavior in the rhesus monkey. Behavior 27: 107-149.

Hinde, R.A. and Rowell, T.E. (1962). Communication by postures and facial expressions in the rhesus monkey (Macaca mulatta). Proc. zool. Soc. Lond. 138: 1-21.

Hines, M. (1942). The development and regression of reflexes, postures and progression in the young macaque. Contrib. Embryol Carneg. Instn. 196: 155-209.

Hockett, C.F. and Ascher, R. (1964). The human revolution. American Scientist 52: 70-92.

Jones, C. and Sabater Pi, J. (1968). Comparative ecology of Cercocebus albigena (Gray) and Cercocebus torquatus (Kerr) in Rio Muni, West Africa. Folia primat. 9: 99-113.

Kaufman, I.C. and Rosenblum, L.A. (1966). A behavioral taxonomy for Macaca nemestrina and Macaca radiata: Based on longitudinal observation of family groups in the laboratory. Primates 7: 205-258.

Kummer, H. (1956). Rang-kriteriem bei mantelpavianen gruppe. Rev. Suisse Zool. 61: 288.

Kummer, H. and Kurt, F. (1965). A comparison of social behavior in captive and wild Hamadryas baboons. In "The Baboon in Medical Research," (Ed. Vagtborg, H.). Austin: Univ. of Texas Press.

Lahiri, R.K. and Southwick, C.H. (1966). Parental care in *Macaca sylvana*. *Folia primat.* **4**: 257-264.

Marler, P. (1961). The logical analysis of animal communication. *J. theoret. Biol.* **1**: 295-317.

Marler, P. (1967). Animal communication signals. *Science* **157**: 769-774.

Mason, W.A., Davenport, R.K., Jr. and Menzel, E.W., Jr. (1968). Early experience and the social development of rhesus monkeys and chimpanzees. In "*Early experience and behavior: The psychobiology of development*" (Ed. Newton, G. and Levine, S.). Springfield, Ill: Thomas.

Rowell, T.E. (1967). A quantitative comparison of the behaviour of a wild and a caged baboon group. *Anim. Behav.* **15**: 499-509.

Simonds, P.E. (1965). The bonnet macaque in South India. In "*Primate Behaviour: Field studies of monkeys and apes*". (Ed. I. DeVore). New York: Holt, Rinehart and Winston.

Southwick, C.H. (1967). An experimental study of intragroup agonistic behavior in rhesus monkeys (*Macaca mulatta*). *Behav.* **28**: 182-209.

Struhsaker, T.T. (1967). Behavior of vervet monkeys and other Cercopithecus. *Science* **156**: 1197-1203.

Van Hooff, J.A.R.A.M. (1962). Facial expressions in higher primates. *Symp. zool. Soc. Lond.* **8**: 97-125.

Van Hooff, J.A.R.A.M. (1963). Facial expressions in higher primates. In "*The Primates*," (Eds. Napier, J. and Barnicot, N.A.). *Symp. zool. Soc. Lond.* No. 10: 103-104.

1. Macaca nemestrina male showing facial expression presumed to be species-specific. Various authors have labeled this: "flehmen face," "Pucker," "LEN" and "jaw-thrust."

2. Cynopithecus niger (Macaca nigra according to Fooden, 1969) male giving presumed species specific "high grin."

3. Macaca maura male showing "snarl," a presumed species specific response readily differentiated from the Cynopithecus niger "high grin" and the Mandrillus "figure eight smile" both with respect to form of the expressions and communication function of the responses.

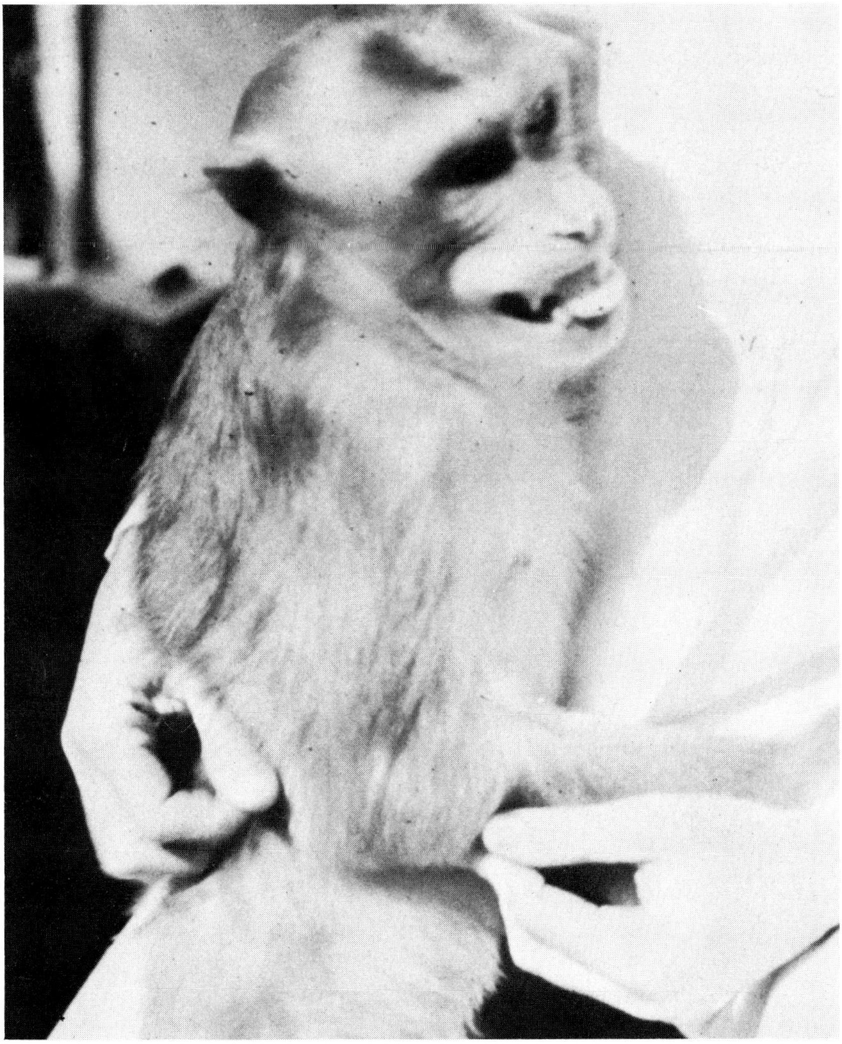

4. Macaca assamensis male showing response labeled as "lipgrin" or clonic jaw movements" or "chatter" by different authors. The response consists of lip retraction and rapid jaw opening with tongue protrusion. Se also, Fig. 5 where the same response may be seen in another species.

5. Macaca arctoides female (with ventrally carried infant in white natal coat) showing "lipgrin" or "clonic jaw movements" or "chatter" of various labeling systems. This response has only been described in four macaque species. It differs from the "grimace" in that jaws are opened and closed rapidly and the tongue is protruded and withdrawn rapidly.

6. *Theropithecus gelada* females giving eyelid display. The pale colored skin of the eyelids is exposed by retracting the scalp and stretching this lightly colored skin over the superorbital arches. This response and the morphological adaptation may be seen in some *Macaca*, *Cercopithecus*, *Cercocebus* and *Papio* species. It is widespread but erratic in that it appears in several genera but may be absent in some species within a genus where other species show it.

7. Macaca nemestrina female showing "grimace" and "crouch," two responses interpreted as submissive and appearing in a wide variety of old world monkeys. The absence, rather than presence of these responses is exceptional in the behavioral repertoire of a cercopithecoid.

8. *Macaca nemestrina* male (subadult) giving "open mouth" response, *usually* described as "threat." This is one of the most widely distributed expressions but may be absent in a few genera such as *Mandrillus* and *Theropithecus*.

ECOLOGY AND BEHAVIOR OF LOWE'S GUENON *(CERCOPITHECUS CAMPBELLI LOWEI)* IN THE IVORY COAST

F. Bourlière, C. Hunkeler, and M. Bertrand

INTRODUCTION

With the exception of Haddow's monograph (1952) on *Cercopithecus ascanius schmidti* and of some data on *C. mitis* contained in Gartland and Brain's paper (1968), very little so far has been published on the ecology and behavior of forest species of the genus *Cercopithecus*. On the other hand, the vervet (*C. aethiops*) which lives in the far more open habitat of the woodland savanna has been extensively studied by Hall and Gartlan (1965), Struhsaker (1967a-d) and Gartlan and Brain (1968). The study of the purely arboreal cercopithecines is in fact very difficult. Hunted almost everywhere, these monkeys are often unapproachable. Even in the most favorable cases, the height of the trees in the mature forest or the density of the vegetation in the second growth makes observation hazardous. It is only in strictly protected areas that one can hope to get good data, but the scarcity of reserves in African forests hardly needs stressing.

In the Ivory Coast we had the good luck to come across a troop of Loew's Guenons (*Cercopithecus campbelli lowei*) which is not hunted and is sufficiently accustomed to man to be observed at close range without apparently modifying its natural behavior. This troop lives in the concession of the Swiss Center of Scientific Research, on the grounds of the ORSTOM campus* 17 km west of Abidjan. Thirty years ago this area was still almost completely covered by a dense forest which spread along the shores of the Ebrié lagoon. Since then much of this forest has been reclaimed for cultivation. On the ORSTOM campus, however, there remain large strips of woodland interspersed with gardens, laboratories and dwellings for the staff. The food resources of this environment modified by man are still sufficient to maintain quite a rich fauna of small mammals and forest birds as well as the troop of Lowe's guenons. Although the guenons supplement their diet with a few bananas provided by the residents of the three closest bungalows, they are not provisioned in the special sense of the word; furthermore the distribution may stop for several months when

---

* ORSTOM: Office de la Recherche Scientifique et Technique Outre-Mer.

the residents are on leave or on field trips.

This troop has been regularly studied by one of us (C.H.) since September 1967 for at least one hour a day. The two other observers visited it on four occasions between September 1967 and January 1970, totalling about 240 hours of observation each. Additional behavioral data were collected by a three weeks study of two tame but freeranging juveniles at the Station of Tropical Ecology of Lamto on the bank of the Bandama (5^o02' West and 6^o13' North). Wild troops and their interactions with other species were observed during two short surveys of one week each in the western Ivory Coast, around Zagné, Keibli and Taï (Fig. 1).

Our study of the Lowe's guenon is still in progress and the present paper, like our earlier one (Bourlière, Bertrand and Hunkeler, 1969), should be considered as an interim report.

ECOLOGY

DISTRIBUTION

Lowe's guenon is one of the most widespread species of the genus *Cercopithecus* in the forested areas of the Ivory Coast. Schwarz (1928), Booth (1955) and Hill (1966) have discussed its taxonomic status and its distribution. The latter coincides almost exactly with the limits of the forest which are not as clear-cut as one might infer from vegetation maps. Gallery forest often penetrates deeply into the savanna, thus permitting forest species to live in zones from which they would normally be excluded. The existence of such "ecological islands" is certainly the explanation for the occurrence of Lowe's guenons in certain spots of the Guinean savannas situated far to the north of the limits of the Western forest bloc (e.g. Béoumi in the Ivory Coast, Bemboi and Kintampo in Ghana).

Modern taxonomists assign the "Mona monkeys" of the Ivory Coast to the subspecies *Cercopithecus campbelli lowei* which spreads from the Volta eastwards to the R. Cavally westwards.

The taxonomic status of the westernmost population living between the R. Sassandra and the R. Cavally is still uncertain; they are possibly intermediate between *C. campbelli lowei* and the nominal race which ranges from Liberia to the R. Casamance in Southern Senegal.

HABITAT AND LOCOMOTION

Booth (1956a, b and 1958), the first primatologist who has tried to define the ecological niches of the West African Cercopithecines, summarizes his observations thus:

"Representatives of the *mona*-group are found throughout the High Forest zone in both primary and secondary forest, and in forest outliners and fringing forest.. The species are.. typical of the lower and middle layer of the forest. They sleep in the middle storey, rarely ascending the tall emergent trees."

Yet, as one can see from Booth's Table I (1956b, p.130), *C. campbelli* also occasionally travels in the canopy and on the ground.

In the Ivory Coast Lowe's guenons may be found at all heights and in all types of forests except marshy and mangrove areas. Although spending most of the time in the middle and lower layers, they often visit the canopy and may occasionally progress on the ground. Forest structure does not seem to concern them unduly. Lowe's guenons are common in the old mature forest between the R. Cavally and R. Sassandra as well as in the second growth of the Basse-Côte and the gallery forest along the R. Bandama.

Extremely agile, Lowe's guenon seems at ease on any type of branch or liana. The trees are the three-dimensional universe which it rarely leaves and where it takes refuge at the first hint of danger.

The arboreal locomotion of *C. campbelli* is a mixture of walking, climbing and leaping which cannot be divided easily into clear-cut categories and which can be much better appreciated in a movie (Bertrand, Hunkeler and Bourlière, 1969) and in photographs (Bourlière, Bertrand and Hunkeler, 1969) than by a written description. On horizontal or slightly sloping branches a more or less rapid quadrupedal walk is used, with hand and feet grasping the support. It is only when the substrate is wide that palms and soles rest flat upon it

The monkeys generally climb up lianas very rapidly, but in climbing down, especially if the liana is slender, they may leap or slide head down, braking with their tails coiled around the support. When the support is very smooth, they may come sliding down feet first like a trapeze artist returning to the circus ring at the end of his performance.

Lowe's guenons usually seem to avoid broad vertical trunks, but when the occasion arises they climb up in a bear-like way, the two arms moving together followed by the two legs which propel the body upwards. On descending sloping trunks, even almost vertical ones, all animals except young infants may run down head first.

Leaps are extremely common and may roughly be divided into three types: "aimed leaps", "free falls" and sideways leaps. The first type is the most frequent and the most impressive. Before springing, the animal briefly marks time, taking off only after a good look at the landing branch which may be 5 to 7 m away horizontally, and sometimes as much as 10 m below the point of take off. If the take off branch has been set in motion by a previous leap, the animal may wait for several seconds for it to settle. Between the starting and landing points the four limbs are partly extended (Fig. 2). At Adiopodoumé the big leaps from the crown of one tree to another are always made at the same points, and only by adults and subadults; younger animals and females in the few weeks prior to and after parturition generally keep to lower layers. In "free falls", the monkey let itself drop on small trees or underlying bushes, sometimes in several stages, steadying itself on the way by catching hold of a branch or twig or even the tail of another monkey. This type of fall is frequent during play. In the sideways leap, the monkey jumps sideways against a vertical support and ricochets at an angle from the direction of impact. This might be equivalent to the running side bounce of the patas monkey (Hall, 1965), but is executed in trees rather than on the ground. It is also used as a means of changing direction and is also frequently seen in play.

Lowe's guenon comes to the ground more often under cover than in the open. At Adiopodoumé all animals except adults frequently hunt insects and play wrestle on the forest floor and sometimes on the forest edge. In order to lure them out onto the lawns of the Center, incentives like bananas are necessary; even then the greatest distance they will move away from the trees is 50 - 60 m. Whole troops may also cross an open space to move from one piece of forest to another (personal observation near Jacqueville).

Locomotion on the ground comprises quadrupedal walking and galloping as well as bipedal running and bipedal hopping. When on the ground movement is generally preceded by and interspersed with bipedal stances during which the monkey looks around. The tail may then be used as a support ("tripod posture") or is raised and curled forwards over the back. The monkeys were observed to maintain a bipedal stance for as long as two minutes when eating bananas on the Center lawn. Lowe's guenons are also very agile when moving bipedally: for example they were seen to jump at least 1.60 m bipedally from the ground to a low branch with both hands full of food and without losing their balance.

The "semi-prehensile" tail of Lowe's guenons is used both in locomotion and at rest. It functions as a counterweight, a brake, a support and a grasping device. Babies anchor their tail to the base of their mothers' tail. Tail twining is frequent both as a means of maintaining balance and as a means of social contact when dozing, sleeping, sitting or grooming (Fig. 3).

HISTORY OF THE ADIOPODOUMÉ TROOP

Lowe's guenons at and near the Swiss Center have been known to the residents for over 10 years. Urs Rahm who was the first director of the Center in 1951 - 1953 and who published the first study on the mammals of the area (1961) writes:

"At Adiopodoumé a band lived in the secondary forest on the shore of the lagoon in a territory of about 500 by 30 m." From 1955 to 1958 heavy poaching resulted in the disappearance of the monkeys (Prof. Huggel, pers. comm.). According to A. Aeschliman (pers. comm.), it was only in 1959 that the monkeys permanently settled in the concession of the Swiss Center and its immediate surroundings. For some three years they were protected, during which the size of the troop never exceeded ten. The maximum number of monkeys counted in 1962 was eight (Mrs. T. Eckert, pers. comm.). At the end of 1963 the population had been reduced to five due to the activities of a local hunter while the directors were on leave. Subsequently protection was more strictly enforced and the troop was able to increase normally, and became more approachable."

The survivors of the 1963 poaching incident formed the nucleus of the present troop (Fig. 4). The nucleus probably consisted of one adult male* and four adult females, since four infants were born in November - December 1964. Three more births occurred in January 1967 and, when we started observing, the troop comprised 13 individuals. With three more births in December 1967 - January 1968, the composition of the troop in 1968 was the following: one large adult male, two large adult females and two smaller "young mothers", two subadult males, two juvenile males in their third year, three juveniles in their second year (2 ♂, 1 ♀) and three

---

* In the genus Cercopithecus females seem to become sexually mature earlier than males (Gartlan and Brain, 1968). Provisionally our age categories are as follows:

	Male	Female
Adult :	Over 4 years	Over 3 years
Subadult :	In the fourth year	In the third year
Juvenile :	12 to 36 months	12 to 24 months
Infant :	0 to 12 months	0 to 12 months

infants (1 ♂, 2 ♀). The adult male and the two largest females certainly belong to the original 1963 troop.

In contrast to the situation in preceding years, only one infant was born in January 1969 and at least one pregnant female miscarried. At the end of February two males, a young adult and a young subadult, together with an adult female, spontaneously left the troop which was thus reduced to 12 individuals. In November 1969 a young adult female spontaneously joined the troop. On December 12, 1969, two males, again a young adult and a young subadult left it. On February 19, 1970, a further splitting of the troop took place as a young subadult male left with an adult female, her two years old daughter, and the young adult female who had joined the troop three months earlier. Two infants were born in December 1969 and January 1970 but one died shortly after birth. Thus on April 1, 1970, the troop was reduced to nine individuals with a single adult male.

HOME RANGE

For the whole duration of the present study the troop has remained in a home range of about 3 hectares comprising mainly a remnant of coastal forest adjacent to the bungalows of the Swiss Center (Fig. 5). This forest remnant still includes many tall trees (± 30 m) towering over a lower stratum (± 15 m). To the north and north-west, this area merges into several hectares of secondary forest surrounding the village school. The monkeys have never been observed at this site, perhaps because of the continual coming and going of the school children and staff. Southwards, the wooded curtain bordering the lagoon changes into a dense bush which has asserted itself following the clear felling which took place at the beginning of 1967. The bush area, which contains a few trees about 10 m tall, acts as a corridor of vegetation communicating with the Botanical Garden and a wooded peninsula where other Lowe's guenons and lesser white nosed monkeys (*C. petaurista*) live permanently.

The Adiopodoumé troop is not tame and the monkeys' reactions to humans depend on the part of the home range where contact occurs. Near the houses of the Swiss Center, the monkeys are "relaxed" and some of them will come within a few metres of the residents, although they do not allow themselves to be touched. In the wood they are always on the alert; they emit alarm sounds ("crr crr") and hide on the approach of humans excepting for the observer (C.H.) whom they see daily.

Different parts of the home range are not equally used, on either a daily or yearly basis. The monkeys visit the northern and southern extremities of the home range only intermittently, whereas every night and the greater part of each day is spent in the most heavily shaded area in Fig. 5. This is the true core area of the home range; its sleeping trees, its main fruit trees, and the favorite playgrounds of the young are all linked by a network of aerial trails. Some limited changes, however, do occur from time to time within the core area. For example the monkeys spent all the nights in some tall trees situated at Point 1 in Fig. 5 in 1967, 1968 and 1969 except for the end of November and beginning of December 1968 when they slept at Point 2. Their daily time-table is also flexible, except for the hours at which they leave the sleeping trees and return; these are more or less constant. During the day the behavior of the troop is variable; the monkeys generally visit the same points of the home range each day, but not always at the same hours or with the same itinerary.

Do Loew's guenons have a territory, that is to say do they actually defend their home range or parts of it against foreign conspecifics? We did not have the opportunity to observe the behavior of two troops coming into contact. Yet unplanned experiments by the Swiss Center staff tend to show that the Adiopodoumé troop does not tolerate foreign conspecifics, unless they happen to be young ones. A pair of Lowe's guenons lived from 1963 to 1968 in a large cage situated near Point 2 of Fig. 5, i.e. at the limit of the troop's home range. The wild monkeys knew the captive ones well; they "acted nervously" when blood samples were taken from the latter. Subadults and juveniles came to play on the roof of the enclosure and the wild adult male often answered the loud calls of the captive male. Yet, when either of the captive monkeys escaped, the troop never accepted it; sometimes the fugitive returned wounded. On the other hand, the troop readily accepted a one-year-old male in February 1969 and, two months later, showed every indication of being willing to accept an 11-months-old female each time we let her free.

The intolerance of adults to foreign adult conspecifics is also suggested by their reactions to their mirror images. On several occasions in 1968 and 1969 the monkeys were attracted by their reflections from the window panes of a house. The adults and subadults would come and threaten their images with head bobbing, whereas younger animals did not show any aggressiveness. Evidence for this behavior was reinforced by experiments using a mirror fixed in a tree.

DAILY ACTIVITY CYCLE

The monkeys leave their sleeping trees at dawn and return to these trees or their vicinity before dark. This was the case both at Adiopodoumé and in the gallery forest of the R. Bandama. Lowe's guenons do not necessarily remain silent and quiet throughout the night. At Adiopodoumé the adult male often emitted loud calls between 8 p.m. and 3 a.m. At Lamto two juvenile males had learned to leave the attic of the laboratory in which they slept as soon as the insect light trap was turned on and stayed there catching insects until 10 or 11 p.m. Often in the middle of the night they could be heard moving when there was no artificial lighting.

Heat and rain are the two components of the weather which seem to have the most influence on the monkeys' activities during the day. The most extensive movements and the main bouts of feeding and play occur mostly in the early morning and in the mid-afternoon. During the warmest hours the monkeys are generally much less active and may take a nap, sitting or lying down with closed eyes, either alone or in small groups.

Moderate rain does not interrupt movement or feeding but thunderstorms compel the monkeys to seek refuge in the foliage of the thickest trees. The adult male often emits loud calls during and after storms, by night as well as by day.

Following Haddow (1952), many authors have suggested that the matings and births of forest *Cercopithecus* occur mostly at night as these activities are very rarely seen during the day. This is simply due to the fact, perhaps, that the monkeys encountered were hunted and unaccustomed to observers. At Adiopodoumé, matings have been observed during the day, and one birth occurred during the hottest part of the day between 10.30 a.m. and 4.00 p.m.

DIET

According to Booth (1956) who examined the stomach contents of a few wild specimens, the diet of Lowe's guenons is almost entirely frugivorous. Such a generalization would appear to be extreme. Both at Adiopodoumé and Lamto, the monkeys also eat many flowers and leaves as well as insects. Table I details the 47 species of food plants which could be identified during a year. These represent moreover only a part of the total plant species eaten by the Adiopodoumé troop. In addition, the leaves of still other species are nibbled at - particularly when young - and then spat out.

The monkeys gobble up certain flowers (e.g. *Ipomoea involucrata*) whereas they eat only the base of the calyx of others (e.g. *Hibiscus*).

Though the staple food of Lowe's guenons is predominantly vegetable, insects constitute, at least qualitatively, an important part of their diet, probably providing the monkeys with the amino acids essential for growth and reproduction. At Adiopodoumé the guenons actively hunt insects each day. They carefully inspect leaves perforated by larvae and may pop the whole leaf into their mouth, spitting out the plant tissue after chewing. Systematically they turn over leaves, flowers and twigs and open the folded pinnae of the dry leaves of the *Elaeis* palm-tree (Fig. 6). They even come down on the lawns to hunt the grasshoppers of which they seem very fond. Lowe's guenons may also pick up or break large pieces of dead wood, to tear off and chew the bark and pick out the larvae. They have also been seen eating whole spiders' webs.

The two juveniles of Lamto showed insect-catching abilities far superior to that of any entomologist. As soon as the light trap was on they would wait in front of the white sheet and grab the big emperor-moths (Saturniidae) and hawk-moths (Sphingidae) as they alighted. The thorax and abdomen would be eaten and the head and wings discarded, a method of eating insects that seems quite general. At Adiopodoumé the remnants of insects in the feces mostly consisted of legs and abdominal segments of Orthopteroids and beetles, whereas the antennae, mandibles and cuticle were absent. Feces also included fragments of wood and many grains of silica.

"Insect" hunting is selective and certain species are discarded or ignored. For example winged individuals of driver ants (Dorylinae), stink-bugs (Pentatomidae), Danaid butterflies (Danaidae) and millipeds (*Spirobolus* sp.) were never touched. All these Arthropods have either repulsive odours and secretions or powerful mandibles.

We never saw Lowe's guenons capturing or eating small vertebrates. When presented with chicks or eggs of various wild species, they would touch or handle them, but they would never bring them to the mouth.

DRINKING

Lowe's guenons seem to drink only sparingly and infrequently. Apparently they obtain sufficient water from their favourite foodstuffs. Arboreal drinking patterns are as follows: (1) after a shower the monkeys often lick the underpart of branches where drops have collected; (2) they visit certain tree-holes where rain water has accumulated, dip in one or the other hand and lick them dry. On the ground they may also suck up water. At Adiopodoumé, for instance, the monkeys sometimes drink from the drinking-trough of captive duikers.

Unlike *C. mona* in Cameroon (Struhsaker, pers. comm.) and in Gabon (Gautier, pers. comm.), *C. campbelli lowei* does not seem to enter water spontaneously in spite of the proximity of lagoons, rivers or stagnant streams. At Adiopodoumé monkeys with wet fur have been seen only three times. Although troop members may run in very shallow water to pass from one tree to another, they do not appear to enter the water just for the sake of it. Juveniles and infants often play in small trees overhanging the lagoon but they were never seen to jump or even fall in.

PLAY

Lowe's guenons, particularly the subadult males and infants and juveniles of both sexes, are very playful animals. In addition to chasing and wrestling, both in trees and on the ground, they play a number of games, both solitary and social. The most original play patterns are "King of the Tree", the "Ant Game" and the "Hornbill Game". The first is an arboreal version of "King of the Castle" played in the middle layer of the forest by two or more (up to five) participants. One monkey forces the other down and off a branch and itself, drops off to the vegetation below; both climb back up and resume the game. The Ant Game takes place on the forest edge. The monkeys repeatedly strike at the nests of tailor-ants (*Oecophylla longinodis*), gradually tearing them open. As shown in our film, the "player" jumps backwards after each strike to avoid a shower of ants and keeps scratching like a madman; nevertheless it cannot avoid being bitten repeatedly. The ant bites are very painful to humans and it is difficult to understand what attracts monkeys to the nests: they were seen only once eating the ant larvae.

The Hornbill Game is described in detail in a later section. Objects commonly found in the habitat are sometimes manipulated for their own sake, without relation to feeding; on several occasions juveniles and subadults of the Adiopodoumé troop were seen playing with twigs and small or large stones. The monkey would carry the object up a tree, drop it, go and fetch it, roll it on a branch, drop it and fetch it again, or perform acrobatics while holding it in one hand or foot.

BIRTH SEASON

According to our observations and to reliable information,

Lowe's guenons give birth at the end of the long rains and the beginning of the main dry season, but never at other periods of the year. The history of the Adiopodoumé troop clearly demonstrates this seasonal breeding:

- in 1964, the first birth occurred in November, and the the three others spread out in the next two or three weeks (Mrs. T. Eckert, pers.comm.).
- in 1965 and 1966, newborn infants were seen at the beginning of the dry season (P. de Rham, pers.comm.).
- in 1967, the three births took place in January.
- in 1968, the first two births were observed between December 22 and December 26, 1967, and the third on January 16, 1968.
- in 1969 the only birth occurred on January 18, during the day, and at least one pregnant female miscarried around that date.
- in 1970, the first baby was born on December 3, 1969, and another one on January 10, 1970.

Lowe's guenons at Lamto also give birth at this period of the year. The two juvenile males whom we studied in August 1967 had been brought into the station in January and in March 1966, the first one with the eyes still closed and the umbilical cord still attached. In 1968-1969, births were also reported in December and January both at Lamto in the West and at Abengourou in the East of the Ivory Coast.

The gestation period of Cercopithecus being about 6 months (Napier and Napier, 1967), the concentration of births in December and January implies that fertile matings take place in June and July, at the beginning of the coolest period of the year (Fig. 7), just after the long rains. Indeed, in 1969 most of the observed matings occurred from June to September (Table II). A relationship between the conception time and declining temperature has been clearly pointed out by Lancaster and Lee (1965) for several species of primates:

"Like those in Japan, monkeys in India appear to conceive in the second half of the year during or after a peak in rainfall and during the annual decline in temperature" (p. 509).

The abrupt fall in temperature is perhaps the "climatic synchronizer" which stimulates spermatogenesis. Furthermore, it is well known that in humans and other mammals with an external scrotum, spermatogenesis is inhibited by a rise in temperature of only a few degrees (Turner, 1966, pp. 443-446 and 457-458).

SOCIAL ORGANIZATION

The following description of the social organization of the Lowe's guenon is based on the study of the Adiopodoumé troop, which has been efficiently protected since 1965 and sufficiently habituated to man to permit observations at close range. All the troop members are individually recognizable and the age of all individuals born since January, 1967 is accurately known. Troops encountered near Lamto, Jacqueville and elsewhere in the Ivory Coast were so wild and the observation conditions were so poor that even a complete count of the monkeys was not possible. Furthermore, hunting pressure is so heavy in the Ivory Coast that the demographic structure and social interactions of these relict troops may not be representative of the group composition and social organization of the species as a whole.

THE ADULT MALE

Although males have been at times more numerous than females in the Adiopodoumé troop, only one of them, the largest and eldest, plays a central role in the group. He has maintained his status since at least 1964. The most obvious role of this "alpha male" is to act as a watch-dog and easy-going patriarch for the whole troop. He is generally to be found at a higher level then the other troop members, watching for potential danger while eating quietly or moving slowly from one look-out to another. The alpha male is the only one to emit the loud calls so typical of the forest *Cercopithecus*. These calls may be given in a variety of circumstances at all times of the day, or even by night. In all cases when the stimulus situation was clearly established, the calls were emitted in response to any situation that was potentially dangerous for the group members, or potentially disruptive to the cohesion of the group. Calls were most often heard in the following circumstances: (1) after loud noises such as thunder, a falling tree, a close gun shot or a low flying plane; (2) when sighting a potential predator such as a large bird of prey, a cobra, an attacking dog, or a cat climbing in a tree, (3) in response to the loud calls emitted by foreign conspecifics; (4) during intra-group squabbles involving infants, or when an infant was in danger; (5) in any unusual circumstance such as the first appearance of a horse or the arrival of the observer wearing a new and bright clothing or holding a strange piece of apparatus. In addition the alpha male regularly calls in the evening when the troop arrives in the sleeping trees, although no danger threatens. In all cases, whether danger is imminent or not, the male calls only from the central part of the home range, never from its periphery. Whenever the loud calls are emitted, group members look in the direction of the sound and, depending on the situation, regroup in a few trees, gather around the male, or approach the stimulus. Sometimes, they may ignore the calls altogether.

The alpha male does not exert a strong dominance over the other troop members. Unlike the situation in most species of baboons and macaques (and even vervets), the social hierarchy in Lowe's guenons is not overt and rigid, but latent and flexible. However, on occasions, it may reveal itself. For instance a subadult may threaten a juvenile which tried to grab its food; animals may spontaneously move away when the alpha male approaches; a female may prevent another from grooming him, and so on. Agnostic interactions are rare. The most common aggressive gestures are head bobbing (or incipient head bobbing with the chin thrust forward), chases and, in males, yawning and tree-shaking. The "jumping around" display described by Gartlan and Brain (1968) for vervets is performed by the alpha male only. The aggression of the alpha male is mainly directed towards maturing males of the troop between three and four years of age, and towards foreign adult males, imaginary (mirror image) or real. As shown in Table II, the male is more aggressive in July, August and September when he is more sexually active. However, the alpha male is not feared by the other group members. Although more often alone than not, he is attractive to adult and subadult females, and to young animals (Fig. 8). Juveniles and infants often play very close to him, sometimes even touching him, while he simply ignores them. He is often groomed by the three adult females whom he may also groom, although less frequently. The alpha male never grooms the other troop members.

ADULT FEMALES

Females of *Cercopithecus cambelli lowei* probably reach sexual maturity when three to four years old. The youngest mature female of the Adiopodoumé troop, f-3, was probably just three years old when she had her first infant on January 18, 1968. In 1969 she was still smaller than the two other females f-1 and f-2 and also had smaller nipples. She did not groom the older females regularly nor was she much groomed by them.

Adult females are rarely aggressive, and even then the threat is limited to head bobbing, chin thrusting or a brief chase. When it occurs, their aggression is generally directed to other females. They may also slap infants or tug their tail to bring them under control.

Adult females are the only members of the Adiododoumé troop who frequently interact with the alpha male. They particularly seek his company during the mating season, often moving across to join him when he sits alone.

During the first weeks following the birth of an infant, mothers are more wary than usual and hide in thick foliage; the alpha male is often seen near them. Adult females, even those having a one year old infant of their own, are often intensely interested in the new born infants. They adopt the aunt-role very readily. Generally, "aunts" are very permissive, even allowing infants to take food from their mouths. Mothers and aunts carry infants ventrally but the infants were never observed to ride on their mother's back.

The mother-infant relationship is probably the strongest bond in the life of Lowe's guenon. A mother utters alarm sounds (crr-crr) as soon as she sees her infant in danger and she may even come to its rescue. On one occasion a three-month-old infant entered a cage but was unable to get out; its mother jumped on the cage and remained there while the other troop members kept away. The alpha male continuously emitted loud calls but did not approach the cage.

SUBADULTS AND JUVENILES

Subadults and juveniles are the most active members of the troop and, with the exception of the subadult females, the most playful. They are also the most inquisitive animals in the group and are the first to investigate strange objects or animals. Subadult and older juvenile males are particularly dauntless. They are the ones who most often steal bananas in the back of the observer and who, on occasions, invade the verandahs of the houses. They were generally the first to approach the experimental animals or objects which we introduced to the troop.

Subadult and older juvenile males interact freely with the adult females, whom they may join, groom and even mount without provoking aggression from the adult male except during the mating season. Unlike the males who only play with infants, juvenile and subadult females show definite signs of maternal care (see below).

Subadults and juveniles have sometimes been seen masturbating with the hand or mouth.

INFANTS

During its first year an infant has several "aunts" of different ages who not only accompany the mother but may "baby sit" for her while she is feeding, resting or being groomed. In some cases one aunt is permitted to cradle - but not to carry - the newborn on the very day of birth, and may also perform perineal cleaning (Fig. 9). The mother may leave the baby with one or several aunts as early as 10 days after birth while she is feeding a few meters away. A few weeks after parturition, mothers begin to leave their babies in secluded places in thick foliage, at first with aunts then, after the second month, more and more frequently alone. There the babies indulge in solitary or social play. Mothers run back to them at the slightest alarm. When the infant is isolated or emits a contact call after playing alone, an aunt will go and retrieve it as readily as will its mother. This pattern may continue until the infant is about 10 months old. An aunt may also carry the infant when it is in difficulty, for instance when climbing a wet tree-trunk after rain. Aunts may also doze with the infant during the mid-day siesta or play with the juvenile. The aunt-infant relationship lasts for at least a year.

When an infant is born not all the females of the troop attempt, or are permitted, to assume the aunt role (Table III). One aunt is often a female sibling of the infant. For instance, the female f-2 delivered an infant, Piagette, on December 3, 1969, and another infant, Septante, on January 10, 1970 (Fig. 10). The aunts of Piagette were: f-1, a mature female who, from the beginning of the study, had always been the closest companion of the mother; a juvenile female, Trudi, just over two years old at the time of birth; and Piagette's sibling, the juvenile Vreni, only 13 months old at that time. The aunts of Septante were, again, the f-1, Trudi, and Vreni, plus the young adult female f-5, a newcomer who joined the troop at the end of November, 1969 but was perhaps a former group member, an inference which was made when she was immediately accepted without threats. Septante's youngest female sibling, Piagette, then just 13 months old, repeatedly tried to touch and handle the newborn but was often threatened or even chased away by her mother. Similarly, Piagette's sibling, Vreni, had repeatedly been threatened away by her mother for about two weeks following parturition, then permitted - in spite of her small size - to cradle and carry her sister. The two other mature females (f-3 and f-4) were only seen twice in the company of Piagette (Table III); and the following year f-3 did not appear to take any interest in Septante. On the other hand, f-3's daughter, the juvenile female Heidi, who was of the same age as Vreni, often attempted to interact with Piagette when the latter was a young infant but was threatened away by the two younger aunts.

In the case of i-3, the daughter of f-3 born on December 3, 1969, the aunts were again f-5 and Trudi, together with the newborn's female sibling, Heidi, at that time almost 23 months old. The two oldest females f-1 and f-2 showed no interest in the newborn. Soon after parturition, the mother began to move with difficulty. On the second day the baby was carried by f-5 while the mother was abandoned by the aunts who started to follow the "new mother". The sick f-3 lagged behind the troop which made no attempts to wait for her, and was even threatened away by f-1 and f-2. The baby died when 4 days old, apparently of starvation, while its mother recovered her health within a few weeks. Thus, adoption by an aunt in the event of the mother's sickness may prove fatal instead of beneficial if the infant is too young. The "new mother" f-5 went on grooming, licking and cradling the corpse for at least 6 hours following death.

During the first days following birth, the alpha male was often seen with the mothers but did not show active interest in the infants. He simply ignored them when they happened to touch him. He began to mildly threaten Piagette away when she was about 5 months old. The other males were very tolerant with infants and played with them, allowing them to jump on their backs or swing from their tails. However they did not show any specifically paternal care.

However one cannot say that paternal care does not exist in Lowe's guenons. In January 1970 one of us adopted an infant lesser white-nosed monkey (*C. petaurista*) which was about 3 months old and whose mother had been killed. During the two weeks following adoption, the infant was repeatedly stolen, for intervals ranging from 30 minutes to several hours, by three members of the troop: namely, f-5 and Trudi, who were decidedly "infant oriented", and the male Coriace who was entering his fourth year and was the only subadult male left in the troop (see next section). Coriace would enfold the infant in his arms and carry it away ventrally to some elevated branch where, like a female, he would groom and cradle it.

Weaning appears to be an abrupt process only when a sibling is born the following year. This was the case for the two infants of f-2, Vreni and Piagette, for each of whom suckling and cradling stopped on the birth of the succeeding infant. Yet weaning is mostly a psychological process, as infants may well feed themselves when 3 or 4 months old. Weaning did not seem to be a particularly traumatic affair as it was started about 2 months before the birth of the sibling and was eased by the presence of aunts. Furthermore, the frustration of a female infant on being rejected by her mother may be compensated by her (the sibling's) attraction to the newborn and the fact that she may be permitted to play an aunt-role. In the absence of younger siblings, "infants" were still occasionally seen on the nipple, sometimes actually sucking, up to 26 months of age (e.g., Boubi, the son of f-1). Mothers and aunts would still occasionally carry infants ventrally when the latter were a year old and extremely agile.

LEAVING THE TROOP

The spontaneous departure of troop members was observed three times in two years. It appeared to be correlated with the maturation of young adult males and the occurrence of the birth season.

At the end of December, 1968 we noticed that one male who was just entering his fifth year tended to isolate himself in tree-crowns like the alpha male. On January 2, 1969, he was observed to emit on several occasions a double "bark" which sounded almost like the loud calls of the alpha male. In mid-January, soon after the birth of the 1969 infant, this male formed a subgroup within the troop with a female of the same age (f-4) and one of the two males who were just entering the fourth year. These three individuals kept apart from the other monkeys during the day but joined them in the sleeping trees for the night. They disappeared at the end of February 1969 and were seen migrating southwards in the direction of the ORSTOM Botanical Garden and the adjacent wooded peninsula.

On December 12, 1969, the male then entering his fifth year

again left with one of the two males entering the fourth year. This time, however, the "secessionists" left suddenly, without isolating themselves for a few weeks before departure. Within a few days they settled at the southernmost extremity of the troop's home range, again in the vicinity of the Botanical Garden (Point 3 in Fig. 5). These were unusually bold individuals and not afraid of people. They confined themselves until January 22, 1970 to the neighbourhood of two houses and of the ORSTOM landing-stage before disappearing permanently. They kept tightly together, often eating or resting side by side, grooming each other, wrestling in play, and sleeping in the same tree. They were never seen to threaten each other. Once, in the evening, we observed their reaction to the loud calls of the alpha male emitted from the troop's sleeping trees, about 300 m away. On hearing the calls both males became tense, motionless and very attentive, facing the direction of the calls; one leaned forward. On another occasion the elder of the two spontaneously emitted a few sounds which again resembled a higher pitched and shorter version of the loud calls.

After the departure of these two males, the oldest male left in the troop, "Coriace", was entering his fourth year. At the end of January 1970 he began to isolate himself frequently at the southern extremity of the home range, and also to form a subgroup with the adult female f-3, her 24 months old daughter Heidi, and the young adult female Berna. As was the case for the splitting in 1969, the members of the sub-group generally kept together during the day but continued to spend the night with the other monkeys. They left the troop on February 19, 1970 and, again, settled at the southernmost extremity of the home range.

What might be the causes of the departure of troop members? And why do these occur at the same period of the year? In all three cases maturing males appear to have taken the initiative, taking with them females of the same age group. One exception is the female f-3 which was probably two years older than Coriace. She had, however, become partly peripheral to the troop, early in December, following her postpartum sickness (see section on "infants"). In two out of the three instances, departures did not seem to be provoked by overt aggression from the alpha male.

In the first instance he was seen briefly to chase the maturing male only once (on January 31, 1969). No aggression was observed in the second instance. It was only in the instance of Coriace that antagonism with the alpha male was observed (January and February, 1970). Coriace would often approach him and would sometimes pull his tail or slap him, as if in play. The adult male would threaten Coriace, either in retaliation or spontaneously. Yet, if antagonism with the alpha male is the main cause of the departures of maturing males, should not these occur during the mating season when the alpha male is at his most aggressive (Table II)? The only noticeable change in the behavior of the alpha male, before and after the departure of the secessionists, was an increase in the emission of loud calls and in the frequency of the "jumping around display" in January and February, 1969 and 1970. The fact that departures always occurred during the birth season does not necessarily imply that the carrying capacity of the home range had been reached. All monkeys were in excellent physical condition and none showed any sign of loss of weight. Furthermore, the first and third departure occurred after the births of the young of the year, but the second took place before the birth of the 1970 generation. Moreover, about two weeks before the final departure of three troop members, in mid-February 1969, the troop readily adopted a young male of about 12 months old who had recently been captured in another area. It appears therefore that departures are due more to the social maturing of males than to a limitation of food resources.

What happened to the secessionists? They all left in the same direction, that is in the direction of the Botanical Garden, which communicates with the wooded peninsula where Lowe's guenons and lesser white-nosed monkeys are living. These wild monkeys are subject to poaching and almost impossible to observe. It was only during the last departure that interactions between the secessionists and the troop were observed. On at least two occasions (17 and 24 days after departure respectively), the two sides approached within 6 - 7 meters of each other at the southernmost extremity of the troop's territory. Females of both sides exchanged contact calls and moved towards one another. In each case the alpha male interfered by threatening the secessionists away and "his" females back to the troop.

INTERACTIONS WITH OTHER VERTEBRATES

The relationship of the Lowe's guenon to other species of vertebrates fall into four categories: interspecific associations with other monkeys, predation, commensalism and interspecific play.

Cercopithecus campbelli lowei often associates with other species of monkeys. On August 30, 1967, near Lamto, we observed a mixed troop of Lowe's guenons and lesser white-nosed monkeys (*C. petaurista*) spending the night in a tall tree in the forest gallery along the R. Bandama. The famous "Holy monkeys" of Zagné, which have been protected by the villagers for over 30 years, also belong to these two species. In August 1967, the mixed troop comprised more than 50 individuals with a predominance of *C. petaurista*.

In the mature forest of the Taï reserve, about 10 km east of Keibli, twice we came across mixed groups within a two-day period. In one group, Lowe's guenons were associated with black colobus (*Colobus polykomos*) and red colobus (*Colobus badius*); in the second, they were associated with *Colobus badius* and *Cercopithecus diana*.

In August 1967 at Lamto we observed a miniature mixed band composed of two 17- and 18-month-old males of *C. campbelli lowei* and a 17-month-old female *C. petaurista*. These juveniles had been captured as very young babies and reared by a member of the staff who had left after one year. The monkeys returned to the attic of the laboratory at night but otherwise roamed freely around the station. They were occasionally visited by wild troops of both species; some wild monkeys were bold enough to follow them as far as the buildings. They would also disappear for a few days in the gallery forest along the R. Bandama, but they did not become assimilated in wild troops. These juveniles formed a closely-knit social unit; each species seemed to understand perfectly the vocalizations of the other and if one member of the trio was attacked, the others would fly to its rescue.

In the Ivory Coast the major predator of Lowe's guenons is undoubtedly man. All monkeys are considered as very palatable by the villagers who actively hunt and trap them the year round. A few other vertebrates may occasionally kill monkeys, mostly infants and juveniles. We have no information on predation in the Taï Reserve where leopards are common. At Adiopodoumé, the monkeys' reactions to birds-of-prey are not stereotyped. They are not afraid of the common black kites, *Milvus migrans*, except the females which appear to avoid them during the birth season. A captive owl, *Bubo poensis*, attracted juveniles and subadults who came to investigate it at close range; a few days later a subadult was seen threatening two young owls of the same species in the forest. On the other hand the approach of an immature palm-nut vulture (*Gypohierax angolensis*) sent the whole troop diving into thick foliage and elicited loud calls from the adult male.

Lowe's guenons seem to be afraid of certain snakes. At Adiopodoumé, the release of three living West African cobras, (*Naja haje*) about 1.5 m long, elicited the loud calls of the adult male who was then about 30 m away. Every evening at Lamto, the three juveniles used to emit alarm calls from the beams of the roof while looking in a certain direction towards the ground. Finally it was realised that thereby they were signalling the passage of a large spitting cobra.

Data on commensalism are still scant. At Adiopodoumé, as elsewhere in the Ivory Coast, the white-crested hornbill, (*Tropicranus albocristatus*), often stays close to the monkeys when they eat and move about in the middle and lower layers. Quite probably hornbills follow the monkeys in order to catch the airborne insects disturbed by them.

At Adiopodoumé the monkeys also play the "Hornbill Game" with an allied species of hornbill, *Tockus semifasciatus*. We first noticed this in December 1968. Every evening, around 5.30 - 6.00 p.m., both the monkeys and a group of 10 - 15 hornbills arrived in a *Ricinodendron heudelotii* which was near the sleeping tree of the monkeys and was, at that time, without leaves. Juveniles waited until the hornbills had settled down to roost, then leaped towards them and shook the branches (Fig. 11). The birds waited until the last moment to take flight and returned soon afterwards. Both "sides" would watch each other quietly for a few moments then the "game" was repeated several times during the evening. The same behavior was observed in different trees on several occasions in 1969, not only with hornbills but also with green pigeons and even common black kites. In the latter instance the game took place in the middle of the afternoon.

Juveniles were also seen playing with wild squirrels (*Heliosciurus gambianus*) and two tame mongoose. Play then was not always one-sided. For instance, a juvenile *mona* and a squirrel would chase each other in a tree, reversing directions, each trying to catch the tail of the other. A mongoose would charge a monkey who would run away on the ground, then come back, pat the mongoose and run away again, apparently in order to elicit a pursuing reaction.

COMPARISON WITH OTHER FOREST SPECIES OF THE GENUS *CERCOPITHECUS*

Comparative data on the ecology and behavior of other forest species of the genus *Cercopithecus* are still scanty. Published information is presently limited to the East African redtail *Cercopithecus ascanius schmidti* (Haddow, 1952), various subspecies of *C. mitis* (Gartlan and Brain, 1968; Aldrich-Blake, pers.comm.) and seven species from North East Gabon and Southern Cameroon: *C. cephus*, *C. nictitans*, *C. mona*, *C. pogonias*, *C. erythrotis*, *C. lhoesti* and *C. neglectus* (Gautier 1969; Gautier and Gautier-Hion, 1969; Struhsaker, 1969). However none of the above mentioned species have been the subject of any long-term field study. Detailed information is only available for the talapoin, a very peculiar animal simetimes included within the genus *Cercopithecus*, but best classified separately in *Miopithecus* (Gautier-Hion, 1966, 1968, 1970).

The only species of the genus *Cercopithecus* which has been adequately studied is *C. aethiops* which never enters the high forest (Hall and Gartlan, 1965; Struhsaker, 1967a, b, c, d; Gartlan and Brain, 1968). Some of its behavioral characteristics have been considered as adaptations to woodland savannas where production varies seasonally. However a better knowledge of its forest relatives is needed to appreciate which characteristics are referable to environmental factors and which are shared with other *Cercopithecus* species.

HABITAT

Among the various forest species of *Cercopithecus* some show a distinct preference for a given type of habitat. For example, in West Africa, *C. diana* is found only in middle and upper layers of both primary and secondary forest. *C. petaurista* shares the same type of habitat, but also enters coastal scrub and swampy areas (A.H. Booth, 1956, and personal observations). In West Cameroon, N.E. Gabon and Rio Muni, *C. nictitans* and *C. pogonias* are apparently more common in mature forest and are seen most frequently in the upper strata. *C. cephus* is more adaptable and is also found in less mature forest and second-growth (Gautier and Gautier-Hion, 1969; Struhsaker, 1969). *C. neglectus* has a distinct preference for the dense vegetation adjacent to rivers (Sabater Pi and Jones, 1967; Gautier and Gautier-Hion, 1969). *Miopithecus talapoin* lives in the lower layers of all forest types provided they are not too far from water. Both the De Brazza monkey and the talapoin sleep near water and are excellent swimmers. *C. lhoesti* is reported by Struhsaker (1969) to have a distinct preference for lower levels and is commonly seen on the ground.

In East Africa, the redtail monkey *C. ascanius* schmidti inhabits virtually every type of lowland rain forest (Haddow, 1952). As for *C. mitis*, it has the widest distribution of all forest *Cercopithecus*. Besides the rain forest of Central West Africa, the species is also found in montane and bamboo forests of the high mountains of East Africa and in a number of isolate evergreen forests scattered from Ethoipia to the Cape Province (Gartlan and Brain, 1968).

DIET

To date the most detailed study of the diet of any forest *Cercopithecus* is that of Haddow (1952) on *Cercopithecus ascanius schmidti*. One hundred stomachs were examined for contents and 42 species of food plants identified. The most commonly recorded foods were leaves and green shoots. The next important group of foods consisted of fruits. The redtails were also "inordinatcly fond" of the yellow flowers of *Markhamia platycelyx*. Only three cases of insect-eating (*Crematogaster*) were reported and no remains of eggs or small vertebrates were found. The Gautiers (1969) have examined a number of stomach contents of *Cercopithecus pogonias grayi*, *C. nictitans*, *C. cephus*, *C. neglectus* and *Miopithecus talapoin* in N.E. Gabon. The bulk of the diet of these species was represented by a great variety of fruits, but buds, young leaves and some flowers were also eaten. Insects were found in the stomach contents of these five specimens, but were particularly abundant in the talapoin.

Forest *Cercopithecus* must not therefore be considered as purely fruit-eating monkeys and never in competition with the sympatric *Colobus* which are reputed to be strictly leaf-eaters. Actually both seem to share the same types of foods, but in different proportions. Species of the genera *Cercopithecus* and *Miopithecus* appear to be more insectivorous than the *Colobus*.

BIRTH SEASON

As long ago as 1952, Haddow reported a definite breeding season for the East African redtail *Cercopithecus ascanius schmidti*. His data (pp. 361-368) show that births occurred only from September to April. Ten years later C. Booth (1962) reported, that for Kenya as a whole, there was no fixed breeding season for *C. mitis*, *C. neglectus* and *C. aethiops*. She pointed out, however, that seasonality existed locally in areas where "the available food fluctuates considerably in the course of each year" (p.482). During their investigations in N.E. Gabon, the Gautiers (in Gautier-Hion, 1968), were surprised to find definite breeding periodicity for most of the forest cercopithecines living in the rain forest areas close to the Equator. All births occurred between December and April in Cercopithecus nictitans, C. cephus and C. pogonias grayi, and in February - March for C. neglectus. As for the talapoin, there appears to be a definite mating season between May and August (the coldest part of the year), all births taking place between December and April.

Although far more data are needed to decide whether there is any reproductive seasonality among forest Cercopithecus, the facts already at hand corroborate our own findings for the Ivory Coast's guenons.

GROUP SIZE

All the forest species of the genus Cercopithecus studies so far live in small groups averaging 3.7 - 18.0 animals

Struhsaker (1969) has given the following values for 5 species in West Cameroon:

Species	Group size (1st number based on actual counts, 2nd number based on estimates)		
	Mean	Mode	Range
Cercopithecus nictitans martini	10.5 (12.6)	7 (13)	7-17 (7-19)
Cercopithecus n. nictitans	9.3 (11.6)	8 and 10 (12)	5-13 (9-15)
Cercopithecus p. pogonias	13.3 (14.2)	10-11 (15)	9-19 (11-17)
Cercopithecus p. grayi	15.0		11-19
Cercopithecus mona	8.8 (10.8)	11 (15)	3-13 (2.5-20)
Cercopithecus cephus	3.7		2-5

Species (contd.)	Group size (1st number based on actual counts, 2nd number based on estimates		
	Mean	Mode	Range
Cercopithecus lhoesti preussi	3.7 (5.6)	2 and 3 (3)	2-7 (2-9)

Gautier and Gautier-Hion (1969) indicate the following average troop size for 3 species in N.E. Gabon:

Cercopithecus pogonias grayi	11.5
Cercopithecus cephus	10.5
Cercopithecus nictitans	18
Cercopithecus neglectus	4

In the Budongo forest, Aldrich-Blake (pers.comm.) found that the average troop size for *C. mitis* was 14 (ranging from 12 - 17).

It appears therefore that most of the forest *Cercopithecus* live in groups of 10 - 15 individuals. The two outstanding exceptions are *C. lhoesti preussi* in W. Cameroon and *C. neglectus* in N.E. Gabon, both having a group size about half that of the other species.

On the other hand talapoins live in much larger groups: 40 - 60 animals in areas distant from human settlements and 74 - 158 near villages where they have become pests (Gautier-Hion, 1970).

GROUP COMPOSITION

Struhsaker (1969) has come to the conclusion that "heterosexual groups of forest *Cercopithecus* are characterized by the presence of only one adult male". The same conclusion has also been reached by Aldrich-Blake for *C. mitis* (pers.comm.). However, some of the observations made by the Gautiers in Gabon show that the situation is not as simple as it first appears. In some troops of *C. pogonias grayi* and of *C. nictitans* the Gautiers have noted that one large adult male remained silent while another individual in the group emitted lout calls (pers.comm.). In 1968 and 1969 the Adiopodoumé troop of *C. campbelli* lowei included, besides the alpha male, two maturing males which would mount the adult females but did not emit loud calls; these younger males subsequently left the group. Thus the concept of the one male group does not necessarily imply that there is only one sexually mature male in the troop or that one male has the monopoly in mating.

The talapoin troop includes many adult males although only one of them normally emits "leader" calls. For instance, in the Ntsi-Belong troop studied by Gautier-Hion (1970), there were 13 adult males and 27 adult females for a total of 115 individuals. Most of the time there was only one principal "caller", but there could be two when the troop split for a few hours into two subgroups.

EXTRAGROUP INDIVIDUALS

Haddow (1952) was the first writer to call attention to the high incidence of solitary individuals in some forest species of genus *Cercopithecus*. For instance, out of a total sample of 143 *Cercopithecus ascanius schmidti*, he found 14 solitary males and one solitary female. In a composite sample of three subspecies of *C. mitis*, he found nine solitary males out of a total of 81 individuals (p. 354). Haddow also points out that "the fact that two-thirds of the present sample of solitary Redtail males were full adults, apparently in good condition, suggests strongly that this state is probably a temporary one".

In his 1969 paper, Struhsaker also mentions that solitary monkeys were commonly observed among the forest *Cercopithecus* of W. Cameroon: "With one exception these solitaries were adult, and when their sex was verified, proved to be adult males" (p. 105). Yet it appears from his data that solitary individuals were not uncommon in *C. nictitans* and *C. erythrotis*, but nevertheless were rare in *C. pogonias*, *C. mona*, *C. cephus* and *C. lhoesti*. Furthermore, most of the extragroup individuals observed by Struhsaker were not sexed. Out of 33 solitaries, only five could be definitely verified as adult males.

Far more observations are needed before one can be certain of the role played by extragroup individuals in societies of forest *Cercopithecus*. It is probably that these individuals are mostly males and adults, at least in certain species. But adult females have also been observed in isolation. Haddow (1952) has reported an isolated adult female redtail in advanced stage of pregnancy. The Gautiers have observed a female *C. neglectus* spending the night in isolation, and on several occasions have seen isolated females and juveniles of both sexes of *C. pogonias grayi* (pers. comm.). On the basis of the observed behavior of the secessionists of the Adiopodoumé troop, one can easily understand how adults, subadults and juveniles of both sexes in several species of *Cercopithecus* may become solitary. The maturing male leaves with a younger male and/or with a female of the same age group and/or with an adult female who has become peripheral. If this maturing male should die or is shot - and he is more likely to be killed than the others because his loud calls and watch-dog behavior makes him more conspicuous - the young animal(s) who accompanied him becomes solitary, at least temporarily. The females may even try to return to their former group. This might have been the case for the young adult female who joined the Adiopodoumé troop in November 1969. These examples show that females may also participate in the gene flow between groups.

The term "solitary" should always be used with caution. Many extragroup monkeys may simply be animals temporarily feeding at the periphery of the group. This is apparently the case in the talapoin, although on 3 occasions single adult males have been found sleeping at 300, 500 and 2100 m from the closest communal "dormitory" (Gautier-Hion, 1970). In heavily poached areas, "solitaries" may be no more than the last remnant of over-hunted troops.

MIXED GROUPS

Forest *Cercopithecus* very often form mixed groups for a variable duration of time. Various species often associate with each other and with talapoins, mangabeys and colobus monkeys. The existence of these mixed groups has been known for a long time (Haddow, 1952). In Ghana A.H. Booth (1957) observed groups comprising *Cercopithecus campbelli*, *C. mona*, *C. petaurista* and *Procolobus verus*. In Rio Muni, Sabater Pi and Jones (1967) and Jones and Sabater Pi (1968) have reported numerous cases of associations between four species of *Cercopithecus*, three species of *Cercocebus*, the black colobus and the talapoin. In Uganda, Chalmers (1968) has noted *Cercopithecus mitis* to be associated with *Cercocebus albigena* and *Colobus badius*.

The most comprehensive study of mixed groups has been carried out in N. E. Gabon by Gautier and Gautier-Hion (1969). Some species of *Cercopithecus* are more often met in association with other species than alone. *C. cephus*, for instance, is encountered in mixed groups in 86% of the cases. Others (*C. neglectus*) are seldom, if ever, observed in association with other monkeys.

The duration of the inter-specific association is very variable. At Ntsi-Belong *Cercopithecus cephus*, *C. pogonias grayi* and *C. nictitans* associate with talapoins only temporarily. At Belinga, on the other hand, mixed groups of *C. cephus*, *C. pogonias grayi* and *C. nictitans* can remain for days, weeks and perhaps years. In such semi-permanent associations one species apparently "dominates" another. When associated with *C. nictitans*, *C. cephus* never assumes the leading role nor does he play in territorial defense; his "loud call" has even lost its spacing function (Gautier, 1969). The tendency of the Lowe's guenon to associate, sometimes permanently, with other species of the genera *Cercopithecus* and *Colobus* agrees therefore with findings derived from other parts of Africa.

REFERENCES

Aldrich-Blake, F.P.G. (1968). A fertile hybrid between two *Cercopithecus* spp. in the Budongo Forest, Uganda. *Folia Primat.* **9**: 15-21.

Bertrand, M., Hunkeler, P. and Bourliere, F. (1969). Ecology and behaviour of the Lowe's guenon (16 mm movie obtainable from the authors).

Booth, A.H. (1955). Speciation in the mona monkeys. *J. Mammal.* **36**: 434-449.

Booth, A.H. (1956a). The Cercopithecidae of the Gold and Ivory Coasts: geographic and systematic observations. *Ann. Mag. Nat. Hist.* **9**: 476-480.

Booth, A.H. (1956b). The distribution of Primates in the Gold Coast. *J. W. Afr. Sci. Ass.* **2**: 122-133.

Booth, A.H. (1958). The zoogeography of West African Primates: a review. *Bulletin IFAN, A* **20**: 587-622.

Booth, C. (1962). Some observations on behavior of Cercopithecus monkeys. *Ann. N.Y. Acad. Sci.* **102**: 477-487.

Bourliere, F., Bertrand, M. and Hunkeler, C. (1969). L'écologie de la Mone de Lowe (*Cercopithecus campbelli lowei*) en Côte d'Ivoire. *Terre et Vie* **23**: 135-163.

Devore, I. and Hall, K.R.L. (1965). Baboon ecology. In *Primate Behavior*, (Ed. Irven DeVore). New York: Holt, Rinehart and Winston.

Gartlan, J.-S. and Brain, C.-K. (1968). Ecology and social variability in *Cercopithecus aethiops* and *C. mitis*. In *Primates, Studies in Adaptation and Variability*, (Ed. Phyllis C. Jay). New York: Holt, Rinehart and Winston.

Gautier, J.-P. (1969). Emissions sonores d'espacement et de ralliement par deux Cercopithèques arboricoles. *Biol. Gabon.* **5**: 117-145.

Gautier, J.-P. Gautier-Hion, A. (1969). Associations polyspécifiques chez les Cercopithèques du Gabon. Terre et Vie, 23: 164-201.

Gautier-Hion, A. (1966). L'écologie et l'éthologie du Talapoin Miopithecus talapoin talapoin. Bio. Gabon. 2: 311-329.

Gautier-Hion, A. (1968). Etude du cycle annuel de reproduction du Talapoin (Miopithecus talapoin) vivant dans son milieu naturel. Biol. Gabon. 4: 163-173.

Gautier-Hion, A. (1970). L'organisation sociale d'une bande de Talapoins (Miopithecus talapoin) dan le Nord-Est du Gabon. Folia Primat. 12: 116-141.

Haddow, A.-J. (1952). Field and laboratory studies of an African monkey, Cercopithecus ascanius schmidti Matschie. Proc. zool. Soc. London. 122: 297-394.

Hall, K.R.L. (1965). Behaviour and ecology of the wild Patas monkey, Erythrocebus patas in Uganda. J. Zool. 148: 15-87.

Hall, K.R.L. and Gartlan, J.-S. (1965). Ecology and behaviour of the vervet monkey, Cercopithecus aethiops, Lolui island, Lake Victoria. Proc. zool. Soc. London. 145: 37-56.

Hill, W.C.O. (1966). Primates. Vol. 6, Edinburgh: University Press.

Jones, C. and Sabater Pi, J. (1968). Comparative ecology of Cercocebus albigena (Gray) and Cercocebus torquatus (Kerr) in Rio Muni, West Africia. Folia Primat. 9: 99-113.

Lancaster, J.-B. and Lee, R.-B. (1965). The annual reproductive cycle in monkeys and apes. In Primate Behavior. (Ed. Irven DeVore), New York: Holt, Rinehart and Winston.

Napier, J.R. and Napier, P.H. (1967). A Handbook of Living Primates. London and New York: Academic Press.

Rahm, U. (1961). Esquisses mammalogiques de basse Côte d'Ivoire. Bulletin IFAN A 23: 1229-1265

Sabater Pi, J. and Jones, C. (1967). Notes on the distribution and ecology of the higher Primates of Rio Muni, West Africa.

Tulane Stud. Zool. 14: 101-109

Schwarz, E. (1928). Notes on the classification of the African monkeys of the genus Cercopithecus Erxleben. Ann. Mag. Nat. Hist. 1: 649-663.

Struhsaker, T.T. (1967a). Behavior of vervet monkeys (Cercopithecus aethiops). Univ. Calif. Publ. Zool. 82: 1-64, 11 pl.

Struhsaker, T.T. (1967b). Behavior of vervet monkeys and other Cercopithecines. Science, 156: 1197-1203.

Struhsaker, T.T. (1967c). Social structure among vervet monkeys (Cercopithecus aethiops). Behaviour, 29: 83-121.

Struhsaker, T.T. (1967d). Ecology of vervet monkeys (Cercopithecus aethiops) in the Masai-Amboseli Game Reserve, Kenya. Ecology, 48: 891-904

Struhsaker, T.T.(1969). Correlates of ecology and social organization among African Cercopithecines. Folia Primat. 11: 80-118.

Turner, D.C. (1966). General Endocrinology: 4th edition. Philadelphia: Saunders.

Table I. Food Plants Utilized by *Cercopithecus campbelli lowei* at *Adiopodoumé*.

Name	Parts Eaten
MONOCOTYLEDONS	
Musaceae	
Musa sapientium	Fruits (ripe and unripe) ++
Palmae	
Elaeis guineensis	Fruits ++
DICOTYLEDONS	
Ampelidaceae	
Cissus aralioides	Fruits
Anacardiaceae	
Spondias mombin	Flowers
Anonaceae	
Anona muricata	Fruits
Cleistopholis patens	Fruits and flowers
Apocynaceae	
Rauvolfia vomitoria	Flowers
Caesalpiniaceae	
Anthonota macrophylla	Fruits and flowers
Caesalpinia pulcherima	Flowers and young pods
Cassia sp.	Flowers
Crudia Klainei	Fruits
Dialium dinklagei	Fruits
Griffonia simplicifolia	Fruits

Combretaceae	
Terminalia catappa	Fruits
Commelinaceae	
Palissota hirsuta	Fruits
Convolvulaceae	
Ipomoea involucrata	Flowers
Euphorbiaceae	
Alchornea cordifolia	Fruits
Macaranga barteri	Young fruits
Microdesmis puberula	Fruits
Uapaca heudelotii	Leaf bases
Hypericaceae	
Harungana madagascariensis	Fruits
Icacinaceae	
Leptaulus daphnoïdes	Fruits
Labiateae	
Hoslundia opposita	Flowers
Linaceae	
Hugonia foliosa	Fruits
Malvaceae	
Hibiscus sp.	Flowers
Meliaceae	
Trichilia heudelotii	Fruits
Menispermaceae	
Rhigiocarya racemifera	Fruits

Mimosaceae	
Albizzia dentifolia	Flowers
Pentaclethra macrophylla	Folioles and leaf exudates
Parkia bicolor	Flowers
Moraceae	
Ficus exasperata	Fruits ++
Ficus vogelii	Fruits ++
Morus mesozygia	Flowers
Myrianthus arboreus	Fruits
Myrtaceae	
Eugenia uniflora	Fruits
Psidium guajava	Fruits
Oleaceae	
Heisteria parviflora	Fruits and flowers
Papilionaceae	
Baphia nitida	Fruits
Centrosoma plumieri	Flowers
Passifloraceae	
Passiflora edulis	Fruits ++
Rubiaceae	
Coffea liberica	Fruits
Mussaendra stigmatica	Fruits
Sapindaceae	
Blighia parviflora	Aril of the seed
Cardiosperumum grandiflorum	Fruits

Simaroubaceae	
Hannoa klaineana	Fruits
Sterculiaceae	
Cola gabonensis	Pericarp of the fruit
Sterculia tragacantha	Flowers

Symbol ++ indicates that plant is heavily used.

Table II. Seasonal variations of mountings and aggressions by the alpha male in 1969.

	Jan.	Feb.	Mar.	Apr.	May	Jun.	Jul	Aug.	Sept.	Oct.	Nov.	Dec.
Mountings with thrusting	0	1	0	0	0	1	2	3	1	0	0	0
Aggressions towards 4 and 3 years old males	2	1	0	0	1	3	5	8	5	2	2	1
Aggressions towards juveniles	0	0	0	0	0	2	2	3	3	3	3	0
Aggressions towards adult and subadult females	0	0	0	0	0	0	0	0	0	0	0	0
Number of observation hours	50	56	50	26	27	29	30	27	28	27	27	52

Aggressions consisted mainly of threats and chases

Table III. Parental and "friendly" interactions of the troop members with female infant (born on January 18, 1969).

Seen in company of:	Jan.	Feb.	March	April	May	June	July	Aug.	Sept.	Oct.	Nov.	Dec.
Mother f-2	33(100)	65(85)	16(50)	15(67)	25(52)	31(39)	10(70)	11(36)	13(31)	5(33)	15(13)	12(8)
f-1	6(100)	12(100)	8(75)	1(100)	2(100)	6(50)	0	1(100)	1(100)	0	6(17)	14(7)
f-3	0	0	0	1(0)	0	0	0	0	0	1(0)	0	0
f-4	0	0	has left the group									
♀ sibling Vreni	0	14(100)	19(79)	7(71)	23(83)	18(28)	14(71)	14(57)	19(21)	13(38)	10(0)	16(0)
♀ subadult Trudi	0	2(100)	11(91)	5(100)	12(50)	18(39)	6(67)	7(71)	8(25)	10(20)	10(20)	8(25)
Alpha male	0	2	1	0	0	0	1	0	0	0	0	0
Subadult and juvenile males	0	5	18	3	19	35	20	22	24	41	33	37
Seen alone	2	22	15	21	50	51	70	34	40	21	31	35

Number of times the female infant has been seen in company of another troop member who would cradle her, carry her (number in brackets, expressed in percentage of the preceding number), groom her, play and "baby sit". The hours of observation are the same as in Table II.

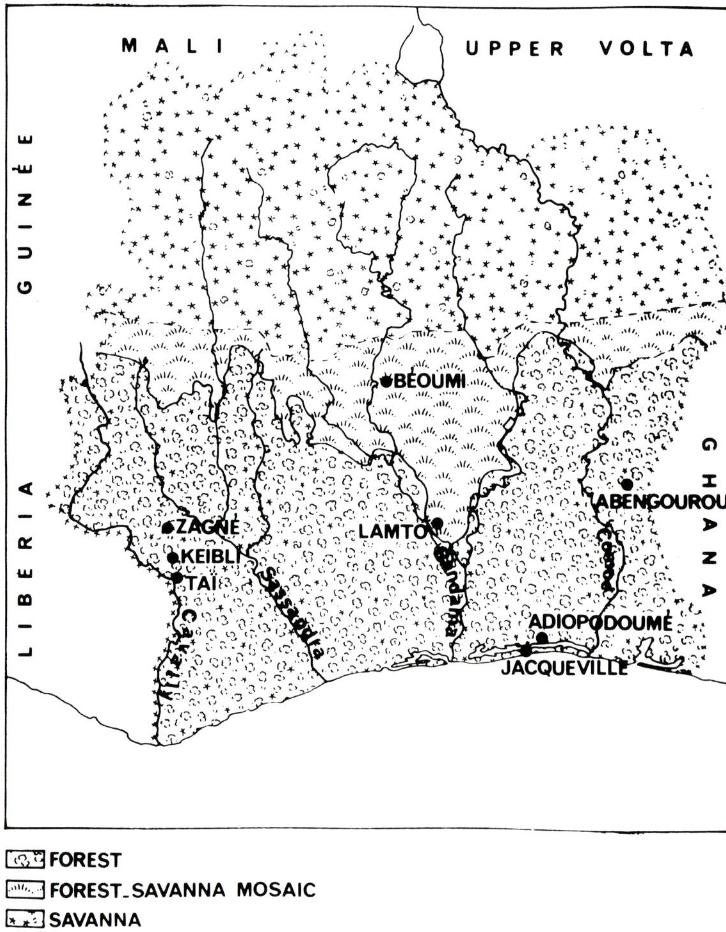

Fig. 1: The major vegetation zones of the Ivory Coast with the localities mentioned in the text.

Fig. 2: Aimed leap of an adult between tree crowns.

Fig. 3: Tail twining between a mother and an "aunt".

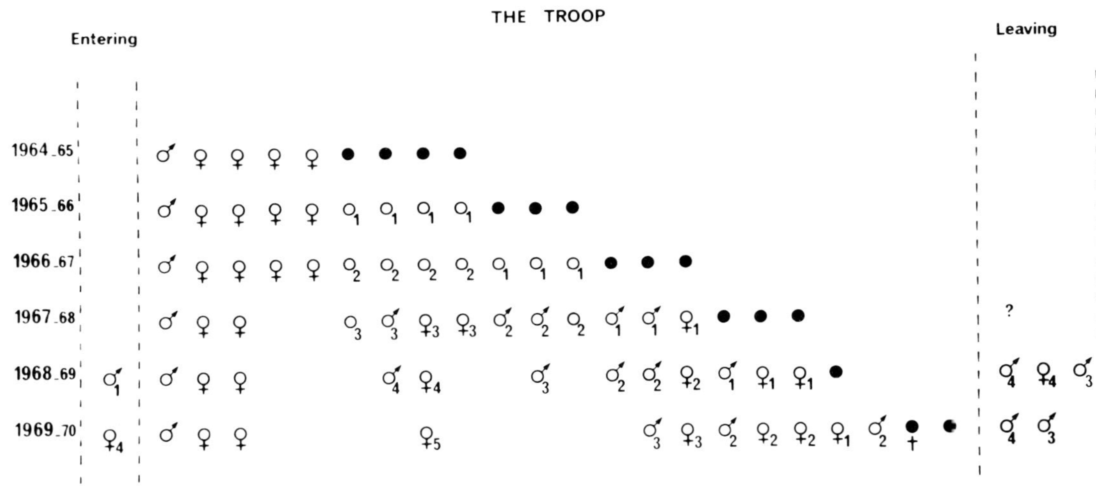

Fig. 4: Demographic structure of the Adiopodoumé troop from 1964 to January 1970. Each year class refers to the period from February in one year to the end of January in the following year. Ages are indicated by numbers (4: Male having just completed the 4th year). Solid circles indicate young of the year.

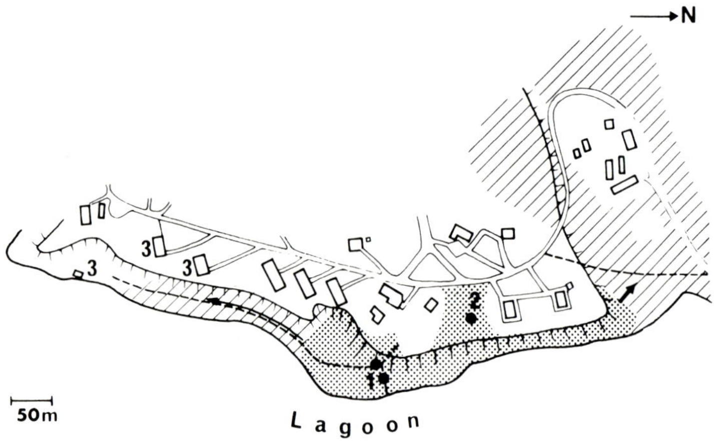

Fig. 5: Study area at Adiopodoumé. Hatching indicates the remnant of coastal forest, stippling corresponds to the home range of the troop.

Fig. 6: Insect hunting: a subadult male opens the folded pinnae of an oil palm-tree.

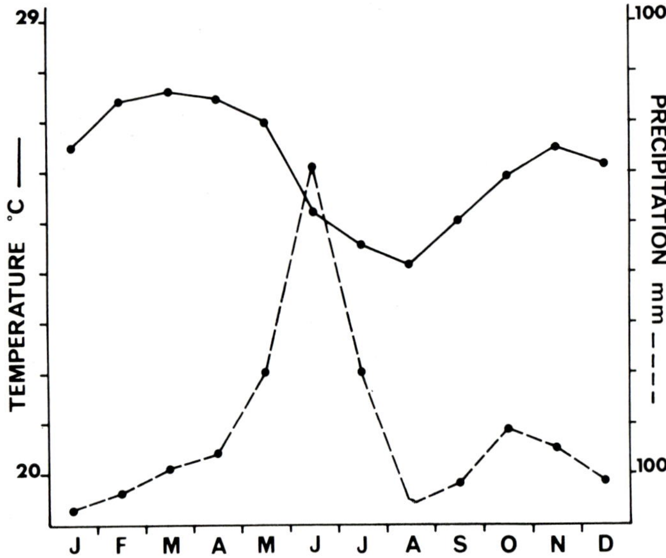

Fig. 7: Average values of the mean monthly temperature and rainfall at Adiopodoumé from 1953 to 1969.

Fig. 8: The adult male, one adult female and her one year old daughter.

Fig. 9: An aunt licks and grooms the perineal area of a young infant lying on its mother's knees.

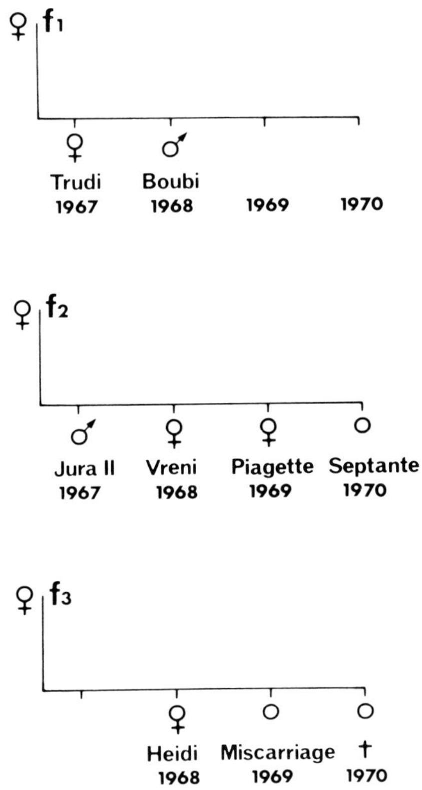

Fig. 10: The progeny of the three adult females of the Adiopodoumé troop from 1967 to 1970, showing the sibling relationships.

Fig. 11: The Hornbill Game, December 1968.

CROSS-SPECIES MODIFICATIONS OF SOCIAL BEHAVIOR IN BABOONS

H. Kummer, W. Goetz, and W. Angst

INTRODUCTION

This paper is based on recent field experiments on the social behavior of Ethiopian baboons. The experiments were designed to identify behavioral characters that bring out different social organizations in different populations or species. For this symposium we selected preliminary results showing how social behavior and the degree of its flexibility can affect gene flow across the common geographical boundary between two closely related, allopatric species.

The subjects were the two species *Papio anubis* and *Papio hamadryas*. Whereas the motor repertoires of social behavior are very similar in these two species, their social organizations differ drastically (Kummer, 1968a). *P. anubis* groups consist of many males, females and young. There are no stable subgroups; all males potentially have access to all females of the group. Only temporary consort pairs are formed during the few days of the females receptivity. The female then pairs with one male at a time, and several males succeed each other as the female's mates. In such a pair, the male follows the female. Whenever she is not receptive the female moves freely and makes contacts throughout the group (Hall and DeVore, 1965). This social organization seems to be typical of all *Papio* except *P. hamadryas*.

In contrast, hamadryas baboons have permanent male-female bonds (Kummer, 1968b). Several females are constantly and exclusively associated with one male. The bond persists during receptive and non-receptive days, pregnancy and lactation. In the hamadryas pair-bond the female follows the male. A varying number of such one-male groups associate with bachelor males to form large troops. This social system has been found in only one other primate species, the gelada. Like hamadryas, geladas form one-male units within large troops (Crook, 1966). It is obvious from the taxonomic position of the two species that their "polygynous" systems evolved independently, and the behavioral evidence shows that their similarity of group structure was achieved by different behavioral means.

The distribution of social systems in the genus *Papio* suggests that hamadryas polygyny derived from an anubis-type organization as a more recent specialization, possibly as an adaptation to an arid habitat. In the present context, however, it is not our concern to identify the selective pressures that favor polygyny but to study the behavioral mechanisms which produce the polygynous pattern once the selective pressures are there. More precisely, we shall examine just how firmly the polygynous pattern is rooted in hamadryas behavior; to what degree, in other words, it has become a fixed species character.

OBSERVATIONS

1. THE HAMADRYAS PAIR BOND

Hamadryas polygyny is a one-sided affair; the one-male group is imposed on the females by the male. He threatens his females when they do not follow him closely enough. Eventually, the male will attack the female and bite her on the nape of the neck. She screams in response and then proceeds to follow the male at a closer distance than before. The male never attempts copulation with any females but his own, but the females do not reject a chance copulation with some sub-adult male. The male's predominant concern is not sex but the pair-bond, the fact that a female "belongs" to him. A hamadryas male is as eager to take over an unreceptive female as a female is ready for copulation; at least the difference, if it exists, is not easily noticed by an observer. Many pair bonds are established even before the female attains sexual maturity: young adult males take over small juvenile females from their mothers. They do not attempt to mate with these juveniles but instead they condition the female to follow by threatening her when she is too far away. At first the female responds to the threats like any "normal" animal would, by running away. Only after days of experience, she learns that immunity from attack comes from staying close to the aggressor. This lesson when learned, however, is learned for life. The following response will be transferred to every new male who may attach the female to his harem when her old male dies or loses her in a fight.

Thus it seems that the hamadryas female must learn the behavior appropriate to the social organization of her species from her male; she does not actively promote or support the harem system even after she has learned it. Is it possible then that polygynous behavior only evolved in the hamadryas male and that - except for the effects of social learning - the hamadryas female still is a less specialized type, such as the anubis female? Clearly, the two sexes must be considered separately.

2. FLEXIBILITY OF FEMALE BEHAVIOR

Our project first focussed on the behavior of the baboon female. What is the hamadryas female's contribution to the polygynous system, and in which respects is she different from a female anubis baboon? If the inexperienced hamadryas female contributes little or nothing to the pair-bond relationship, then an anubis female should do as well under the same conditions. The answers were sought in a series of field experiments conducted around the Awash National Park in Ethiopia. Here, the species border between anubis and hamadryas baboons cuts across the R. Awash, the anubis occupying the western, up-river areas while the hamadryas occur down-river to the east. For the first experiment, adult anubis females were trapped, marked, and transported across the species border to a sleeping-cliff of a hamadryas troop. There, the famales were exhibited in wire cages for one or two days in order to familiarize them both with the locality and the troop, and to introduce the troop to its new member. On the second day, the anubis females were released singly in the presence of the hamadryas troop. The experiment was repeated eight times with different females and different troops. For each anubis transplant a hamadryas female was released into the same host troop as a control.

There is as yet no systematic field research on the relationship of monkey groups with strangers of both sexes and all ages, but there is good evidence that some groups reject strangers. Our hamadryas transplants, far from being rejected, were eagerly accepted by the resident hamadryas males. Even during the initial exhibition of the strange females, hamadryas males approached the cage and tried to induce the caged female to follow them. When a female was released, a male usually rushed towards her, clasped or mounted her and then walked away, checking the female's following-response by looking back at her. The released hamadryas females followed immediately and soon settled into the routine of harem life with their new males.

The anubis females also attracted male suitors during the introductory period. At the moment of release it became evident that they know nothing of a following response. The respective hamadryas male naturally responded with threat and attack, which provoked an escape reaction in the anubis female and, thus, more attacks by the male. Some males finally caught their anubis females and restrained them physically. Most males soon learned not to press their female too hard. The harassed anubis females gradually learned not to run away. The "following-score" of the anubis females rose and reached the level of the hamadryas controls by the end of the first hour after the release. Pair formation with anubis females was marked by a high frequency of aggressive male herding behavior and of escape behavior by the females. In the process some hamadryas males developed conditioning techniques of unusual subtlety; for example they often refrained from attacking when the female walked away. Instead they passed her and walked in front of her carefully adjusting their route to the female's prospective path. Not all anubis experiments were successful. Several females escaped and were not seen again. Others escaped and were seen in neighboring troops weeks or months later; but none of the females managed to remain in a hamadryas troop without adapting to the harem system. The social organization of the host troop prevailed. The results suggested:

(1) that hamadryas males establish pair bonds with anubis females nearly as readily as with females of their own species;

(2) that the hamadryas and anubis system of communication are similar enough to permit cross-species interaction without obvious "misunderstandings";

(3) that the unusual response of keeping close to an aggressor needs no particular genetic basis but that it even forms part of the behavioral potential of a species that knows no herding behavior.

In the female, the hypothetical evolutionary shift from a social system of the anubis type to one of the hamadryas type can thus be re-enacted with moderate success within <u>one hour</u>. The hamadryas females' contribution to the social structure of their species is not elaborate. The experiments show quantitatively that their response to a given degree of male aggression is less intense that the response of anubis females to the same degree of male aggression. It was also observed that captive hamadryas females adapted much faster to our presence than anubis females and that their escape reactions to standardized human behavior were significantly less intense than those of their anubis counterparts. This difference may be explained by assuming a simple selection process. In the course of hamadryas evolution, females with a high escape response were probably more often driven from their troops to become the prey to predators than more threat-resistant females.

The main cause of hamadryas polygyny must be the hamadryas male. This conclusion was confirmed by the reverse experiment, in which hamadryas females were released into anubis groups. The females at first tried to follow a particular anubis male, but they soon learned that they received neither protection nor herding from the anubis males. By the second day they had already learned to move independently through the anubis group and, thus, they contacted many individuals in succession, a behavior that is typical of a female anubis baboon.

3. ONTOGENY OF MALE BEHAVIOR

To understand the social organization of the hamadryas baboon we must turn to the hamadryas male as the central figure of the harem system. In what respects does his behavior differ from the behavior of an anubis male? It is profitable to follow the comparison of the development of male behavior in the two species.

Two year-old males of both species carry out intromission in a technically correct manner. Their mates are receptive females of three to four years. Thus the juvenile male copulates with females that are larger than himself. These relationships may last for only a few minutes or for as long as an hour. In neither species does the male try to induce a following response in his sexual partner nor does he claim exclusive access to her. Two or three juvenile hamadryas males may mount and penetrate the same female in quick succession without antagonism from other juveniles or any other indications of pair-bonding behavior.

The first observed difference between the males of the two species appears at about four years of age. In both species, occasional sexual behavior is still directed at older females, i.e. at subadults and adults, without any evidence of a permanent bond. At the same age, the hamadryas males show the first signs of herding behavior. The object, however, is not a potential sexual partner but any small infant or juvenile which the male can "borrow" from its mother. Two years later, as a young adult the hamadryas male tends to adopt and herd orphaned infants and juveniles of both sexes. Eventually he establishes his own harem by adopting and herding a juvenile female which is destined to become his sexual partner when she reaches puberty. None of these early forms of herding and pair behavior is found in anubis males.

Sexual behavior and pair behavior in the hamadryas male thus develop independently and with two different types of partner. In the early stages, the sexual partner is older than the male and is not herded; in contrast, the object of herding is a much younger, immature animal with whom no copulation occurs. It is only in the fully adult leader of a one-male group that sexual and pair behavior converge on the same object. We therefore assume that the hamadryas pair-bond is not primarily based on sexual motivation. It may stem from a hypothetical motivation which we can, tentatively, label as social possessiveness.

Social possessiveness is no evolutionary novelty but merely an extreme expression of a tendency found in many Old World monkeys. The anubis male claims exclusive access to his sexual consort during her estrus. Baboon and macaque mothers are intolerant of another female's attempt to handle their infants. A gelada famale may aggressively prevent another female from approaching her male. But most of these assertions can be overcome by patient and persistent rivals even if they are of inferior rank and strength. The macaque mother eventually allows an "aunt" to mother her baby, and the gelada female eventually accepts a new female to her one-male group. The anubis male, once sexually satisfied, allows an inferior male to take over his mate. The hamadryas male is unique in that his claim for exclusive access is permanent and inflexible. He yields only to a rival of superior fighting power and persistence and, perhaps, to the limitations of his own old age.

4. INFLEXIBILITY OF MALE BEHAVIOR: EVIDENCE OF ITS EFFECT AT THE SPECIES BORDER

Our recent study did not include systematic cross-species transplantations of males, but a few accidental observations give the impression that baboon males are less likely to modify their social behavior than females. In each of two pure hamadryas troops, situated several kilometers from the species border, we observed single adult anubis males. Neither of the two anubis males was followed by hamadryas females, nor were any interactions with females observed. In a fight with a hamadryas male over an estrous female, one of the males chased his rival away, but the hamadryas nevertheless retrieved the female. The anubis male seemed to lack persistence, but obviously he had not acquired the herding techniques of the host species.

5. AN ANUBIS-HAMADRYAS HYBRID ZONE

When we decided to work near the species border, we hoped to study the natural encounters of *P. hamadryas* and *P. anubis* groups. The task was undertaken by one of our students (Nagel, in preparation). Nagel investigated the baboon troops along the Awash River, which corsses the border at a right angle from west to east. He found that a zone of 15 to 20 kilometers is inhabited neither by P. anubis nor P. hamadryas, but by hybrids of various degrees. Considering the capacity for travel of baboons, the hybrid zone is relatively narrow in extent. In a sequence passing down-river, the last anubis group was replaced by three or four hybrid groups with an increasing proportion of hamadryas characters and, finally, by troops of pure P. hamadryas. Each hybrid group included a broad spectrum of morphological combinations. The same group contained, for instance, almost pure anubis and three-quarter hamadryas individuals. The morphological characters seemed to re-combine freely, producing specimens with extra-long brown mantles and others with short gray coats. Nagel found various intermediate forms of social structure corresponding roughly with these physical distinctions.

Two observations which hint at the social conditions that produce hybrids, and confirm the impression that baboon males do not modify their social behavior as readily as the females, arose by chance. One day we found two anubis females in one of the P. hamadryas troops to the west of the hybrid belt. No anubis baboons had been seen there in the previous ten months; even now there were no anubis males. One of the anubis females could be observed on two evenings following a young adult hamadryas male as his only consort. A few weeks later, one of us inspected the sleeping cliff of an almost pure P. anubis group at the up-river edge of the hybrid zone and discovered a single almost pure P. hamadryas male who had not been seen there before. The male was conditioning an estrous anubis female of the resident group. At intervals the hamadryas male watched a resident anubis male who was running about the cliff in an agitated manner. The anubis male finally attacked the hamadryas male and drove him off. The hamadryas, however, returned immediately and once again placed himself close to the female. Surprisingly, the anubis male did not attack. Next morning, the hamadryas male prevented his conquest from following her group. When the group was safely out of sight, he led her in the opposite direction. The following night he was observed alone with the anubis female on a cliff further downriver in the direction of the hamadryas area. The female P. anubis was now following him obediently. Once again it was observed that the visiting male did not modify his behavior to conform with the social organization of the host group.

These two observations suggest that our transplantation experiments were not as remote from reality as we had thought. Apparently some of the young *P. hamadryas* males obtain their first females from *P. anubis* groups across the species border. The procedure is perhaps easier than extricating a hamadryas female from a harem in the home troop. In terms of reproductive success, the observations suggested a clear advantage of the hamadryas males, regardless of whether an anubis male visited a hamadryas troop or vice versa. If this is typical of the border situation in general, it is likely that the hybrids are produced mainly by anubis ♀ - hamadryas ♂ crosses; the reverse cross-mating situation appears to be less likely.

Thus, the social organization of two closely related and neighboring species could affect the geographical location of their common border. If hamadryas males routinely annexed anubis females into their population, the border would be shifted to the disadvantage of their own species. As the hamadryas groups on the border gradually became transformed into hybrid groups, they would themselves become the raiding grounds of hamadryas males farther inland. In reality, different ecological adaptations may prevent an actual loss of area: perhaps some hamadryas males may balance the gene flow by taking up residence in the anubis groups permanently.

CONCLUSIONS

The behavior underlying the different social organizations of two species of baboons was investigated. *Papio anubis* forms no permanent pairs, *Papio hamadryas* is polygynous. Experiments in the field showed that females of both species are able to modify their behavior to fit the social organization of a host group of the other species. There is some evidence that the males are incapable of such cross-species modification. If the two social systems are genetically controlled to any significant degree, then such control must act mainly through the behavior of males.

This sex difference is related to the fact that, at least in *P. hamadryas*, the social organization is patterned by male behavior. The sexual behavior of anubis and hamadryas males is essentially the same, and does not account for the difference in social structure. Hamadryas polygyny, instead, is caused by the strong tendency of the males to herd smaller animals and to prevent all interactions with rivals. Sexual behavior and herding develop independently and, in the earlier stages of ontogeny, with different classes of partners.

The possessive behavior of the hamadryas males appears to be an advantage in their competition with anubis males for females of either species. The hybrid zone at the species border therefore is more probably an effect of ♀ anubis - ♂ hamadryas cross-breeding than vice versa.

The evidence on the hybrid zone is preliminary but does not, in our opinion, justify the inclusion of both *P. anubis* and *P. hamadryas* in a single species. A detailed analysis will be required to explain why the hybrid zone is only about three home-ranges wide. As Bigelow (1965) pointed out, selection against hybrids may "Inhibit gene flow between two differently well-integrated gene pools despite interbreeding". The occurrence of hydrids thus does not exclude a high degree of reproductive isolation.

ACKNOWLEDGEMENTS

This research was supported by grant GB-6167 from the National Science Foundation and by a grant from the Wenner-Gren Foundation for Anthropological Research. The study was carried out while the principal author was a research associate at the Delta Regional Primate Research Center, Covington, La. We thank the Ethiopian Wildlife Conservation Department for permitting and facilitating the field research.

REFERENCES

Bigelow, R.S. (1965). Hybrid zones and reproductive isolation. *Evolution* **19**: 449-458.

Crook, J.H. (1966). Gelada baboon herd structure and movement - A comparative report. *Symp. zool. Soc. Lond.* **18**: 237-248.

Hall, K.R.L. and I. DeVore. (1965). Baboon social behavior. In *Primate Behavior*. (Ed. DeVore, I.). New York: Holt, Rinehart and Winston.

Kummer, H. (1968a). Two variations in the social organization of baboons. In *Primates. Studies in adaptation and variability*. (Ed. Jay, Phyllis). New York: Holt, Rinehart and Winston.

Kummer, H. (1968b). *Social organization of hamadryas baboons. A field study*. Chicago: University Press.

PHYLOGENETIC IMPLICATIONS OF SOME VOCALIZATIONS OF *CERCOPITHECUS* MONKEYS

Thomas T. Struhsaker

INTRODUCTION

Species of the genus *Cercopithecus* are usually distinguished by taxonomists on the basis of pelage coloration. Karyology and osteology provide some limited data, although they are inadequate in themselves for diagnosing species. The genus *Cercopithecus* has more species and forms than any other genus in Africa, perhaps more than any other living primate genus. The extreme diversity in pelage coloration within this genus and within many of its so-called species has resulted in considerable disagreement over basic problems of taxonomy and phylogenetic affinity. Various schemes attempting to depict these phylogenetic relationships have ranged from extreme "lumping" to extreme "splitting" of the genus (Hill, 1966). This paper presents new data on the behavior and ecology of several rain forest *Cercopithecus* species about which little was previously known. Greatest emphasis is placed on the vocalizations of these monkeys. Some of these calls prove to be extremely stable characters and are among the best indicators of phylogenetic affinity within the genus *Cercopithecus*. The use of sounds in taxonomic and phylogenetic studies has been advanced most in studies of orthopterans (Alexander, 1968), anurans (Littlejohn and Oldham, 1968) and birds (Lanyon, 1969). All of the new data presented in this paper were collected under field conditions during a 19-month study in Cameroon (17 months), Rio Muni (circa 1 month), and Ghana (circa 1 month), West Africa.

RATIONALE

The polymorphic *Cercopithecus aethiops* group demonstrates little, if any, interpopulational variability in the basic structure of its vocalizations, as concluded from study of populations in Kenya, Uganda, Tanzania, Cameroon and Ghana. *C. mona* which is distributed over a smaller range in Africa than *C. aethiops* demonstrates a similar uniformity between populations. Likewise two subspecies of *C. nictitans* lacked differences in vocalizations. Given these three examples in which the data on vocalizations conform with the phylogenetic conclusions based on morphological data, it is argued that certain kinds of vocalizations can be used to elucidate the phylogenetic affinities of Cercopithecus monkeys whose relationships are less clearly understood on morphological grounds.

Different kinds of vocalizations may be subject to dissimilar kinds of selective pressures consequently evolving at disparate rates and in different directions. An understanding of function permits a more realistic decision as to which types of vocalizations can best contribute to an understanding of phylogeny. For example, the development of similar alarm calls by sympatric congeners having the same predators can readily be understood as a case of parallelism. Such calls may be less reliable indicators of phylogenetic affinity than are calls whose specific physical structure is less dependent on environmental pressures. The loud calls given by adult males appear to be an example of the latter. They are given in response to a variety of stimuli, all of which seem to have in common the potential to disrupt the coherence of the social group (Struhsaker, 1969). In the rain forests of West Africa where visibility is limited and temporary associations of several congeneric species of similar size and habits are common, one could readily understand strong selection for some kind of audible behavior that keeps conspecifics together with fleeing from alarming stimuli. This is expecially critical among the forest Cercopithecus where the majority of heterosexual groups have only one adult male member. Limited observations (Strusaker, 1969) indicate that the loud calls of adult male Cercopithecus monkeys living in rain forests provide an acoustical focal point for fellow group members and thus serve this function. Of course this central focus function may not be the only one performed by the loud calls, but it is probably the critical one in terms of selection. During alarming situations conspecifics of mixed species associations seem to maintain their group coherency through these loud calls. If such a behavioral device were not present, it is suggested that the conspecific social groups would be continually breaking-up and hindering differential reproduction within the species, reproductive isolation between species, and any ecological advantages of intraspecific cohesion, such as the location of food suitable for that particular species. Although these loud calls are surely modified by selective pressures and are undoubtedly modified faster or more radically in area where more species are sympatric than in areas of less sympatry, their detailed physical structure bears less direct relationship to the environment than does that of others such as alarm calls. In the evolution of adult male loud calls, natural selection presumably favored differences between those of sympatric species and that these differences could be of any sort, regardless of the physical environment. Similar kinds of selective pressures presumably influence the evolution of Cercopithecus cohesion calls. In contrast, it is plausible that alarm calls evolved toward similarity with sympatric congeners. The data for Cercopithecus species support this latter hypothesis less well than they do the former two. Better examples of characters whose details have apparently evolved in direct response to the social and physical environment are the shape and size of teeth and limb proportions (see Jolly, this volume).

Four forms were studied within the mona superspecies, C. mona, C.p.pogonias, C.p.grayi, and C.campbelli lowei (Booth, 1955). Study of their vocalizations substantiates Booth's conclusions on their phylogenetic affinities. The loud calls of adult males of all three species have a basic spectrographic and acoustical similarity, but are readily distinguished. The calls of the two subspecies of pogonias are the same. Analysis of the alarm calls or so-called sneeze calls and 000 (cohesion) calls revealed similar patterns. In my opinion, these results, combined with observations of sympatry and with the conclusions from Booth's study, dispel all doubts regarding the relationships of these four forms. Reference to the grayi and campbelli forms as subspecies of mona are no longer justified (Schwarz, 1928; Rode, 1937; Rahm, 1961; Gautier & Gautier-Hion, 1969).

The erythrotis and cephus froms have long been considered valid species. Even as recently as 1966 (Hill) they have been aligned with different superspecies groups. The hacks of the adult male of these two forms are indistinguishable. Combined with my observations of an area containing monkeys with coloration intermediate to these two forms, the similarity of all of their vocalizations supports the conclusion that they are conspecific. However, the correspondence of their adult male loud calls is most meaningful because some of the other vocalizations are virtually identical with calls of the otherwise very different *C. nictitans*. For example, some kinds of chirp calls given by adult females and juveniles are identical in *nictitans* and *erythrotis-cephus*. This could be interpreted as a case of parallelism between two species that commonly form mixed foraging associations. However, *mona* and *pogonias* also associate frequently with these two species and yet their alarm calls (sneeze calls) are very distinct. This suggests a closer relationship between *nictitans* and *erythrotis-cephus* than between either of them and members of the *mona* superspecies. Grunts, apparently used to enhance intragroup cohesion, are given both by *nictitans* and *erythrotis-cephus*. Again they are indistinguishable. No such calls are given by members of the *mona* superspecies. But, *C. aethiops* which is rarely sympatric with any of these forms also gives cohesion grunts identical with *nictitans* and *erythrotis-cephus*. Although the so-called cohesion calls enhance intraspecific cohesion, they would seem not to have functions identical with those of the adult male loud calls nor are they exposed to the same selective pressures. Vocalizations having different functions evolve at different rates. Calls serving to alarm or to coordinate groups may evolve more slowly than those enhancing reproductive isolation through the maintainance of conspecific intragroup cohesion around one adult male. If such reasoning is correct, it would appear that *nictitans*, *erythrotis-cephus*, and *aethiops*, although readily distinguished, are more closely related to each other than to members of the *mona* superspecies.

C. lhoesti was less well studied than the aforementioned species, but does have chirps resembling *nictitans* and *erythrotis-cephus* and hacks superficially like *erythrotis-cephus*. The adult male loud call of this species is totally unlike that of the others.

Limited experience with *C. petaurista* demonstrates its calls to be more like those of *erythrotis-cephus* than other species, but are, none-the-less, very distinct.

METHODS

All tape-recordings and observations considered in this study were made under field conditions in Africa. Most of the work on vervets (Cercopithecus aethiops) was done in Kenya and Uganda in 1962-64 (Struhsaker, 1967a, b, c, d). The other species were studied in Cameroun (17 months), Rio Muni (1 month), and Ghana (1 month), West Africa from November 1966 through May 1968. During this period in West Africa a total of 2, 803 hours was spent searching for, observing, or tape-recording primates. Of these 2, 803 hours, 1, 095 were spent in contact with primates. During this time the animals were in view or within 30 yards and the primary goal was to observe them in an undisturbed state. An additional 346 hours was spent tape-recording monkeys and conducting playback experiments with them.

The majority of recordings were made with a Nagra III recorder at $7\frac{1}{2}$ ips and a Sennheiser MKH unidirectional microphone. A few recordings were made with an Uher Report S. The tapes were analyzed with a Kay-Sonagraph 6061B 85-16, 000 Hz Spectrum Analyzer and a Sanborn 151 Recorder Oscillograph.

The gazetteer of the localities mentioned in this study is as follows:

Atimpoko Island, Ghana 6° 14' N., 0° 6' E.

Bifos, W. Dja Reserve, Cameroun 3° 15' N., 12° 37' E.

Boa, Cameroun 4° 27' N., 8° 54' E.

Bomana, Cameroun 4° 16' N., 9° 3' E.

Dipikar Island, Cameroun 2° 9' N., 10°3' E.

Idenau, Cameroun 4° 15' N., 9° 0' E.

Kade, Ghana 6° 5' N., 0° 52' W.

Kindonge, S. Bakundu Reserve, Cameroun 4° 32' N., 9° 25' E.

Makweli, Cameroun 6° 6' N., 9° 23' E.

Manmenyin, Campo Reserve, Cameroun 2° 22' N., 10° 16' E.

W. of Meyébot, E. Dja Reserve, Cameroun 3° 0' N., 13° 30' E.

W. of Nazam, E. Dja Reserve, Cameroun 3° 7' N., 13° 37' E.

Nsepe (Nyoungou-Toko), Cameroon 3° 38' N., 9° 57' E.

Obenika, Cameroun 4° 22' N., 8° 54' E.

Ongat, Campo Reserve, Cameroun 2° 21' N., 10° 9' E.

Seventeen-kilometer camp, W. Dja Reserve, Cameroun 3° 16' N., 12° 40' E.

Tinaso, Cameroun 3° 37' N., 9° 57'E.

Lac Tisongo, Cameroun 3° 34' N., 9° 53' E.

Volta Dam, Ghana 6° 18' N., 0° 5' E.

RESULTS

INTERMEDIATE FORMS OF *CERCOPITHECUS* MONKEYS IN CAMEROUN

On five occasions, monkeys were seen in the forest near Idenau with coloration intermediate between that of *C. mona* and *C. p. pogonias*. Four of the sightings were made within an area of about 10 hectares and were probably of one individual. The fifth sighting was made several kilometers away. The first intermediate was seen on 17 May 1967. I could not determine its sex, but it was nearly adult size. It had a reddish-brown back like *mona*, but the white hip-patches that are well-developed in *mona* were only faintly visible. A black median forehead stripe was conspicuous, much like *C. p. pogonias*, and its ventral surface was faintly washed with yellow. In typical *C. p. pogonias*, the venter is a richer orange-yellow, whereas in *mona*, it is white. This monkey was in a mixed association of *C. p. pogonias*, *C. erythrotis*, and *C. nictitans*. No *mona* was present. The next intermediate was seen eleven months later on 10 April 1968 and in the same area as the latter. This was coincident with the seasonal return of the *C. p.* pogonias to this area. This monkey was of approximately adult female size. The lateral surface of its arms was black, whereas the lateral surface of its legs was dark gray. In typical *mona*, the lateral surfaces of both arms and legs are black. In *C. p. pogonias*, the lateral surface of the arms varies from gray with a yellowish-brown tinge to black. The legs are always gray. The back of this monkey was deep chestnut on the anterior half (like mona), but black on the posterior half (like p. pogonias). Its tail was black and the inner surface of its arms was white like mona and p. pogonias. There were no prominent white hip patches at the base of the tail, but instead there were white areas there which blended into the gray of the lateral surface of the thighs and were more or less continuous with the white on the inner surface of the legs. The face was not clearly seen. This monkey was also with a mixed association of C. p. pogonias, C. erythrotis, and C. nictitans. A monkey of similar description was observed again on the 14th and 28th of April 1968 in the same area and with the same species as the latter. The 28th of April was the last date of study in this area, but there is little doubt that all four sightings were of the same individual.

The other monkey having coloration intermediate between mona and p. pogonias was seen on 26 January 1968. It was in a group of at least ten typical p. pogonias. This monkey had features reminiscent of the grayi and nigripes forms of pogonias. It was about 2/3 to 3/4 the size of an adult male. The shoulders and lateral sides of the anterior half of the body were reddish-brown, but less brilliant than in mona. There was a broad, black, dorsal stripe over the posterior half of the back and the lateral sides of the posterior half of the body were light gray, like p. pogonias. Ther lateral surface of the legs was light gray and those of the arms were black, also like p. pogonias. The medial surface of the arms and legs and the under surface of the body were pale yellow and not brilliant orange-yellow like p. pogonias. The head was not clearly seen, but there was a black sagittal crest of moderate development.

Only one monkey was seen having coloration intermediate between mona and pogonias grayi. This was on 12 February 1968 in the vicinity of Tinaso. It was in a mixed association of mona, pogonias grayi, n. nictitans, erythrotis-cephus, and Cercocebus albigena. This monkey, of nearly adult size, had a predominance of mona characters including a reddish-brown back, arms with a black lateral surface, absence of a well-developed sagittal crest and light-colored parasagittal lines on the head, and a white ventral surface. The white hip-patches on the base of the tail were only slightly developed. In addition to this latter feature, there was one other which qualified this monkey as an intermediate. The lateral surface of its legs was gray, typical of pogonias grayi. C. pogonias grayi always have gray legs and mona always have black, those of p. grayi can vary from gray to black. In other ways, mona and pogonias grayi are very similar in color (Booth, 1955 and Hill, 1966) and it would seem that the black sagittal crest and light-colored parasagittal lines on the head of pogonias grayi and the white rump patches of mona provide important visual cues for species recognition.

Monkeys with facial color patterns intermediate to C. erythrotis and C. cephus were observed only along the left bank of the R. Sanaga below Edéa, Cameroun. Six of these intermediates were seen in the vicinity of Tinaso and one was seen about 10 kilometers away near Lac Tisongo (Fig. 1). Only one typical erythrotis was seen on the left bank (near Tinaso). On the right bank only typical erythrotis were seen. Although I have seen typical cephus further south in the Campo Reserve, none were seen in the vicinity of Tinaso and Lac Tisongo. Previously (Booth, 1958), the R. Sanaga was believed to separate erythrotis and cephus, but these sightings invalidate that conclusion.

On 12 February 1968, five intermediate forms were seen in a mixed association of C. n. nictitans and C. mona. Their body color was indistinguishable from C. erythrotis camerunensis seen on the right bank of the Sanaga. However, their facial color and patterning were variable. An adult male had a very faint white moustache (fainter then typical cephus), broad white side whiskers with a faint yellowish tinge, and a very light blue nose (lighter than typical cephus). Another monkey of adult female size had a pinkish nose and upper lip, but no moustache and very little yellow on the white side whiskers. Of similar size to the latter, the third intermediate had a nose and upper lip of a more rusty color, but was unlike typical erythrotis and lacked a moustache characteristic of cephus. The fourth monkey was only 2/3 adult size, but had a facial pattern and color like the latter. The final intermediate seen in this association was of adult male size and had a white moustache that was more or less well developed and a very light blue nose.

Ten kilometers away near Lac Tisongo, another intermediate erythrotis-cephus was observed on 15 February 1968 with a group of n. nictitans. This was an adult whose nose was rusty-pink like a typical erythrotis, but whose upper lip was whitish blending into the nose color. Its white side whiskers had a yellow tinge. The final intermediate was seen on 19 February 1968 near Tinaso, but in a different part of the forest from those described on the 12th. It was an adult whose nose and upper lip were very light pink and lacked a white moustache. The side whiskers were broad and white. Nearby, and a few minutes later, a typical erythrotis was seen in an association of n. nictitans, mona, and Cercocebus albigena. This was a juvenile about 2/3 to 3/4 adult size. Its nose and that part of the upper lip immediately ventral to the nose were rusty colored, typical of C. erythrotis camerunensis. It had no moustache.

Seven out of the eight erythrotis-cephus type monkeys who were clearly observed along the left bank of the Sanaga had intermediate facial patterns. The high incidence of intermediacy in this area is further indicated by the fact that these were seen during only 40 contact-hours with monkeys in this area. Although the data are few and certain details are lacking, these intermediate forms seem best interpreted as a case of secondary intergradation (Mayr, 1969). This is also referred to as allopatric hybridization of two subspecies. In this zone of renewed contact, intermediate forms occur and if the variation in form is great and the zone of hybridization is relatively narrow, it is clearly a case of hybridization between two subspecies. Mayr states: "If two taxa that were previously recognized as two allopatric species completely intergrade in a zone of secondary contact, it proves that they are not reproductively isolated and that they should be considered subspecies of a single polytypic species." True, it has not been established that erythrotis and cephus forms completely intergrade and interbreed randomly, or that this area represents a zone of secondary contact; but in conjunction with their identical vocal repertoires and similar ecology, this seems the most reasonable interpretation. Clearly, this zone of intermediacy has not stabilized enough to warrant taxonomic distinction.

Only three intermediate forms between *mona* and *pogonias* were observed during 672 contact-hours with monkeys in areas where *mona* and *pogonias* are sympatric. These cases clearly fall into the category of the occasional breakdown of isolating mechanisms between two sympatric species (Mayr, 1969). Their specific distinction is further substantiated by their different vocal repertoires and ecology.

CERCOPITHECUS AETHIOPS GROUP

The species *C. aethiops* is extremely polymorphic. There is considerable variation in pelage coloration and patterning within and between populations. However, this variation does not justify the formation of a superspecies divided into three species as proposed by Dandelot (1959). For example, the color of the scrotum, along with other characters, is used by Dandelot to distinguish species and subspecies. Scrotal color varies not only with the age of the individual, but also between the adult males within any given social group (personal observation). The extent of these age and individual variations are equivalent to the variation between Dandelot's proposed species. Study of many populations in E. Africa, including three of these subspecies (*C. aethiops johnstoni*, *C. a.* centralis, *C. a. callidus*), revealed no qualitative differences in the more common vocalizations (Struhsaker, 1967). I was unable to distinguish any differences in the so-called threat-alarm-barks, woof-woof call, waa call, or the two kinds of infant-lost calls. Subtle differences in temporal patterning may exist between these populations, but this would not invalidate the point that vocalizations are a relatively stable group of characters within this species. This is further substantiated by limited observations on a population of *C. aethiops tantalus* in the Waza Reserve of Cameroun, W. Africa. Dandelot (1959) considers this form to be a species distinct from those mentioned above. Again, the preceding four calls were heard in this population and were distinguishable from those in E. Africa. Tape recordings were made of a young juvenile *C. aethiops tantalus* (estimated as six months old) in Ghana, W. Africa. Spectrographic analysis revealed that its lost rrr calls had the same basic structure as those recorded in Kenya, E.Africa. The variation was no greater than that between individuals of the same social group in Kenya. A possible exception to this stability of vocalizations would appear to exist on Lolui Island, Lake Victoria where J.S. Gartlan (pers.comm.), in a year-long study of *C. aethiops johnstoni* never heard the woof-woof and waa calls. However, he did hear many of the vocalizations recorded from other populations, and the absence of these two calls seems less significant than if they had been structurally altered.

LOUD CALLS OF ADULT MALE *CERCOPITHECUS SPP.*

The details of the stimuli evoking and the responses evoked by adult male loud calls have already been described (Struhsaker, 1969). These calls are given under a great variety of conditions and it has been suggested that they serve as a central focus which enhances intragroup cohesion. They probably affect intergroup spacing as well (also see Gautier, 1969). These calls were given whenever the group was likely to become dispersed, such as during intergroup agonistic encounters, after the roll of thunder or the crash of a falling tree, after an attack by a crowned-hawk eagle, prior to a group progression, and in response to loud calls from conspecifics of neighboring groups.

Diurnal Periodicity: Although the loud calls of *C. lhoesti* show a very marked diurnal periodicity (Fig. 2, and Struhsaker, 1969), those of other *Cercopithecus* are less periodic or are aperiodic (Fig. 2). All bouts of loud calls were tabulated during the sample periods indicated below. Those call-bouts which were obviously given in response to our presence have been eliminated from these data. At all of the localities, except Idenau, we were camped in the forest and whenever loud calls were heard while we were in camp, they were tabulated. These data cannot, however, be compared with those collected during the much more careful sampling done while we were actually searching for or observing monkeys during the day. In addition to those data presented in Fig. 2, *C. n. nictitans* loud call-bouts were heard twice between 0531 and 0600 hours and 4 times between 0601 and 0630 hours at Bifos and the 17

Kilometer Camp. At Lac Tisongo, one *mona* loud call was heard at 0337 hours, 3 *n. nictitans* loud calls, one *pogonias grayi*, and 4 *mona* between 0631 and 0700 hours and a *mona* loud call was heard at 2122 and 2141 hours.

The localities from which these data were collected, the dates, number of days, and hours spent systematically tabulating loud calls are as follows: Bifos and 17 Kilometer Camp 10-16 December 1967, 4.5 days, 58.6 hours; Dipikar 21-24 August 1967, 3 days, 31.4 hours; Idenau 1-28 April 1968, 18 days, 183.0 hours; Lac Tisongo 13-15 February 1968, 2.5 days, 27.8 hours; Mboni Bush near Bomana 18-19 January, 4 March, 2-3 April, 22-23 April 1968, 6 days, 74.5 hours; Obenika to Boa January 1968, 3/4 days, 9.4 hours. Those days on which tape recording or activities other than systematic observations were done are not included in this sample. Of the *Cercopithecus spp.* only the loud calls of *lhoesti*, *mona*, *nictitans*, and *pogonias* could be heard over distances of 100 m. or more. Consequently, these were the only *Cercopithecus* species for which quantitative data were collected. At Bifos and the 17 Km. Camp, *n. nictitans* and *pogonias grayi* were common. Many *n. nictitans* were present at Dipikar but *pogonias* were few. In April at Idenau, both *nictitans martini* and *p. pogonias* were common, but *mona* were only locally common in the younger secondary vegetation. *C. lhoesti* were sporadically seen. At Tisongo, *n. nictitans* and *pogonias grayi* were common, but *mona* were slightly less so. Mboni Bush had many *nictitans martini*, *p. pogonias*, and *lhoesti*, but very few *mona*. The river from Obenika to Boa is lined with *Rhizophora sp.* mangrove and only *mona* were seen and heard here, but in great abundance.

Although these data are not amenable to statistical analysis, certain trends seem apparent. Both subspecies of *nictitans* seem to lack diurnal periodicity in the frequencies with which they give these loud calls. This is in contrast to the findings of Gautier (1969) which indicate a morning and evening peak for *C. n. nictitans*. With the exception of the data from Bifos and 17 Km. Camp, *nictitans* in the other localities had a fairly high level of calling just before sunset between 1731 and 1900 hours, but only at Idenau, L. Tisongo, and Mboni did this level appear as a peak. Cycles in the loud calls of nictitans might be apparent if one were to segregate the different variations on the basis of temporal pattern, giving special consideration to the time interval between phrases. The pogonias loud calls also lack a consistent and marked diurnal periodicity. Although their rates of calling are usually lower than those of nictitans, they often follow the same trend. A notable exception to this was the sample taken at L. Tisongo. The similar trends of these two species might be related to the fact they often form mixed associations and that adult males often seem to respond reciprocally to the loud calls of the other species. The data for mona at Idenau indicate a greater periodicity than either nictitans or pogonias, with a pronounced evening peak just before sunset. However, this could be an artifact of sampling in that mona are most common at Idenau in the younger secondary vegetation. In a typical day of sampling, we always began and ended in this kind of vegetation and usually spent the period from about 0830 to 1730 hours in more mature forests where mona were relatively uncommon. It does seem significant, however, that in the very brief sample taken between Obenika and Boa, a similar pattern was attained with a high peak of calling from mona monkeys at sunset. Between 1831 and 1900 hours, a peak of 20 call-bouts per 100 minutes of listening was attained. This high rate of calling probably reflects the greater population density of mona in this area compared with that at Idenau. Although this area was not sampled between the hours of 1231 and 1500, the evening peak of calling is real and was accentuated even more by the fact that no mona calls were heard during the preceding hour.

Differences in the rates or frequencies of calling bouts for a given species between different localities seem to be the function of at least two important variables: population density and sample time. The differences in the frequency of mona calling bouts between Idenau and Obenika-Boa are undoubtedly due to the higher population density at Obenika-Boa. Differences in sampling time must also contribute to the magnitude of this difference however. The difference in the rate of calling bouts of pogonias grayi at L. Tisongo and the western Dja (Bifos and 17 Km. Camp) is most certainly due to higher population densities at the latter locality, because it was here that more time was spent sampling and yet higher rates attained. Differences in the frequency of calling bouts due to differences in sampling time can best be demonstrated by considering a species with a relatively small and stable home range in one locality. From the April 1968 sample for *C. nictitans martini* at Idenau, it can be seen that the highest rate of calling bouts for any given 30-minute period on any one day is 6.66/100 minutes. In contrast, the highest rate obtained for any 30-minute period averaged over the 18 days sampled in April is 1.45/100 minutes. A difference of similar magnitude is seen when the same comparison is made within the Mboni Bush sample. For any given day and 30-minute period the highest rate of calling bouts for *C. nictitans martini* is 13.33/100 minutes. The highest rate for any 30-minute period averaged over the entire 6-day sample is only 5.98/100 minutes.

LOUD CALLS (HACKS AND POWS) OF ADULT MALE NICTITANS

The loud calls of adult male *Cercopithecus nictitans* were typically given as single-unit phrases in a series that lasted for several minutes, although single phrases were infrequently heard. A series of phrases is referred to as a bout. During the actual field study, it was apparent that the temporal pattern and quality of these calls were quite variable. Two extreme kinds of phrases were recognized early in the study and referred to as pows and hacks. Intermediate forms were variously referred to as hack-pows and pow-hacks. Spectrographic analysis revealed a wide range of variation in the duration of these phrases from 0.05 to 0.32 seconds (Table 1). A selection of 28 loud calls from eight to nine different individual *C. nictitans martini* were subjectively classified before sonagrams were made of them as pows, hacks, or pow-hacks and then their duration was measured from sonagrams. Although all three classes overlapped in duration, the longest phrases were all pows and the shortest all hacks. Only one pow phrase was within the range of variation of hack phrase duration. As was expected, the pow-hacks were all intermediate in duration. It seems justifiable to conclude that there is a continuum of phrase types, based on duration, from the shorter hacks to the longer pows. Often, but not invariably, pow phrases occurred early in the series and were followed by hack phrases. Gautier (1969) recognized a gradation in the development of harmonics in the loud calls of *C. n. nictitans*.

The interval between phrases varied considerably from very rapid series in which the phrases were separated by tenths of seconds to considerably slower ones in which the phrases were separated by several seconds. These different temporal patterns may correlate with the stimulus situation. For example, the most rapid series of hacks seemed to be given in response to the presence or attack of a predator such as a crowned hawk eagle, and in some cases during an intergroup conflict. The bouts of pows and hacks given near sunrise and sunset were usually of slower temporal patterns. Unfortunately, data on the exact situations evoking these calls are difficult to obtain and are too few to permit confident conclusions (see Gautier, 1969).

Variation in these loud calls seems to be as great within an individual's repertoire as that within and between populations of the same subspecies (Fig. 2). This does not exclude the possibility of individual recognition on the basis of statistical differences in the loud calls of adult males within the same population. However, this problem is extremely difficult to solve because of apparent variation in these calls according to the stimulus situation. How is one to distinguish between individual variation having significance in individual recognition, such as between neighboring groups, and the variation dependent on stimulus circumstances?

There appears to be no significant difference in the loud calls of the two subspecies of *C. nictitans* studied (Table 1 & Fig. 2). *C. nictitans nictitans* has a black chest and is thus most readily distinguished from *C. nictitans martini* whose white chest contrasts with the black of its sides and back. I am not aware of any controversy over the conspecificity of these two forms (Hill, 1966). These findings, therefore, support the general thesis that among *Cercopithecus* monkeys, vocalizations are relatively stable characters and good indicators of phylogenetic affinities.

Most nictitans loud calls have phrases with obvious tonality.

This tonality is usually modulated, beginning with a higher frequency and then descending to a lower frequency (Fig. 2 & Gautier, 1969). Several harmonics are developed and sometimes sidebands* develop in the higher frequencies above the harmonic system. In 44.8 percent of the 67 nictitans nictitans phrases and 86.4 percent of the 88 nictitans martini phrases analyzed, these tonal portions were followed by a shorter non-tonal portion in which a harmonic system was not developed. In the other phrases the non-tonal portion was absent. The Wilcoxon two-sample test (two-tailed) for the unpaired case was used to compare the average measurements of six different nictitans nictitans and eight different nictitans martini. There was no significant difference in the total duration of the phrases (0.07 level), in the duration of tonal portions (o.22 level), nor in the non-tonal portion present more often than n. nictitans and often have longer phrases, this difference is not statistically significant. The distribution of energy in these loud calls does not appear to differ between the two subspecies (Table 1).

Comparison of certain isolated phrases of n. nictitans hacks with single-unit phrases of pogonias hacks reveals some basic similarities (Fig. 2). The two examples were tape recorded in the same locality (about 10 kms. west of Meyébot on the east side of Dja Reserve). The energy distribution of these selected examples is very similar. However, in spite of selecting the very shortest of nictitans phrases available, they are consistently longer in duration than the pogonias single-unit phrases. Frequency modulation is more pronounced, even in these very short nictitans phrases (Fig. 2), than in pogonias and tonality is either prominent or at least faintly present in all nictitans phrases, whereas it is absent from all pogonias phrases. In addition to these differences in phrases, there are differences in temporal pattern that are probably more important for species recognition. Although one can selectively locate portions of a bout of nictitans hacks that resemble a similarly selected portion of pogonias hacks, the calling bout of each species taken as a whole cannot be confused. The pogonias series is usually preceded by booms and always contains some two-unit phrases, neither of which are present in nictitans bouts. Nictitans

*See Watkins, 1967 for a discussion of sidebands.

usually have some pows in their bouts, which cannot be confused audibly or spectrographically with pogonias hacks. Aside from an obvious indication of generic affiliations, the full phylogenetic implications of the aforementioned similarities are not apparent.

LOUD HACKS OF ADULT MALES OF C. ERYTHROTIS, C. CEPHUS, AND C. PETAURISTA

The loud hacks characteristic of these forms are given only by adult males and possibly by large subadult males. They are often given within tenths of seconds after the loud hack of nictitans or mona males (Fig. 3 & Gautier, 1969), but rarely preceed them.* They were also given in the absence of these other species and typically in alarming situations as when disturbed by a human observer.

Two kinds of hacks were tape recorded from erythrotis. The

---

*The significance of this correlation is not clearly understood, but because it occurs with such regularity in mixed associations of nictitans and cephus or nictitans and erythrotis it seems worthy of conjecture. The available data, few as they are, indicate that nictitans groups are territorial (Struhsaker, 1969 & Gautier, 1969), but that cephus and erythrotis are not. This may be related to what appears as a lower threshold for giving the loud calls among nictitans adult males than among the adult males of cephus and erythrotis. Nictitans give these loud calls during intergroup conflicts and they may also function in intergroup spacing. One of the other functions of these adult male loud calls is to facilitate intragroup cohesion of conspecifics. It is reasoned, therefore, that because erythrotis and cephus frequently associate with nictitans and because nictitans has a lower threshold for giving loud calls following stimuli that tend to disperse the monkeys, selection has favored adult male cephus and erythrotis who give their loud calls immediately after those of another species and thus enhance the coherency of their conspecific group members. The adaptive features of maintaining intraspecific cohesion within polyspecific groups are presumed to be reproductive isolation and the location of food supplies most suitable for each species.

multilunit phrases were tape recorded from only one individual *erythrotis*, whereas the more common single-unit phrases were tape-recorded from four *erythrotis*. The basis for this difference was probably a function of the stimulus situation, for the same individual gave both multi- and single-unit hacks in the same bout. The single-unit phrases of three *cephus* were indistinguishable from those of *erythrotis* (Fig. 3 & Table 2). The average duration of single-unit phrases and the upper frequency of major energy of these *erythrotis* and *cephus* hacks were compared using the Wilcoxon two-sample test for the unpaired case and found to be not significantly different with a probability of 1.00. The lower frequency of major energy for all hacks was 0.085 kHz. Sidebands were present in the narrow band analysis of all hacks and ranged in number from 2-14 for *erythrotis* and 3-19 for *cephus*. These data support the hypothesis that *erythrotis* and *cephus* are conspecific.

Although considerably less time was spent observing and tape-recording *petaurista* than *erythrotis* and *cephus*, it is worthwhile to compare the loud hacks or klunks given by adult male *petaurista*. They too were given in alarming situations, usually toward the observer. Acoustically these calls are readily distinguished from the hacks of *erythrotis* and *cephus*. Spectrographically they bear a striking similarity to the multi-unit hacks of *erythrotis* (Fig. 3). Their units are alike in having a rapid series of short subunits or spikes. All ten of the *petaurista* klunks than were recorded consisted of three units. The first two units are shorter than the last one, whereas in *erythrotis*, the multi-unit hack phrases consist of alternativing shorter and longer units (Table 3). *Erythrotis* multi-unit phrases are longer than those of *petaurista*. The first unit in *petaurista* phrases consistently reaches higher frequencies than the last two units. In contrast, there is no conspicuous difference in the in the energy distribution among units of *erythrotis* multi-unit hacks. The spectrographic similarity of *erythrotis* multi-unit hacks and *petaurista* klunks lends support to the hypothesis that these two species are closely related phylogenetically and that it is justifiable to place them in the same superspecies.

ADULT MALE BOOMS OF THE *MONA* SUPERSPECIES

Four forms belonging to this superspecies (Hill, 1966) and

distinct in pelage coloration were studied: *C. mona*, *C. campbelli lowei*, *C. p. pogonias*, and *C. p. grayi*. Adult males of all four forms gave deep resonant calls of high amplitude that have been termed booms. These booms typically preceed a series of hacks. Just as hacks are sometimes given without booms, so are booms sometimes given without subsequent hacks. Booms can be heard by humans over greater distances than can hacks and this is, presumably, a function of the concentration of the major energy of this call in the lower frequency ranges between 0.1 and 0.2 kHz.

Audibly the booms of *mona* always occur in pairs referred to as 2-unit phrases. Those of *p. pogonias*, *pogonias grayi*, and *campbelli lowei* are indistinguishable and occur as single-unit phrases. The number of boom phrases occurring in a bout ranged from 1-4, but was typically two. Tape-recordings were made of all except *campbelli lowei* and, although this latter species was never actually observed giving booms, on five different occasions two single-unit boom phrases preceded hacks typical of this species.

Spectrographically, these boom phrases sometimes appeared differently than was anticipated from impressions formed by hearing them (Table 4). Often the boom phrases consisted of more units than they seemed to have audibly (Fig. 4). The interval between these units makes it unlikely that the "unexpected" units are echoes, but this still remains a possibility. Occasionally units that one might have suspected as echoes had a slightly different distribution of energy than the unit preceeding it and thus provide further evidence that the latter were not echoes (Fig. 4). However, even when tape-recordings of these booms were played at half-speed, I was still unable to hear these "extra" units.

The energy distribution of the booms for the two species is identical and the two outstanding and consistent differences are phrase duration and the number of units per phrase which are greater in mona than *pogonias*. There is no obvious difference in the booms of the two subspecies of *pogonias* that were studied.

LOUD HACKS OF ADULT MALES OF THE *MONA* SPECIES

The hacks which follow booms are summarized for these four forms in Table 5.

Hacks are invariably given in a series ranging in this sample from 7-107 phrases. In a bout of *mona* hacks, the impression is gained that the interval between hack phrases becomes progressively shorter until a crescendo of rapidly repeated hack phrases is reached in approximately the middle of the bout after which the inter-phrase interval becomes progressively longer until the series or bout is terminated. A similar though less conspicuous temporal pattern sometimes exists in bouts of *pogonias* hacks. Only complete tape-recordings of complete hack sequences were considered in this analysis. The sample consisted of one complete series of hacks for each of three adult male *mona* monkeys, one for each of four *pogonias grayi*, one for one and two for two different *p. pogonias*. Each series or bout of hacks was divided into five parts and the mean or average inter-phrase interval was calculated for each of these five parts. It was hypothesized that the mean inter-phrase interval would be shortest in the second 20 percent of the series, next longest in the first 20 percent of the series, and then becoming progressively longer from the 3rd through the last 20 percent of the series. Accordingly, the five parts were ordered and ranked with the first 20 percent being given the rank of 2, the second 20 percent given the rank of 1, and the third, fourth and fifth parts given the ranks of 3, 4, and 5 respectively. The mean inter-phrase intervals for each fifth of the series were also ranked and ordered with the shortest means receiving the rank of 1. These data were tested statistically with the Spearman Rank Correlation Coefficient. The three *mona* bouts of hacks showed significant correlations as hypothesized ($0.05 > p > 0.01$, $p=0.01$). One of the four *pogonias grayi* was also significantly correlated ($p=0.01$) in the same manner, but the other three were not, although they too demonstrated a similar trend with the shortest inter-phrase intervals occurring in the second or third 20 percent of the series and the longest intervals always in the last 20 percent. Three of the five *p. pogonias* series were significantly correlated as postulated ($p=0.05$ once and 0.01 twice). The two which were not positively correlated were given by two different monkeys, one of which also gave one of the series that was positively correlated. The hypothesis as stated above is apparently valid for mona and valid for pogonias about half the time. Although, it is also evident that an individual pogonias is capable of giving hack series with variable temporal patterns.

Cercopithecus campbelli lowei, C. mona, and C. pogonias give similar, but consistently different hacks that are readily distinguishable audibly and spectrographically (Fig. 5). C. campbelli lowei has longer phrases and units than the others. The hacks of campbelli lowei and mona are similar in that they often have two portions with marked differences in energy distribution. Typically the first portion (type A) reaches higher frequencies (about 3 kHz) than the second (type B reaching to about 0.50 kHz). All but one of the mona and 57 percent of the campbelli lowei phrases had type B portions. In campbelli lowei, the type B occurred as a separate unit, but with mona hacks, it was not separated from the type A portion. Virtually all type A portions had sidebands developed (97.7 percent for mona and 100 percent for campbelli lowei). The number of sidebands ranged from 2-25 for mona and 5-27 for campbelli. Sidebands were in most of the type B portions (75.8 percent for mona and 94.7 percent for campbelli). The number of sidebands in the type B portions was also similar for the two species, ranging from 1 to 5 for mona and 2 to 5 for campbelli. Further comparison reveals that the type A portion of campbelli hacks are about twice as long in duration as those of mona, but the type B portions of the two species are of virtually the same duration.

The two subspecies of pogonias have hacks whose two-unit phrases are only slightly longer than half the duration of those given by campbelli males. The units, whether occurring in single-, double-, or triple-unit phrases, are relatively uniform in duration being comparable to the type B portions of mona and campbelli hacks, but distinct from them in their energy distribution which is more like that of type A portions (Table 5). Pogonias hacks invariably have sidebands developed ranging in number from 1-15 per phrase.

Individual and intrapopulational variation in hacks for all parameters measured was a great as interpopulational variation. The best example of this is provided by mona. Geographically, it is the most widely distributed of the four forms studied. Monkeys were sampled from three areas in Cameroun (Kindonge, Nsépé, and Makweli) and two areas in Ghana (Atimpoko and Volta River Dam). These samples represent the most Westerly and Easterly populations of mona. An adult male mona monkey in the vicinity of Nsépé, E.Cameroon gave hacks whose phrases varied in duration from 0.14 to 0.27 seconds. The range of variation in phrase duration for all nine mona monkeys sampled, regardless of population locality was 0.10 to 0.27 seconds. Mona monkeys giving hack phrases with durations as short as 0.10 seconds were found at Makweli, W.Cameroon and at Atimpoko, Ghana. These two localaties are more than 600 miles apart. Phrases as long as 0.21 seconds were also recorded from a mona at Atimpoko. These two widely separated populations also showed similar ranges of variation in the duration of the A and B type of unit portions.

The equality of intra- and interpopulational variation can be further demonstrated with a comparison of the upper frequencies of mona hacks. Two different mona monkeys from the region of Nsépé, E. Cameroon gave hacks that varied in their upper pitch from 1.00 to 2.25 kHz and 2.25 to 6.75 kHz respectively. Similar patterns of variation were found in two mona along the R. Volta in Ghana. Their upper frequencies varied from 2.50 to 7.25 and 1.75 to 2.50 kHz respectively. Only one bout of calls was recorded for each monkey and consequently this variation also represents intrabout variation. Clearly, the individual and intrapopulational variation equals that between populations.

Subspecific variation in the adult male loud hacks of C. pogonias appeared to be negligible. The hacks of four different males of C. pogonias grayi from two localities (three from the west side of Dja Reserve and one from 85 miles away on the east side) were compared with those of four different C. p. pogonias from three localities (two from Idenau, one from Kindonge, and one from Nsépé). The average measures for each of these males were tested with the Wilcoxon 2-Sample Test for the Unpaired Case (2-tailed). No significant difference was found between the samples of these two subspecies in the duration of hack units or in the duration of two-unit phrases. However, it was demonstrated that the mean number of units per hack phrase in pogonias grayi was significantly lower (0.026 level) than in p. pogonias. In this latter analysis seven different pogonias grayi males from the Dja Reserve were compared with seven different p. pogonias from three localities (Idenau, Kindonge, and Nsĕpĕ). The mean number of units per phrase for each male was used in this analysis. The sample sizes for each of these meles are summarized in column one of Table 5.

Of seven p. pogonias sampled, one gave some single-unit phrases, two gave some three-unit phrases, and all gave two-unit phrases. And of the seven pogonias grayi sampled five had some single-unit phrases, all had two-unit phrases, and none had three-unit phrases. The extreme case of variability was a p. pogonias who gave all three kinds of phrases in one bout.

The energy distribution in the hacks of these two subspecies also appeared identical (Table 5). With the exception of the significant difference in the number of units per phrase, the comparison of the hacks of these two subspecies is consistent with the thesis that vocalizations within the genus Cercopithecus are relatively stable in the course of evolution.

Comparison of the measures (Table 5) and sonagrams (Fig. 5) of hacks for these three species and two subspecies demonstrates that they have enough in common to suggest a closely related evolutionary history. The details and direction of this history, must of course be highly speculative. But, if one accepts the general concept that the morphologically simpler forms evolve into more complex ones, it is reasonable to suggest that the hacks of C. pogonias grayi, which have fewer units per phrase, whose units are shortest and demonstrate least variability in duration and are least elaborate in structure are most like the ancestral form of this superspecies because their hacks are the most simple in form. C. campbelli and mona have the most complex, longest and most variable hack phrases and are therefore presumed to be more recently evolved than pogonias (Fig. 5). This is in contrast to Booth's (1955) conclusion based on pelage patterns in which pogonias is elaborate than mona and campbelli and therefore considered more recently evolved. This conflict sheds doubt on the basic concept that morphological elaboration, as perceived by humans, is an indication of more recent evolution and that simplicity is an indication of "primitiveness".

CHIRPS AND TWITTERS OF CEPHUS, ERYTHROTIS, AND PETAURISTA

Calls included in this category were given only by adult female and immature monkeys. Primarily given in alarming situations, such as toward an observer or a crowned hawk-eagle, they were sometimes given in circumstances lacking an obvious alarm-evoking stimulus. In the latter situation these calls were seldom repeated or, if so, at relatively long intervals compared with those given in alarm. In these cases the calls may have facilitated group cohesion.

Four types were distinguishable in cephus and erythrotis (Fig. 6). In single-unit chirps the elements were separated by greater time intervals than were the units of two- and three-unit chirps. Single-unit twitters were separated by intervals less than those between single-unit chirps, but greater than those between the units of two- and three-unit chirps. Two-unit twitters were the rarest of the four groups with only two examples occurring in the 123 phrases sampled. They occurred only in the sample from Bifos, W. Dja and their units were closer together than the single-unit twitters occurring in the same bout of calls. The interval between the units of two-unit twitters was similar to that separating the units of two- and three-unit chirps, but were distinguished by their occurrence within a twitter bout (Fig. 6).

No consistent trend is apparent when comparing the duration of single-unit chirps with the units of two- and three-unit chirps Tables 6 & 7). It is therefore concluded that the slight differences are not significant. Thus in some populations, the units of single-unit chirps are longer, and in other populations of the same subspecies, they are shorter than the units of multi-unit chirps. Twitter units may be slightly longer than chirp units, but the difference is not great (Tables 6, 7 & 8).

Comparing the distribution of energy of the call types within the two subspecies reveals little, if any, difference (Tables 6, 7 & 8). The slight differences between call types do not form a consistent pattern. In some populations one call type has its upper limit of major energy at a slightly higher frequency than another call type, but other populations within the same subspecies demonstrate a reverse trend. It seems probable that these differences are not significant. There is some suggestion that *erythrotis* calls reach slightly higher frequencies than *cephus* (Tables 6, 7 & 8). Although the major energy usually occurred below 8 kHz, some of the *erythrotis* chirps had ultrasonic components (24 and 32 kHz, Fig. 6N), whereas the highest frequency measured in a *cephus* call (twitter) was about 16 kHz.

There is a consistent pattern for both subspecies in the distribution of energy in the units of a two- or three-unit chirp. The first unit had its energy distributed over a shorter frequency range than did the subsequent unit or units (Fig. 6). The lowest frequency of the first unit was higher than the subsequent units in all of the *cephus* multi-unit chirps (N=17) and in 91 percent (N=11) of *erythrotis* multi-unit chirps. The upper frequency of major energy was lower in the first unit than in subsequent units in 88.2 percent of *cephus* and 81.8 percent of *erythrotis* multi-unit chirps.

The variation in the spectrographic shape of these calls was the most difficult feature to analyze. It was, however, demonstrated that calls virtually indistinguishable in shape occur in both subspecies (Fig. 6). In addition to the slightly higher frequencies of *erythrotis* calls, there is one other apparent difference between *erythrotis* and *cephus*. Twitters definitely appear to be a more common element in the repertoire of *cephus* than in *erythrotis* (Table 8). Nonetheless these differences are slight; of greater importance are the remarkable similarities in the calls of these two forms further supporting the hypothesis that *erythrotis* and *cephus* are conspecific.

Cercopithecus petaurista has been aligned with *erythrotis* in the superspecies C. petaurista (Hill,1966). Although very few observations were made of C. petaurista, it seems pertinent to compare the recordings of some of their calls with those of cephus and and erythrotis. The so-called squeak calls were given by adult females and juvenile petaurista in alarming situations, e.g. when disturbed by the observer, and would appear to be homologous with the chirps and twitters of cephus and erythrotis. The squeaks of petaurista resembled the chirps of erythrotis and cephus in being bird-like, but were always audibly distinct from them. Petaurista squeaks most typically have three units per phrase. Of the ten examples for which sonagrams were made, three had two units and seven had three units. The first unit or units are always non-tonal and of relatively short duration (Fig. 6 & Table 9). The last unit of petaurista squeaks was always of longer duration and tonal in structure. Although there were several obvious differences between the petaurista squeaks and erythrotis and cephus chirps, e.g. the phrase duration was longer in petaurista, they were most readily distinguished by differences in the kind of units and their sequential regularity.

NICTITANS CHIRPS

The circumstances in which these calls were given are the same as those described for erythrotis and cephus. As with these latter two forms, only adult female and immature nictitans gave chirp calls. No distinction was apparent in the chirps given during alarm and non-alarm situations other than in the frequency of repetition which was greater and faster in cases of alarm.

Single-unit chirps were considerably more common than two-unit chirps (Tables 10 & 11). Unlike erythrotis and cephus, chirp phrases with more than two units were not given by nictitans. Measurements of unit duration, the lower limit at which energy occurred in the call, and the upper limit of the major energy of the call as determined from relative darkness on the sonagrams are summarized in Tables 10 and 11. The intra-populational variation in these measures of single-unit chirps is greater than the variation between populations and between subspecies. It is reasonable to assume that the single-unit chirps are not different between the two subspecies of *nictitans* studied.

The smaller and, presumably, less reliable sample of two-unit chirps would indicate slight differences between the two subspecies. Although uncertain and not amenable to statistical tests because of the uncertainly of statistical independence, these data indicate that *nictitans martini* has longer units than *n. nictitans* (Table 11). Furthermore, the major energy of two-unit chirps may reach higher frequencies in *n. martini* than *n. nictitans*. However, the differences in the means of these measurements are not great, and those of the modes are even less, thus further undermining the importance of these differences.

It is difficult to distinguish audibly the chirps of *nictitans* and *erythrotis* or *cephus*. Comparison of the measurements of single and two-unit chirps reveals no apparent differences between the two species (Tables 6, 7, 10 & 11). No calls comparable to the twitters of *erythrotis* and *cephus* were given by *nictitans*, which represent a difference between these two species in the range of variation of the temporal pattern of these calls (see section on *erythrotis* and *cephus* chirps). The two-unit chirps of *nictitans* resembled those of *cephus* and *erythrotis* not only in the duration of units and distribution of energy, but in the differences between the first and second units of such phrases. In the 12 two-unit *nictitans* chirp phrases available for comparison the first unit in all but one phrase had a higher low-frequency than the subsequent unit in the same phrase. In 50 percent of these phrases the second unit had a higher high-frequency or one that was equal to that of the first unit in the same phrase. Although this percentage is not as high as that for *erythrotis* and *cephus* (81 percent and 88 percent respectively), the relationship is the same, namely that the second unit has energy distributed over a greater range of frequencies than the first unit. This is also true for those phrases in which the first unit has a higher high-frequency than the second unit. Although both subspecies of *nictitans* consistently demonstrated this relationship, the scant data would indicate that *n. nictitans* more often has a higher upper-frequency in the first unit than in the second unit, whereas the converse seems to be so with *n. martini*.

Comparison of selected sonagrams (Fig. 6 & 7) reveals remarkable similarities in the single-unit chirps of nictitans, erythrotis and cephus. Two-unit chirps of nictitans were similar but less so than single-unit chirps in spectrographic shape to those of erythrotis and cephus (Figs. 6 & 7). However, these differences were not great enough to permit species distinction audibly or spectrographically. Some of the nictitans chirps had elements reaching 16 kHz, but none as high as those of erythrotis.

The major distinction between chirps of nictitans and the erythrotis-cephus group is the extreme variation in temporal pattern exhibited by twitter bouts. The remarkable similarity in this class of calls between two otherwise very different forms of Cercopithecus can be explained as a case of parallel evolution or as the retention of a character common to the ancestry (synapomorph character of Remane, 1961).

The advantage of similar alarm calls in sympatric species having common predators is apparent. However, this similarity implies more than parallelism, because other sympatric species of the same genus have alarm calls very distinctive from nictitans and erythrotis-cephus. An alternative hypothesis to parallelism is that erythrotis-cephus and nictitans are more closely related to each other than to other sympatric Cercopithecus species, e.g. mona and pogonias. This hypothesis is further supported by the fact that the grunts given by nictitans, cephus, and erythrotis, apparently to facilitate group cohesion, are extremely similar and unlike any calls given by other sympatric congeners.

CHIRPS OF C. LHOESTI

Adult females and juvenile lhoesti gave chirps in alarm situations. These chirps were acoustically reminiscent of those given by erythrotis, cephus, and nictitans. However, they differed from these latter three species in at least three ways. The chirps of lhoesti always consisted of one unit and were never repeated as rapidly as those of erythrotis, cephus and nictitans. The six lhoesti chirps of which sonagrams were made all had their energy distributed over a shorter range of frequencies than did those of the other species (Table 12). The terminal portion of lhoesti chirps attained lower frequencies much more abruptly than did the others (Fig. 8). In spite of these differences, one is more impressed with the acoustical and spectrographic similarities of the chirps than with the subtle differences. Because these four forms are from three different superspecies it is possible to interpret these similarities as a case of parallelism, or conservatism in indicating a closer phylogenetic affinity among them than with members of the mona superspecies.

Adult male lhoesti gave hacks in alarm situations. Although tape recordings were not made of these hacks, they sounded very much like the low-intensity form of nictitans hacks. Their repetition rate was slower than most hack sequences of nictitans.

Only one tape recording was made of the adult male lhoesti evening call (Struhsaker, 1969). This call is referred to as the 'oop-uuh' call and each phrase consists of two units. Spectrographically it bore a superficial resemblance to the booms given by adult males of the mona superspecies (Fig. 4). It differed, however, in having a considerably longer phrase (1.46 secs.), longer units (0.47 and 0.44 secs.), and in having a slightly higher frequency (upper frequency was at 0.4 kHz, Fig. 8). Acoustically it was very distinctive and never confused with any other call.

SNEEZE CALLS AND POGONIAS STACCATO CHIRPS

These calls were given in circumstances of alarm, usually toward the observer. They were most commonly given by adult females and juveniles. Adult males gave them less frequently. The adult male versions of these calls were of lower frequency, but had a similar temporal pattern. In fact the temporal pattern was very similar to that of the adult male bouts of hack calls of each species. All phrases of mona sneeze calls consisted of one unit, like the hacks of this species (Fig. 9). The phrases of pogonias staccato-chirps usually consisted of two units like their staccato hacks. Fewer had one or three units, also like the staccato-hacks. The major difference was in pitch (Tables 5, 13 & 14). The sneeze calls of campbelli differed from their adult male hacks in that only single-unit phrases were recorded. However, campbelli sneeze calls did have a frequency distribution similar to the hacks of campbelli males. The first portion covered a greater range of frequencies, whereas the second and contiguous portion was restricted to very low frequencies. It certainly seemed that the adult males of all three species gave gradations of calls ranging from sneezes or staccato-chirps to their hacks.

The mona sneeze calls also seemed related in acoustical and temporal structure to their rapid grunts, which were not common and seemed to be given in certain kinds of alarm situations. The sneeze calls were also similar to the pants and grunts given during copulation. A gradation or continuum in temporal pattern seems apparent in this group of call types. The sneeze calls are usually of shorter duration and always separated by longer time intervals than are the pants and grunts given during copulation. The rapid series of grunts sometimes given in alarm situations are intermediate in their temporal pattern, but have similar unit durations to those of sneeze calls.

Intra-populational variation in these calls was a great as that between populations. There were no apparent differences between populations in phrase duration, unit duration, no in lower or upper pitch (Tables 13 & 14). The data presented in these tables would indicate that the Volta Dam population of mona might have sneeze calls reaching higher frequencies than mona from Cameroon. However, one mona sneeze call recorded from Makweli, West Cameroon had major energy at 5.5 kHz, higher than that of any other sneeze call recorded elsewhere. In other parameters this call was like those of other areas and was acoustically indistinguishable.

The sample of staccato-chirps for p. pogonias was too small to warrant tabulation, but all measures were within the range of variation demonstrated by pogonias grayi. Spectrographically the staccato-chirps of these two subspecies were also similar (Fig. 9).

A comparison of the alarm calls (sneeze and staccato-chirps) within the mona superspecies reveals a greater spectral or structural complexity in those of _campbelli_, a complexity of intermediate degree in pogonias, and a very simple structure in _mona_ sneeze calls. This relationship of complexity is different from that existing among the adult male hacks where _mona_ has the most complex call. This further demonstrates the invalidity of using structural complexity as an indicator of relative "primitiveness".

Because several monkeys usually gave these calls simultaneously, it is not known how many individuals were sampled for each species. It is certain that in some cases more than one call from a particular individual was included in the sample. In others this is less certain and consequently statistical independence cannot be assumed. This prevents statistical analysis of the differences and similarities of these calls.

GRUNTS AND CHUTTERS

Cercopithecus nictitans, _C. erythrotis_ and _C. cephus_ all gave grunts that were acoustically similar. They were certainly given by adult females and juveniles, but it is not certain if adult males gave them. On several occasions these grunts clearly seemed to facilitate group cohesion (Struhsaker, 1969). Acoustically similar to grunts, were the so-called chutters. These were given during intra- and intergroup agonistic encounters by _nictitans_ and _erythrotis_, and chutters from _n. martini_ demonstrate that these calls have phrases of similar duration (Table 15 & Fig. 10). All the _nictitans_ grunts sampled consisted of single-unit phrases. Two of the seven _erythrotis_ grunts consisted of three units; the others were single-unit phrases. In contrast, only one of the 14 _nictitans_ chutter phrases into multiple units may account for their rather staccato-like acoustical effect which distinguishes them from grunts. This limited sample also suggests that the energy of _nictitans_ grunts is concentrated into a lower frequency range than their chutters and the grunts of _erythrotis_.

Although there was little difficulty in distinguishing the chutters from grunts, the same cannot be said for the distinction of _erythrotis_ and _nictitans_ grunts. Even after considerable contact with these two species in the field I was never very confident of the identity of the vocalizer until it was actually seen. One wonders if the monkeys have equal difficulty in distinguishing these grunts. Combined with the similarity of their chirp calls, this similarity in grunts and chutters would imply a closer phylogenetic affinity between nictitans, erythrotis and cephus than between any of them and members of the mona superspecies.

These grunts and chutters are very much like those given by Cercopithecus aethiops (Struhsaker, 1967a). One is tempted to suggest that grunts and chutters are conservative vocalizations and consequently widespread throughout the genus much like the various screams and screeches. However, their notable absence from the mona superspecies would indicate that the problem is more complex than that. Perhaps the presence or absence of chutters and grunts is an indicator of phylogenetic proximity of superspecies within the Cercopithecus. It seems unlikely that these similarities in grunts are the result of parallel evolution. Although nictitans and erythrotis-cephus often occur together in mixed associations and the possibility exists that the adaptive significance of these associations is great enough to allow selection for similar cohesion calls, it seems infeasible, in view of the fact that pogonias does not have such grunts in spite of the fact that it often associates with nictitans and erythrotis-cephus. Furthermore, aethiops who rarely associates with any of these species, and then probably only after the breakdown of ecological barriers by human activities, has grunts used for intragroup cohesion like those of nictitans and erythrotis-cephus. Similarly, the use of chutters in aggressive encounters by nictitans, erythrotis-cephus and aethiops and their notable absence from the mona superspecies begs an interpretation, but lends support to the preceding hypothesis.

OOO CALLS OF THE MONA SUPERSPECIES

Calls of this category were recorded from C. mona, C. p. pogonias and C. pogonias grayi. Information is not available on the presence or absence of this call in C. campbelli lowei. These calls were most typical given as monkeys fed, foraged, or as they progressed through the trees. They were sometimes, but not invariably, heard after an alarming situation, such as presumed attacks by crowned-hawk eagles. The impression was gained that these calls served as contact calls between group members and thus facilitated intra-group cohesion.

Four kinds of calls were distinguished and are fererred to as types A, B, C and D (Fig. 11). Type A is a low frequency non-tonal call that occurs infrequently (Table 17). Only one sonagram of the mona version of this call was available and its duration (0.13 secs.), energy distribution (0.085-0.75 kHz), and structure were indistinguishable from the non-tonal portion of the mona version of the type B call. Only three examples of the type A call were recorded from C. pogonias and these were given by the same group at virtually the time. They were of longer duration (0.51, 0.64, & 0.82 secs.) than the typical non-tonal portion of type B calls for this population, but were within the range of variation (Table 18). Their structure and energy distribution were indistinguishable from the non-tonal portions of type B calls.

The type B call is characterized as having a non-tonal portion followed without temporal separation by a tonal portion of a higher frequency and of equal or greater duration. Harmonics are sometimes present (Fig. 11 & Table 18).

Type C calls are tonal and similar in structure, energy distribution, and duration to the tonal portions of type B calls (Table 19).

Type D calls are similar to type C, but always about 1.5 or more times longer. They are given only by mona and most commonly by the group sampled on the left bank of the R. Volta near the Volta Dam, Ghana. It is not at all certain if the type D calls serve the same function or are given in the same circumstances as the other three types. The eight examples of this call recorded from the Volta Dam group ranged in duration from 0.36 to 2.0 seconds with a mean duration of 0.86 seconds. The two examples from Kindonge were 0.34 and 0.47 seconds and the one from Nsépé was 0.5 seconds long. The energy distribution of all these calls was similar, ranging at the lower levels from 0.75 to 2.0 kHz and at the upper levels from 2.0 to 3.5 kHz. The second harmonic was developed in only one of the 11 examples.

The comparable call types of _mona_ and _pogonias_ are audibly and spectrographically distinct. The type A call of _pogonias_ is considerably longer than that of _mona_. _Pogonias_ usually give longer type B calls than _mona_, but some of those recorded from the _mona_ group at the Volta Dam are exceptions. This difference in duration is due primarily to the longer tonal portion in type B calls of _pogonias_ than in most of those of _mona_. The exceptionally long tonal portions in some of those calls recorded at the Volta Dam are discussed below. The most consistent difference, and one that is equally applicable to the Volta Dam group, is the structure or spectrographic shape of the tonal portion. With the _pogonias_ this tonal portion ascends from about 0.5 kHz to about 1.5 kHz and then descends again to about 0.5 kHz (Fig. 11). The tonal portion of _mona_ type B calls ascends as does that of _pogonias_, but, in contrast to _pogonias_, it terminates without descending (Fig. 11). No exceptions were found in the 13 specimens of _mona_ typ B calls examined. Of the 61 examples of _pogonias_ type B calls measured, two did not have the ascending portion, one had a tonal portion which neither ascended nor descended in pitch but was held at a nearly constant pitch, and three phrases had only a very slight descent following the ascent and consequently might possibly have been confused with _mona_. However, it should be noted that these latter three exceptions were of considerably longer duration than any type B calls recorded from _mona_ living in the same locality. One final _pogonias_ phrase was exceptional in that its tonal portion first ascended, then descended only slightly and finally ascended slightly again.

The structure or spectrographic shape of type C calls is virtually the same as that described for the tonal portion of type B calls and is the most reliable character distinguishing the _mona_ and _pogonias_ versions of this call type. Type D calls were given only by _mona_.

There are no apparent differences in the OOO calls of _mona_ from Kindonge and Nsépé (Tables 18 & 19). Relatively little time was spent observing and recording the mona of Ghana. Consequently, it is difficult to understand some of the apparent differences that exist between them and the _mona_ of Cameroon. With the exception of the type A call, examples can be selected from the three localities to demonstrate remarkable similarity (Fig. 12). The tonal portion of the type B calls seems to be of longer duration in the Volta Dam group than in those from Cameroon. The longest type D calls from Ghana may also represent an interpopulational difference. An alternative explanation exists. It is possible that these differences are a reflection of situational and functional differences, i.e., the longer tonal portion and longer type D calls of the Volta Dam group may not be functionally related to the other OOO call types. It may be pertinent to mention that these mona monkeys (Volta Dam group) demonstrated a greater tolerance for human proximity than any other _mona_ tape-recorded either in Ghana or Cameroon. Militia who were guarding the small forest in which this group of _mona_ lived informed me that this forest had been under military guard for several years because it was next to a chalet formerly used by ex-President Nkrumah. Such protection was undoubtedly of significance in the tolerance shown by the _mona_. Paucity of observations and information limit further elucidation of this difference.

Variations in type B calls within the subspecies of _pogonias_ are as great as those between subspecies (Table 18). The type C calls examined for the two subspecies were few, but not apparently different. Those four recorded from _pogonias grayi_ ranged in duration from 0.19 to 0.34 seconds, whereas the two from _p. pogonias_ were 0.18 and 0.31 seconds long. The distribution of energy in the fundamental harmonic was also similar in the two forms. The lower frequency varied from 0.75 to 1.25 kHz in _pogonias grayi_ and was 0.5 kHz in both _p. pogonias_ calls. The upper frequency varied from 1.75 to 2.25 kHz in _pogonias grayi_ and was 1.0 kHz in the two _p. pogonias_ calls. The second harmonic was developed in only one of the four _p. grayi_ calls, but was present in both of the _p. pogonias_. The absence of type A calls from the sample of _p. pogonias_ may be an artifact of the small sample size.

Several individual monkeys were usually giving OOO calls at the same time. Consequently, it is not possible to determine how many individuals from each locality are included in the sample forming the basis of the preceding description. This deficit prohibits an evaluation of individual and interindividual variation in OOO calls from this sample. This uncertainty regarding statistical independence also invalidates any statistical testing of the data.

SUMMARY

New data are presented on vocalizations, ecology, distribution, and intermediate color forms which elucidate certain phylogenetic relations among *Cercopithecus* species. Stability in the vocalizations of several widely dispersed populations of monkeys belonging to the superspecies *C. aethiops* support the conclusion that this is one species polymorphic for pelage color and pattern and is not composed of three species as concluded by Dandelot (1959). Two subspecies of *C. nictitans* with minor but distinct color differences have identical vocalizations. My field data support Booth's (1955) conclusions that *C. mona* is monotypic and distinct from *C. campbelli lowei* at the specific level. Furthermore, *C. p. pogonias* and *C. p. grayi* are clearly conspecific. Intermediate forms between *pogonias* and *mona* indicate that they are closely related and justifiably placed in the same superspecies, as concluded by Hill (1966). Similar data strongly indicates that *C. erythrotis* and *C. cephus* are conspecific in contrast to Hill (1966) who placed them in different superspecies. It is concluded that study of the vocalizations of *Cercopithecus* monkeys can contribute to an understanding of the phylogeny within this genus. These studies are most meaningful when several vocalizations for each species are available and when something is known of their function. A multivariate approach to the study of phylogeny within this genus still seems best, but of all the single classes of characters so far used, vocalizations seem the most stable and thus the single most reliable indicator of phylogenetic affinities within this genus.

ADDENDUM

Single-unit booms and wa-hoom type hacks are also given by adult males of *Cercopithecus campbelli campbelli*. Tape-recordings made in January 1970 of the wa-hoom type hacks from *C. c. campbelli* in Senegal, although not yet analyzed spectrographically, are audibly indistinguishable from calls of *Cercopithecus campbelli lowei* in Ghana, nearly 1,800 kms. to the southeast. These data further strengthen the thesis that vocalizations among *Cercopithecus* species are extremely stable characters and, consequently, reliable indicators of phylogenetic affinities.

ACKNOWLEDGEMENTS

I thank Drs. Peter Marler and J.S. Gartlan for constructive criticism of the manuscript. Mr. T.S. Jones of the Cameroon Development Corporation, the staff of the West Cameroon Forestry Department, and Mr. Ferdinand Namata are thanked for their co-operation during the field work. This study was financed by NSF grants GB 5792 and GB 8476 and sponsored by the New York Zoological Society and The Rockefeller University.

REFERENCES

Alexander, R.D. (1968). Arthropods. In *Animal Communication.* (Ed. Sebeak, T.A.). Bloomington: Indiana University Press.

Booth, A.H. (1955). Speciation in the *Mona* monkeys. *J. Mammal.* **36** (3): 434-449.

Booth, A.H. (1958). The zoogeography of West African primates: a review. Bulletin de l'I.F.A.N. **20**: ser.A.2: 587-622.

Dandelot, P. (1959). Note sur la classification des *Cercopitheques* du groupe aetiops. Mammalia. **23**: 357-368.

Gautier, J.P. & A. Gautier-Hion. (1969). Associations polyspecifiques chez les Cercopithecidae du Gabon. *La Terre et la Vie,* **23**: 164-201.

Gautier, J.P. (1969). Emissions sonores d'espacement et de ralliement par deux Cercopitheques arboricoles. *Biologia Gabonica,* **5**: 118-145.

Hill, W.C.O. (1966). *Primates - Comparative Anatomy and Taxonomy. VI. Catarrhini, Cercopithecoidea, Cercopithecinae.* New York: Interscience Publishers.

Lanyon, W.E. (1969). Vocal characters and avian systematics. In *Bird Vocalizations.* (Ed. Hinde, R.A.). Cambridge: University Press.

Littlejohn, M.J. & R.S. Oldham. (1968). *Rana pipiens* Complex: Mating call structure and taxonomy. Science, **162**: 1003-1004.

Mayr, E. (1969). *Principles of Systematic Zoology.* New York: McGraw-Hill.

Rahm, U. (1961). Esquisses mammalogiques de basse Côte d'Ivoire. *Bulletin de l'I.F.A.N.* **22**: 1229-1265.

Remane, A. (1961). Probleme der Systematik der Primaten. Z.wiss.Zool. 165: 1-34.

Rose, P. (1937). Les Primates de l'afrique. Paris: Librairie Larose.

Schwarz, E. (1928). Notes on the classification of the African monkeys in the genus Cercopithecus Exleben. Ann.nat.Hist. 1: 649-663.

Struhsaker, T.T. (1967a). Auditory communication among vervet monkeys, (Cercopithecus aethiops). Social Communication Among Primates. (Ed. Altmann, S.A.) Chicago University Press.

Struhsaker, T.T. (1967b). Social structure among vervet monkeys. Cercopithecus aethiops). Behaviour, 29: 2-4.

Struhsaker, T.T. (1967c). Ecology of vervet monkeys (Cercopithecus aethiops) in the Masai-Amboseli Game Reserve, Kenya. Ecology, 43: 6.

Struhsaker, T.T. (1967d). Behavior of vervet monkeys (Cercopithecus aethiops). Univ. of California Publ. Zool. 82.

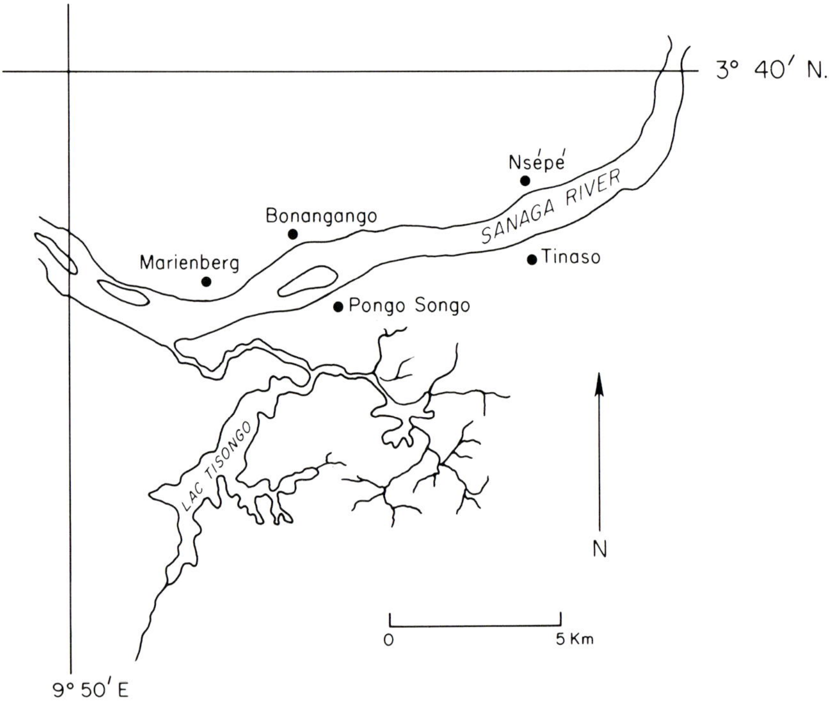

Map 1: monkeys intermediate to erythrotis and cephus were seen near Tinaso and Lac Tisongo. (modified from the Mouanko, NA-32-XXII 1/200,000 map, Institut Géographique National, Paris.)

TABLE 1

ADULT MALE POWS, POW-HACKS, & HACKS OF <u>NICTITANS</u>

measurements: mean, range, mode

subspecies	N*	phrase duration (secs.)	duration of tonal portion (secs.)	duration of non-tonal portion (secs.)	lower limit of energy range (kHz)	upper limit of major energy (kHz)
nictitans nictitans	10, 6, 21 10, 10, 10	0.10 0.060- 0.28 0.090	0.079 0.060- 0.11 0.070	N**: 4, 6, 9, 8, 3 0.047 0.010 - 0.17 0.030& 0.040	0.25	2.44 1.50 - 3.00 2.00
nictitans martini	10, 15, 10, 12, 10, 5, 10, 6, 10	0.16 0.050- 0.32 0.090	0.11 0.050- 0.28 0.070	N**: 9, 12, 7, 10, 10, 5, 8, 5, 10 0.060 0.020 - 0.14 0.030	0.23 0.085 - 0.25 0.25	2.44 0.75 - 3.50 3.00

N*: number of phrases sampled for each individual monkey, i.e., a total of 67 phrases from 6 different *n. nictitans* and 88 phrases from 9 different *nictitans martini* were sampled.

N**: number phrases that had non-tonal portions for each different monkey. One *nictitans nictitans* gave no non-tonal portions.

TABLE 2

ADULT MALE HACKS

species	N*	duration of 1-unit phrases (secs.)	lower limit of energy range (kHz)	upper limit of major energy (kHz)
		measurements: mean, range, mode		
erythrotis	5, 3, 6, 21	0.050 0.020 - 0.090 0.060	0.085	2.72 0.50 - 4.50 3.00
cephus	12, 2, 1	0.033 0.020 - 0.070 0.020	0.085	3.01 1.40 - 5.00 1.40, 3.50, & 4.00

*: number of phrases sampled for each different monkey

TABLE 3

PETAURISTA KLUNKS (HACK) FROM KADE, GHANA

measurements: mean, range, mode

N	phrase duration (secs.)		unit duration (secs.)	lower frequency of energy range (kHz)	upper frequency of major energy (kHz)
10	0.21 0.18 - 0.31 0.22	1st unit:	---- 0.010 ----	---- 0.085 ----	3.03 2.25 - 3.75 3.50
		2nd unit:	0.020 0.010 - 0.030 0.020	---- 0.085 ----	---- 1.00 ----
		3rd unit:	0.10 0.070 - 0.21 0.090	---- 0.085 ----	---- 1.00 ----

TABLE 4

ADULT MALE BOOMS

species	no. booms per bout (N)	no. units per phrase	phrase duration (secs.)	N**	unit duration (secs.)	lower limit of energy range (kHz)	upper limit of major energy (kHz)
			measurements: mean, range, mode				
mona	2.85 2 - 4 2 (7)	(N*: 4, 3, 2, 2, 4, 2, 2) 3.21 2 - 5 3	0.69 0.47 - 1.020 0.47 & 0.68	61 (11, 10, 4, 6, 17, 6, 7)	0.17 0.040 - 0.37 0.12	---- 0.10 ----	0.20
p. pogonias	1.62 0 - 3 2 (8)	(N*: 3, 2, 3) 1.37 1 - 3 1	0.34 0.20 - 0.54 0.27	11 (4, 4, 3)	0.24 0.070 0.53 0.27	---- 0.10 ----	0.22 0.20 - 0.30 0.22
pogonias grayi	1.85 0 - 3 3 (7)	(N*: 2, 2, 2, 1, 3) 1.30 1 - 2 1	0.32 0.21 - 0.42 0.23, 0.36, & 0.42	13 (3, 2, 3, 1, 4)	0.23 0.090 - 0.42 0.21, 0.22, & 0.23	0.11 0.10 - 0.13 0.10	0.21 0.20 - 0.30 0.20

N*: number of phrases sampled for each different monkey
N**: total number units sampled followed by number of units sampled for each different monkey.

TABLE 5

ADULT MALE HACKS

species	N*	no. units per phrase	phrase duration (secs.)	N**	unit duration (secs.)	lower limit of energy range (kHz)	upper limit of major energy (kHz)
					measurements: mean, range, mode		
campbelli lowei	12, 11, 10	1.57 1 - 2 2	0.30 0.12 - 0.53 0.40	A: 33 B: 19	A: 0.22 0.13 - 0.37 0.13 & 0.15 B: 0.070 0.050- 0.17 0.060 & 0.070	A: 0.15 0.085 - 0.40 0.085 B: 0.14 0.085 - 0.25 0.085	A: 3.29 2.25 - 4.7 2.50 B: 0.65 0.50 - 1.5 0.50
mona	7, 22, 10, 19, 18, 18, 13, 8, 14	1	0.16 0.10 - 0.27 0.16	subunits A: 129 B: 128	subunits A: 0.095 0.040- 0.16 0.10 B: 0.066 0.020- 0.15 0.050	A: 0.090 0.085 - 1.00 0.085 B: 0.085	A: 3.040 1.00- 7.25 2.50 B: 0.50 0.25 - 1.5 0.50
p. pogonias	26, 107, 7 15, 24, 14, 72	2.02 1 - 3 2	N+: 6, 10, 9, 6 0.17 0.13 - 0.22 0.16	94	0.063 0.040 - 0.11 0.060	0.12 0.085 - 1.25 0.085	2.92 0.50 - 7.25 2.00
pogonias grayi	27, 7, 16, 90, 23, 22, 62	1.58 1 - 2 2	N+: 8, 3, 1, 4, 2, 1, 1, 1 0.17 0.14 - 0.20 0.16	75	0.064 0.050- 0.10 0.060	0.29 0.085 - 1.25 0.085	2.24 1.30 - 4.50 1.50

N*: number of phrases sampled for each different monkey, e.g., 3 campbelli lowei males were sampled, 12 phrases from one, 11 from another, and 10 from the third

N**: number of units sampled for each different monkey; in the case of campbelli lowei this is further divided into A & B type units. Mona units are divided into 2 kinds of subunits (A with sidebands developed and B without sidebands). Only one kind of unit is recognized in pogonias

N+: number of 2-unit phrases included in this sample

TABLE 6

SINGLE UNIT CHIRPS

Species	Locality	N	Unit Duration (secs.)	Lower Limit of Energy Range (kHz)	Upper Limit of Major Energy (kHz)
			measurements: mean, range, mode		
cephus	W. of Meyébot, E. Dja	8	0.033 0.020 - 0.040 0.040	1.76 1.40 - 2.50 1.75	3.62 3.00 - 4.00 4.00
	W. of Nazam, E. Dja	7	0.041 0.030 - 0.060 0.040	1.41 1.25 - 1.75 1.25	4.28 3.75 - 5.00 3.75
	17 km. Camp, W. Dja	11	0.049 0.030 - 0.070 0.050	1.77 1.50 - 2.00 2.00	3.95 3.00 - 4.50 4.00
	Bifos, W. Dja	6	0.037 0.030 - 0.050 0.030	1.41 1.25 - 1.50 1.50	4.08 2.25 - 7.50 3.50
erythrotis	Kindonge	14	0.043 0.030 - 0.12 0.040	1.86 1.25 - 2.75 2.00	4.50 3.50 - 6.00 4.50
	Idenau	3	0.027 0.020 - 0.030 ----	1.58 1.00 - 2.25 ----	4.08 2.75 - 6.00 ----

TABLE 7

2 AND 3 UNIT CHIRPS

Species	Locality	N phrases (units)	Unit Duration (secs.)	Lower Limit of Energy Range (kHz)	Upper Limit of Major Energy (kHz)
			measurements: mean, range, mode		
cephus	W. of Meyébot, E. Dja	2 (4)	0.043 0.020 - 0.050 0.050	1.87 1.50 - 2.50 1.50	3.68 3.50 - 4.00 3.50
	17 km. Camp, W. Dja	2 (4)	0.038 0.020 - 0.050 0.040	2.12 1.50 - 2.75 ----	3.50 3.25 - 3.75 3.50
	Bifos, W. Dja	13 (28*)	0.035 0.010 - 0.070 0.030	2.14 1.00 - 3.50 1.50 & 3.00	4.03 2.50 - 6.75 4.00
erythrotis	Kindonge	10 (21**)	0.037 0.020 - 0.050 0.040	2.04 0.75 - 3.00 2.00 & 2.50	4.38 3.50 - 7.00 4.50
	Idenau	1 (2)	---- 0.040 ----	---- 1.50 ----	7.37 3.75 -11.00 ----

* includes two 3-unit phrases
** includes one 3-unit phrase

TABLE 8

SINGLE UNIT TWITTERS

Species	Locality	N	Unit Duration (secs.)	Lower Limit of Energy Range (kHz)	Upper Limit of Major Energy (kHz)
			measurements: mean, range, mode		
cephus	W. of Meyébot, E. Dja	6	0.037 0.020 - 0.050 0.040	1.37 0.75 - 2.50 1.00 & 1.50	3.08 2.25 - 3.50 3.00 & 3.50
	Bifos, W. Dja	33	0.041 0.020 - 0.060 0.040	1.70 1.00 - 2.25 1.50	4.26 2.50 - 8.00 3.50
erythrotis	Idenau	5	0.032 0.020 - 0.040 ----	2.30 2.00 - 3.50 2.00	---- 8.00 ----

TABLE 9

PETAURISTA SQUEAKS FROM VOLTA DAM, GHANA

measurements: mean, range, mode

phrase duration (secs.) (N)	duration of non-tonal units (N)	duration of tonal units (N)	lower frequency of non-tonal units (kHz)	lower frequency of tonal units	upper frequency of non-tonal units	upper frequency of tonal units
0.29	0.011	0.067	3.21	3.49	6.22	4.55
0.26 - 0.34	0.0050 - 0.020	0.050 - 0.080	3.00 - 3.50	3.25 - 3.75	4.00 - 8.00	3.50 - 4.50
0.26	0.010	0.070 & 0.080	3.25	3.50	7.50	4.50
(10)	(17)	(10)				

TABLE 10

SINGLE UNIT CHIRPS

Subspecies	Locality	N	Unit Duration (secs.)	Lower Limit of Energy Range (kHz)	Upper Limit of Major Energy (kHz)
			measurements: mean, range, mode		
nictitans martini	Makweli	7	0.039	2.03	4.46
			0.030 - 0.050	1.25 - 2.50	2.75 - 5.50
			0.040	2.50	4.25 & 5.50
	Idenau	9	0.033	1.68	5.36
			0.020 - 0.040	0.75 - 2.50	3.50 - 14.00
			0.040	2.00	3.75 & 4.50
	Nsépé	4	0.038	1.53	4.37
			0.030 - 0.040	1.40 - 1.75	4.00 - 4.50
			0.040	1.50	4.50
	Kindonge	16	0.044	1.58	5.21
			0.030 - 0.060	0.080 - 2.50	2.00 - 7.00
			0.050	1.50	---------
nictitans nictitans	Manmenyin, Campo	3	----------	1.080	---------
			0.040	1.00 - 1.25	4.00
			----------	---------	---------
	Ongat, Campo	6	0.050	1.52	4.29
			0.040 - 0.070	1.25 - 2.00	4.00 - 4.50
			0.040	1.50	4.50
	Bifos, W. Dja	8	0.038	1.84	6.56
			0.030 - 0.050	0.75 - 2.50	4.00 - 8.50
			0.040	2.25 & 2.50	---------

TABLE 11

TWO UNIT CHIRPS

Subspecies	Locality	N phrases (units)	measurements: mean, range, mode Unit Duration (secs.)	Lower Limit of Energy Range (kHz)	Upper Limit of Major Energy (kHz)
nictitans martini	Kindonge	3 (6)	0.043 0.030 - 0.060 0.030 & 0.050	1.52 1.25 - 1.75 1.50	5.00 4.00 - 7.00 4.00
nictitans nictitans	Manmenyin, Campo	4 (8)	0.031 0.020 - 0.050 0.020	1.62 0.50 - 4.00 0.50	3.68 2.00 - 4.50 4.00
	Ongat, Campo	5 (10)	0.028 0.020 - 0.040 0.020 & 0.030	1.88 1.00 - 2.75 1.50 & 2.75	3.68 3.00 - 4.00 4.00

TABLE 12

L'HOESTI CHIRPS FROM IDENAU, CAMEROON

measurements: mean, range, mode

N	unit duration (secs.)	lower frequency of energy range (kHz)	upper frequency of major energy (kHz)
6	0.045 0.040 - 0.050 0.040	2.04 2.00 - 2.25 2.00	3.67 3.25 - 4.00 ----

TABLE 13

SNEEZE CALLS

species	locality	N	phrase duration (secs.)	lower limit of energy range (kHz)	upper limit of major energy (kHz)
			measurements: mean, range, mode		
mona	Nsépé	12	0.10	----	1.29
			0.070 - 0.16	0.085	1.00 - 2.50
			0.090 & 0.10	----	1.00
	Kindonge	12	0.095	----	0.93
			0.060 - 0.11	0.085	0.75 - 1.25
			0.10	----	1.00
	Volta Dam	12	0.095	----	2.71
			0.060 - 0.12	0.085	1.00 - 3.50
			0.10 & 0.11	----	2.50
campbelli lowei	Kade	12	0.15	----	6.04
			0.090 - 0.20	0.085	4.25 - 7.50
			0.16 & 0.19	----	4.50 & 7.00

TABLE 14

STACCATO CHIRPS

species	locality	no. units / phrase (no. phrase sampled)	phrase duration (secs.)	unit duration (secs.) (no. units sampled)	lower limit of energy (kHz)	upper limit of major energy (kHz)
		measurements: mean, range, mode				
pogonias grayi	Meyébot, E. Dja	2.08 2 - 3 2 (12)	0.16 0.14 - 0.18 0.16	0.057 0.020 - 0.070 0.060 (25)	0.40 0.085 - 2.00 0.085	3.28 1.25 - 5.00 3.00
	Bifos, W. Dja	---- 2 ---- (12)	0.16 0.13 - 0.18 0.17	0.058 0.040 - 0.080 0.060 (24)	---- 0.085 ----	2.98 1.00 - 5.50 3.00
	Campo	1.90 1 - 3 2 (10)	0.14 0.030 - 0.22 0.14	0.046 0.020 - 0.090 0.050 (20)	0.73 0.085 - 2.00 0.085	3.83 0.75 - 5.75 4.50

TABLE 15

GRUNTS

species	locality	no. units per phrase (N)	phrase duration (secs.)	unit duration (secs.) (N)	lower frequency of energy range (kHz)	upper frequency of major energy (kHz)
		measurements: mean, range, mode				
n. nictitans & *n. martini*	Bifos, W. Dja and Kindonge	---- 1 ---- (9)	0.14 0.11 - 0.23 0.12	----	0.12 0.085 - 0.25 0.085	0.91 0.75 - 2.00 0.75
erythrotis	Kindonge	1.57 1 & 3 1 (7)	0.14 0.090 - 0.22 ----	0.086 0.030 - 0.22 0.040 (11)	---- 0.085 ----	1.27 1.00 - 2.50 1.00

TABLE 16

AGONISTIC CHUTTERS OF NICTITANS MARTINI FROM KINDONGE, CAMEROON

measurements: mean, range, mode

no. units per phrase (N)	phrase duration (secs.)	unit duration (secs.) (N)	lower frequency of energy range (kHz)	upper frequency of major energy (kHz)
2.50	0.16	0.055	----	1.18
1 - 5	0.11 - 0.27	0.030 - 0.22	0.085	0.75 - 3.25
2	0.13	0.040	----	0.90
(14)		(35)		

TABLE 17
FREQUENCY DISTRIBUTION OF OOO CALL TYPES
WITHIN THE SUPERSPECIES *MONA*

species	locality	CALL TYPE A	B	C	D	total
mona	Kindonge	1	5	13	2	21
	Nsépé	0	2	10	1	13
	Volta Dam	0	6	2	8	16
pogonias pogonias	Kindonge	0	8	0	0	8
	Nsépé	0	9	0	0	9
	Makweli	0	1	2	0	3
	Idenau	0	9	0	0	9
pogonias grayi	E. Dja	0	6	2	0	8
	Campo	0	1	0	0	1
	Bifos, W. Dja	3	21	2	0	26
	17 km. Camp, W. Dja	0	6	0	0	6

TABLE 18
B TYPE OOO CALLS OF THE MONA SUPERSPECIES

measurements: mean, range, & mode
L=lower U=upper limit of energy range (kHz)

species	locality	N	phrase duration (secs.)	duration non-tonal (secs.)	range of energy of non-tonal L	range of energy of non-tonal U	dur. of tonal (secs.)	range of energy of f_o L	range of energy of f_o U	energy of $2f_o$ & (N*) L	energy of $2f_o$ & (N*) U	energy $3f_o$ & (N**) L	energy $3f_o$ & (N**) U
mona	Kindonge	5	0.20 0.17 - 0.24 ----	0.12 0.10 -0.14 0.14	0.12 0.085 to 0.25 0.085	0.70 0.50 to 0.75 0.75	0.10 0.060-0.17 0.080	0.58 0.40 to 0.75 0.50& 0.75	1.24 1.00 to 1.5 1.10& 1.50	1.00	2.00 (1)	(0)	
	Nsépé	2	0.19 & 0.14	0.10& 0.090	0.085	0.75	0.090 & 0.050	0.40 & 0.75	1.25 & 1.40	0.75	1.75 (1)	1.50	2.00 (1)
	Volta Dam	6	0.67 0.25 - 1.13 ----	0.15 0.060- .24 ----	0.18 0.085 to 0.50 0.085	0.79 0.75 to 1.00 0.75	0.52 0.15 - 1.00 ----	0.75 0.50 to 1.00 0.75	2.02 1.60 to 2.50 2.00	1.35 1.00 to 1.50 1.50	2.12 1.50 to 2.50 2.50 (5)	(0)	
pogonias pogonias	Kindonge	8	0.63 0.46 - 0.80 0.69	0.24 0.12 -0.54 ----	0.085 0.085	0.50 0.50	0.38 0.26 - 0.52 ----	0.72 0.50 to 1.00 0.75	1.22 1.00 to 1.75 1.25	1.63 1.25 to 2.50 --	2.35 2.00 to 2.50 2.50 (4)	3.00	3.50 (1)
	Nsépé	9	0.37 0.33 - 0.40 0.37 - 0.40	0.14 0.10- 0.18 0.11, 0.15, & 0.16	0.085 0.085 0.085	0.79 0.60 to 1.00 0.75	0.23 0.18 - 0.27 0.22, 0.23, & 0.26	0.44 0.25 to 0.75 0.50	1.18 1.00 to 1.50 1.00	0.96 0.75 to 1.00 1.00	1.77 1.50 to 2.00 1.75 & 2.00 (6)	1.25	2.00
	Makweli	1	0.44	0.20	0.085	0.50	0.24	0.50	1.00	1.00	1.75	1.75	2.50

f_o = fundamental harmonic
$2f_o$ = 2nd harmonic
$3f_o$ = 3rd harmonic

N* = no. units with $2f_o$
N** = no. units with $3f_o$

TABLE 18

B TYPE OOO CALLS OF THE MONA SUPERSPECIES

measurements: mean, range, & mode
L=lower U=upper limit of energy range (kHz)

species	locality	N	phrase duration (secs.)	duration non-tonal (secs.)	range energy of non-tonal L	range energy of non-tonal U	dur. of tonal (secs.)	range of energy of f_o L	range of energy of f_o U	energy of $2f_o$ & (N*) L	energy of $2f_o$ & (N*) U	energy $3f_o$ & (N**) L	energy $3f_o$ & (N**) U
	Idenau	9	0.43	0.18	0.085	0.52	0.25	0.61	1.31	1.50	2.00	(0)	
			0.28 - 0.50	0.090-0.26	0.085	0.50	0.12 - 0.37	0.50	1.00	(1)			
			0.46 & 0.48	0.13		to	0.24	to	to				
						0.60		0.75	1.50				
					0.085	0.50		0.50	1.50&				
									1.25				
pogonias grayi	E. Dja	6	0.51	0.15	0.085	0.67	0.36	0.79	1.92	1.25	2.25	(0)	
			0.37 - 0.55	0.070-0.23	0.085	0.50	0.22 - 0.60	0.50	1.75	(1)			
			0.55	----		to	----	to	to				
						0.75		1.00	2.50				
					0.085	0.75		0.75	1.75				
	Campo	1	0.33	0.17	0.085	0.50	0.16	1.00	2.00				
	Bifos, W. Dja	21	0.57	0.26	0.11	0.57	0.30	0.61	1.57	1.20	2.00	(0)	
			0.24 - 1.18	0.090-0.96	0.085	0.50	0.060-0.56	0.40	1.00	0.75	1.25		
			----	0.10 - 0.22	to	to	0.24	to	to	to	to		
					0.50	0.90		1.00	3.00	1.75	2.50		
					0.085	0.50		0.50	1.50	1.25	2.00		
										(12)			
	17 km. Camp, W. Dja	6	0.47	0.21	0.085	0.53	0.26	0.40	1.25	0.98	2.25	(0)	
			0.36 - 0.44	0.16 - 0.26	0.085	0.50	0.18 - 0.43	0.25	1.00	0.75	2.00		
			0.44	0.20		to	0.24	to	to	to	to		
						0.60		1.00	1.50	1.50	2.50		
					0.085	0.50		0.25	1.25	0.75	2.25		
										(6)			

f_o = fundamental harmonic
$2f_o$ = 2nd harmonic
$3f_o$ = 3rd harmonic

N* = no. units with $2f_o$
N** = no. units with $3f_o$

TABLE 19

C TYPE OOO CALLS OF CERCOPITHECUS MONA

measurements: mean, range, mode

L=lower U=upper limit of energy range

locality	N	phrase duration (secs.)	range of energy of f_o (kHz) L	range of energy of f_o (kHz) U	range of energy of $2f_o$ & (N)* L	(N)	range of energy of $2f_o$ & (N)* U	$3f_o$ L	(N)	$3f_o$ U	$4f_o$	$5f_o$
Kindonge	13	0.16 0.080- 0.25 0.13	0.60 0.25 to 1.00 0.50	1.44 1.00 to 1.75 1.50	1.14 0.75 to 2.00 1.00	(7)	1.50 1.00 to 2.00 1.50	1.50	(1)	2.00		
Nsépé	10	0.12 0.070- 0.16 0.090	0.45 0.25 to 0.75 0.50 & 0.25	1.53 1.00 to 1.75 1.50	1.19 0.75 to 1.50 1.25	(8)	2.19 1.75 to 2.75 2.00	1.63 1.50 to 1.75	(4)	2.19 2.00 to 2.50	3.00 (1)	3.50- 4.00 (1)
Volta Dam	2	0.23 & 0.26	2.00 1.00	2.75 2.25	1.75	0	2.50					

f_o= fundamental harmonic
* = No. of units with a $2f_o$, $3f_o$, $4f_o$, $5f_o$

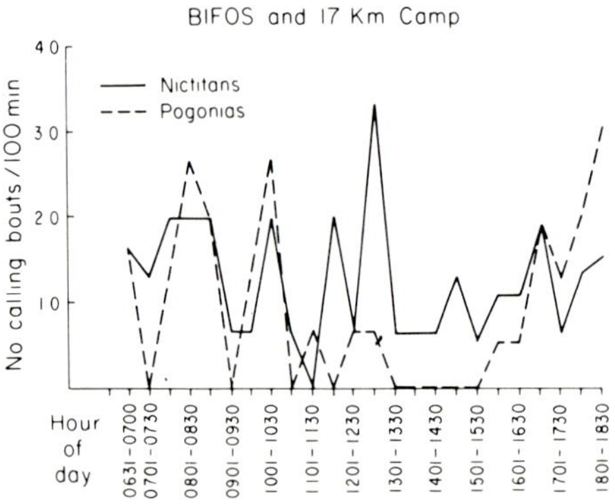

BIFOS and 17 Km Camp
No calling bouts / 100 min
40
30
20
10
Nictitans
Pogonias
Hour of day
0631 - 0700
0701 - 0730
0801 - 0830
0901 - 0930
1001 - 1030
1101 - 1130
1201 - 1230
1301 - 1330
1401 - 1430
1501 - 1530
1601 - 1630
1701 - 1730
1801 - 1830

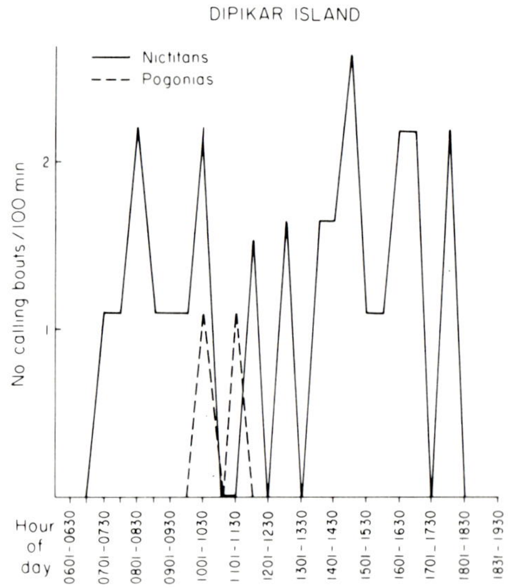

DIPIKAR ISLAND
No calling bouts / 100 min
2
1
Nictitans
Pogonias
Hour of day
0601 - 0630
0701 - 0730
0801 - 0830
0901 - 0930
1001 - 1030
1101 - 1130
1201 - 1230
1301 - 1330
1401 - 1430
1501 - 1530
1601 - 1630
1701 - 1730
1801 - 1830
1831 - 1930

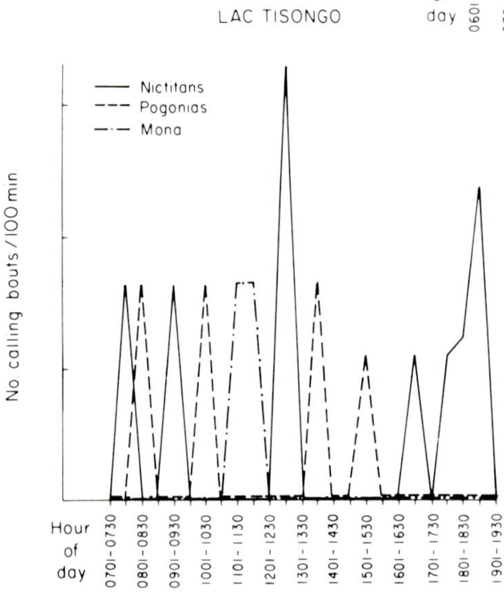

LAC TISONGO
No calling bouts / 100 min
Nictitans
Pogonias
Mona
Hour of day
0701 - 0730
0801 - 0830
0901 - 0930
1001 - 1030
1101 - 1130
1201 - 1230
1301 - 1330
1401 - 1430
1501 - 1530
1601 - 1630
1701 - 1730
1801 - 1830
1901 - 1930

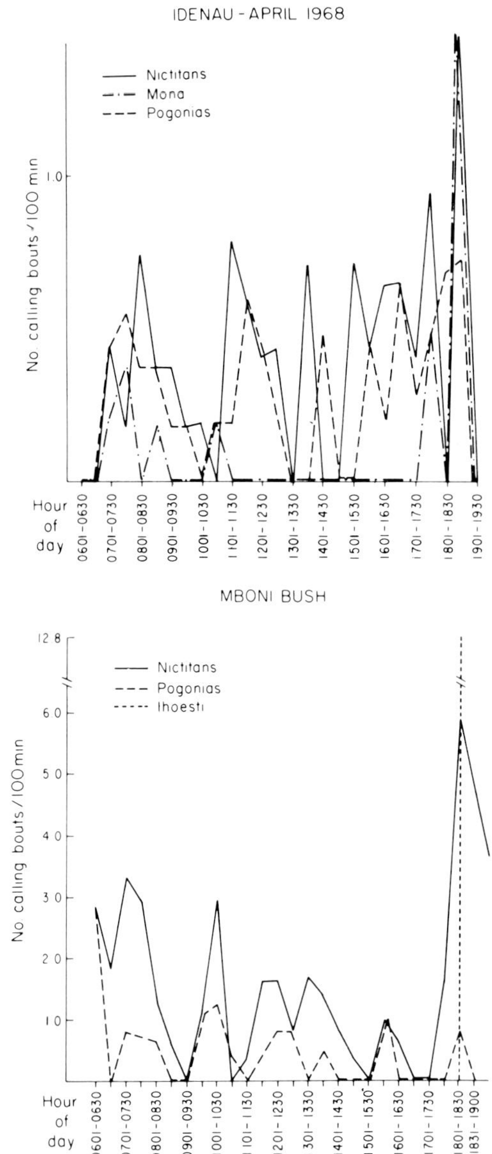

Fig.1: Diurnal distribution of bouts of adult male loud calls in five localities (see text for details).

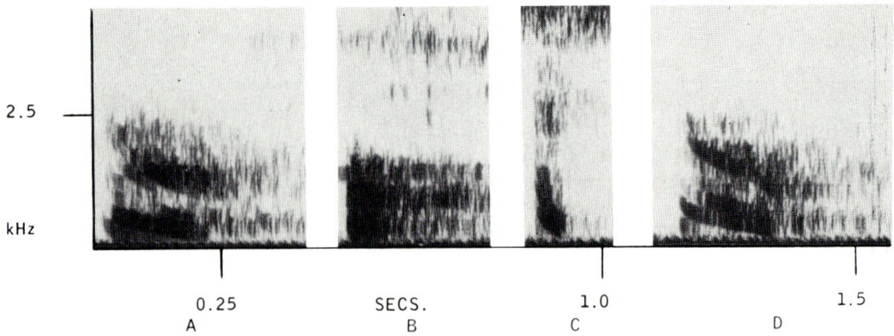
2.5
kHz
0.25
SECS.
1.0
1.5
A
B
C
D

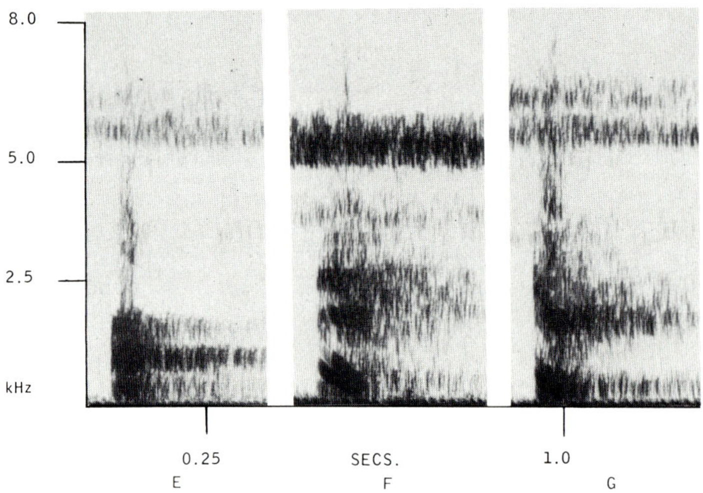
8.0
5.0
2.5
kHz
0.25
SECS.
1.0
E
F
G

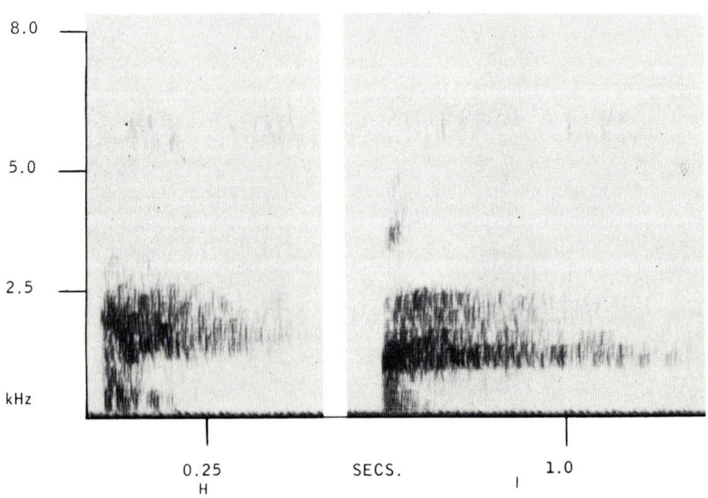
8.0
5.0
2.5
kHz
0.25
SECS.
1.0
H
I

Fig.2: A) _nictitans martini_ pow from Kindonge; B) _n. martini_ hack, same monkey as A; C) _n. martini_, Kindonge, a different monkey from A) & B); D) _n. martini_ pow, Nsépé; E) _n. nictitans_ hack, W. Dja; F) _n. martini_ hack, Kindonge; G) _n. nictitans_ hack, W. Dja; H) _n. nictitans_ hack, E. Dja; I) _pogonias grayi_ hack, Mayebot, E. Dja. Ignore background noise occurring at 3.5 kHz in B) and C) and from 5-6 kHz in E), F), & G).

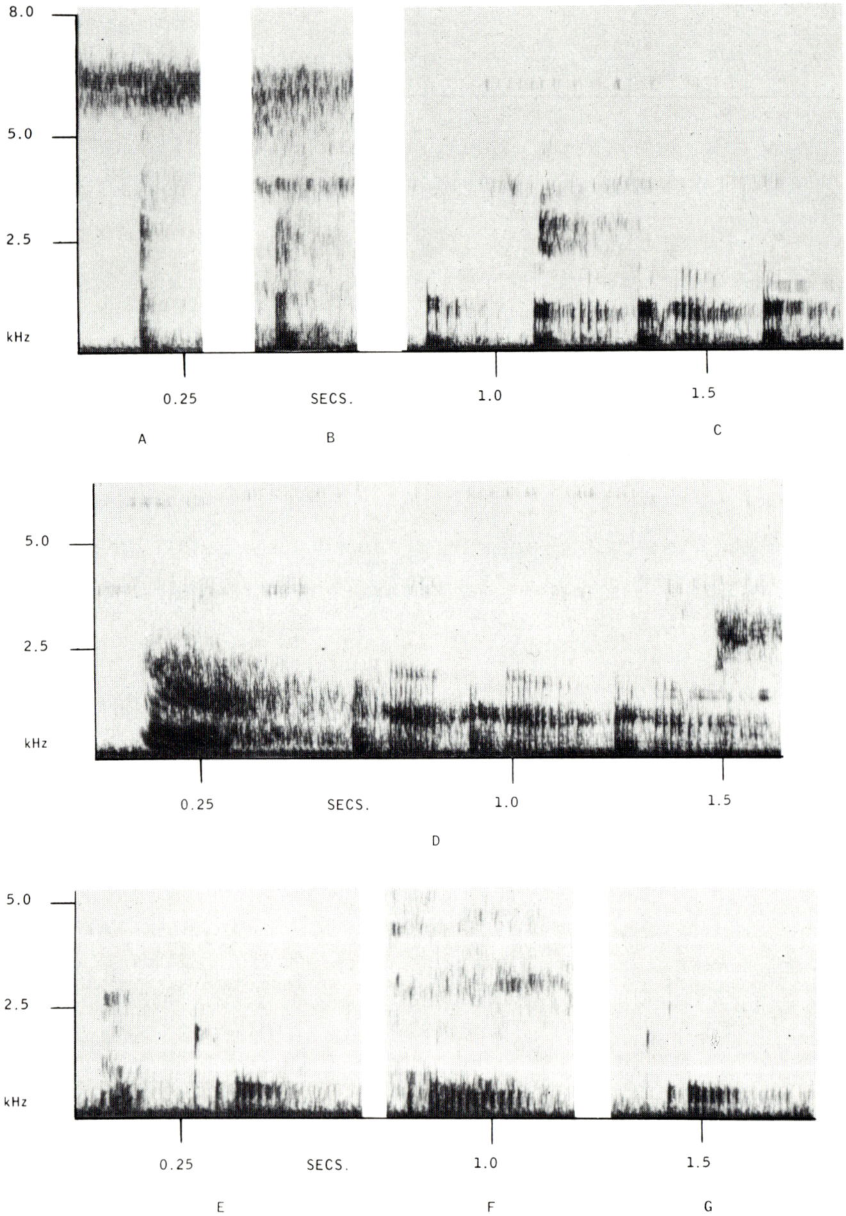
8.0
5.0
2.5
kHz
0.25
SECS.
1.0
1.5
A
B
C
5.0
2.5
kHz
0.25
SECS.
1.0
1.5
D
5.0
2.5
kHz
0.25
SECS.
1.0
1.5
E
F
G

Fig.3: A) *cephus* single-unit hack, Bifos; B) *erythrotis* single-unit hack, Kindonge; C) 4 multi-unit *erythrotis* hack phrases, Kindonge, a chirp from another monkey is present above the first part of the second phrase; D) one *n. martini* pow centered at 0.25 secs. followed by 3 multi-unit *erythrotis* hack phrases, Kindonge, a chirp from a third monkey occurs above the last phrase at 1.5 secs.; E) *campbelli lowei* sneeze followed by a multi-unit *petaurista* klunk (hack) phrase, the first unit of this latter phrase is a spike with a frequency range from 0 to 2.5 kHz, Kade; F) & G) two multi-unit *petaurista* klunk phrases, Kade. Ignore the background noise at 6-7 kHz in A) & B) and also at 4 kHz in B).

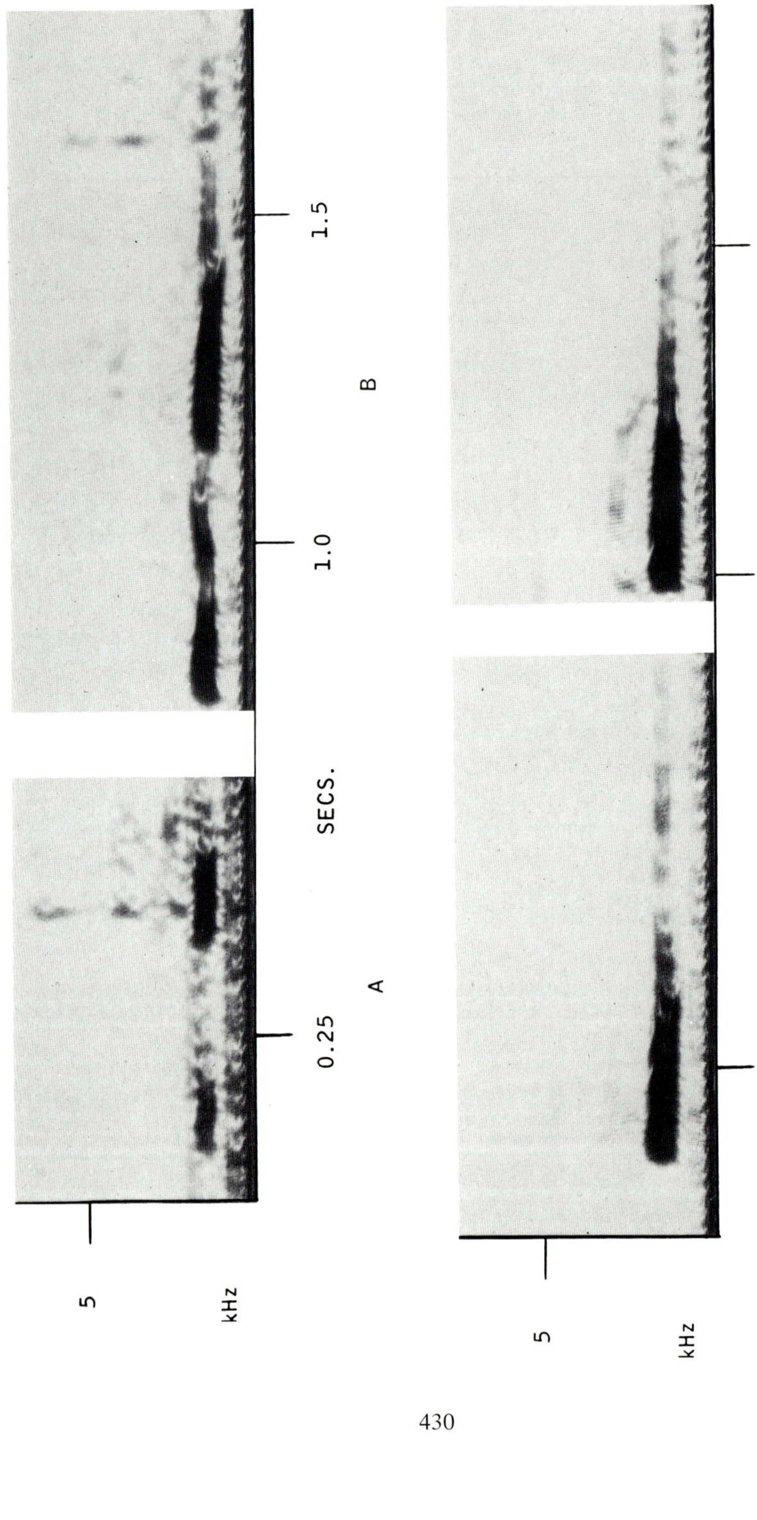
5
kHz
0.25
SECS.
1.0
1.5
A
B
C
D

Fig. 4: A) mona 2-unit boom, Kindonge, ignore background noise occurring as a spike in the middle of the 2nd unit; B) mona "2-unit" boom with 2 "extra" units, Kindonge: C) p. pogonias single-unit boom, Kindonge; D) p. grayi single-unit boom, W. Dja.

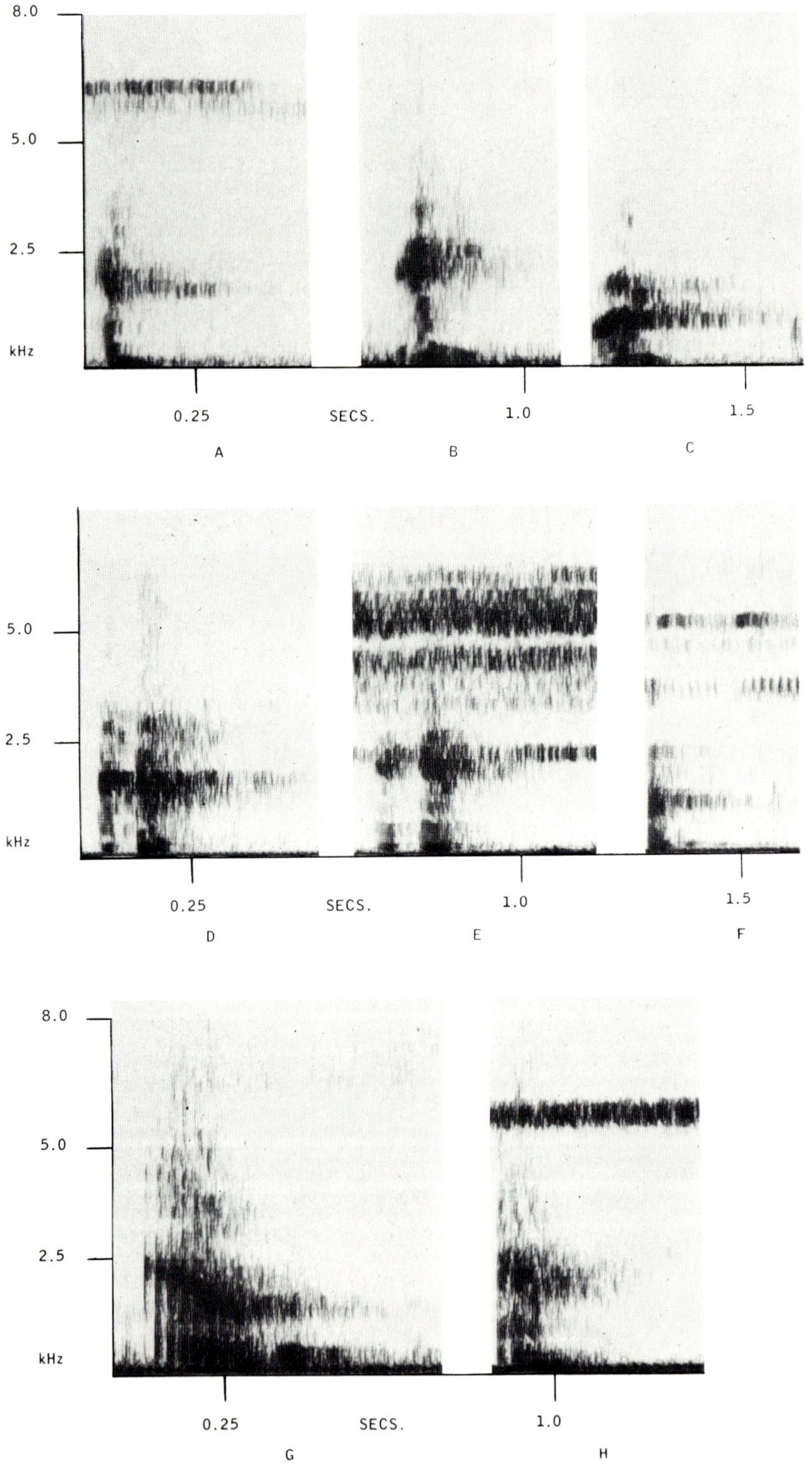
8.0
5.0
2.5
kHz
0.25
SECS.
1.0
1.5
A
B
C
5.0
2.5
kHz
0.25
SECS.
1.0
1.5
D
E
F
8.0
5.0
2.5
kHz
0.25
SECS.
1.0
G
H

Fig.5: A) mona hack from early in a hack series, Kindonge; B) mona hack early in series, Atimpoko; C) mona hack from middle of series, same bout as A); D) p.pogonias 2-unit hack, Idenau; E) p.grayi 2-unit hack, Bifos; F) p.grayi 1-unit hack, Bifos; G) campbelli lowei 2-unit hack, Kade; H) campbelli lowei 1-unit hack (Type A unit), Kade. Ignore background noise at 7 kHz in A), 3-6 kHz in E, 3-5 kHz in F), 6 kHz in H).

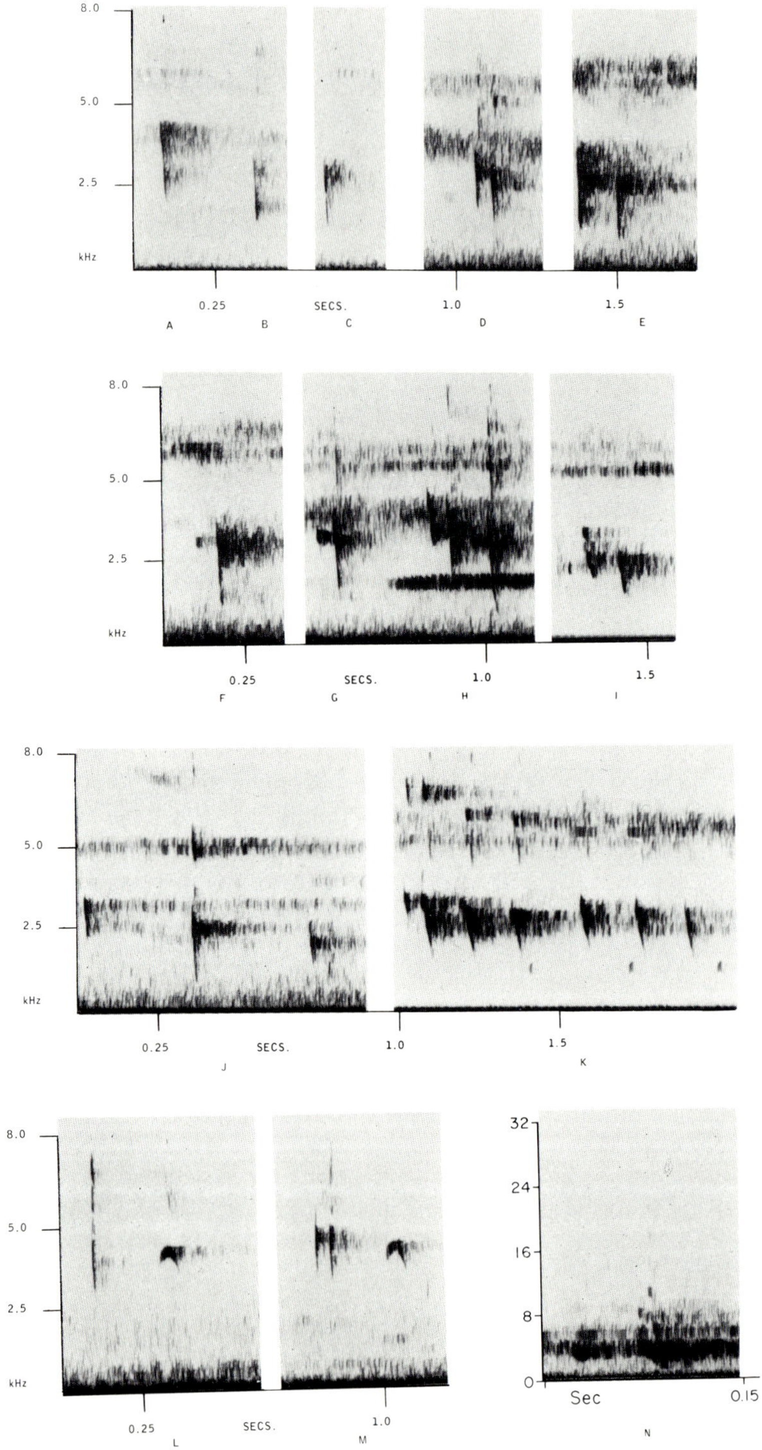

8.0
5.0
2.5
kHz
0.25
SECS.
1.0
1.5
A
B
C
D
E
F
G
H
I
J
K
L
M
32
24
16
8
0
Sec
0.15
N

Fig.6: A) & B) *erythrotis* single-unit chirp, Kindonge; C) *cephus* single-unit chirp, W.Dja; D) *erythrotis* 2-unit chirp, Kindonge; E) *cephus* 2-unit chirp, W.Dja; F) *cephus* 2-unit chirp, W.Dja; G) *erythrotis* 2-unit chirp, Kindonge; H) *erythrotis* 3-unit chirp, Kindonge; I) *cephus* 3-unit chirp, W.Dja; J) 3 single-unit *erythrotis* twitters, Idenau; K) one 2-unit followed by five single-unit cephus twitters (all from one monkey), Dja; L) *petaurista* 2-unit squeak, Volta Dam; M) *petaurista* 3-unit squeak, Volta Dam; N) this is D) at a different scale to demonstrate the ultrasonic components. Ignore the continuous band of background noise at about 6 kHz in A) - K), at 3.5 kHz in D), H), & J), and at about 2 kHz in H).

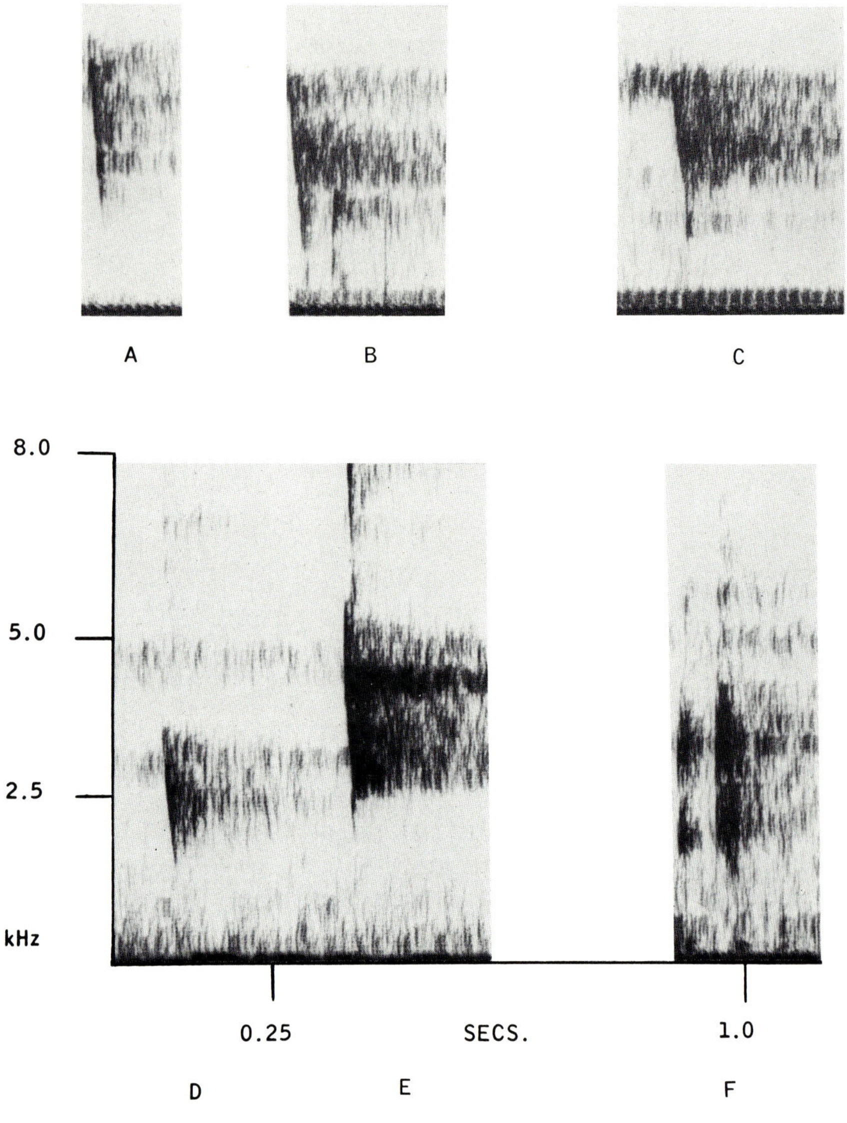

Fig.7: A) n. nictitans single-unit chirp, Ongat; B) n. nictitans 2-unit chirp, Manmenyin; C) n. nictitans 2-unit chirp, Manmenyin; D) & E) n. martini single-unit chirps, Idenau; F) n. martini 2-unit chirp, Kindonge.

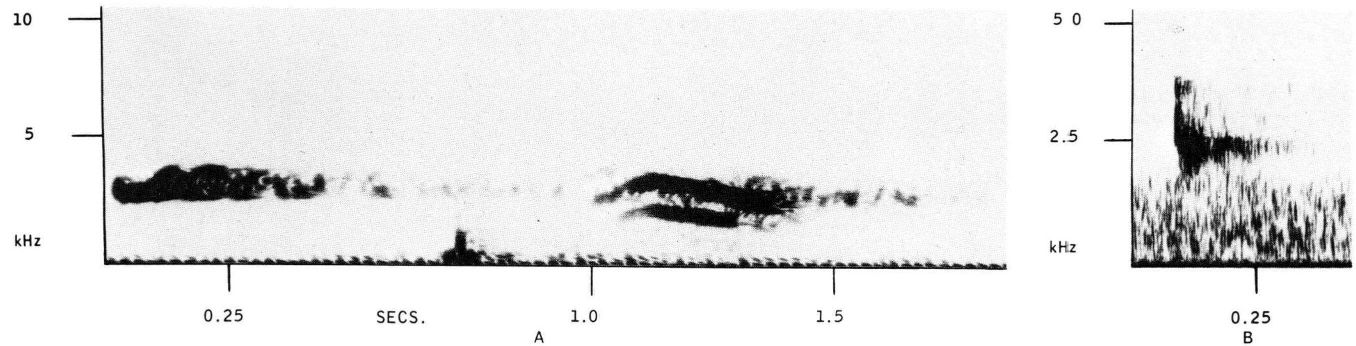

Fig. 8: A) adult male lhoesti Oop-uuh loud call, Bomana; B) lohesti chirp, Idenau, ignore background noise below 2 kHz.

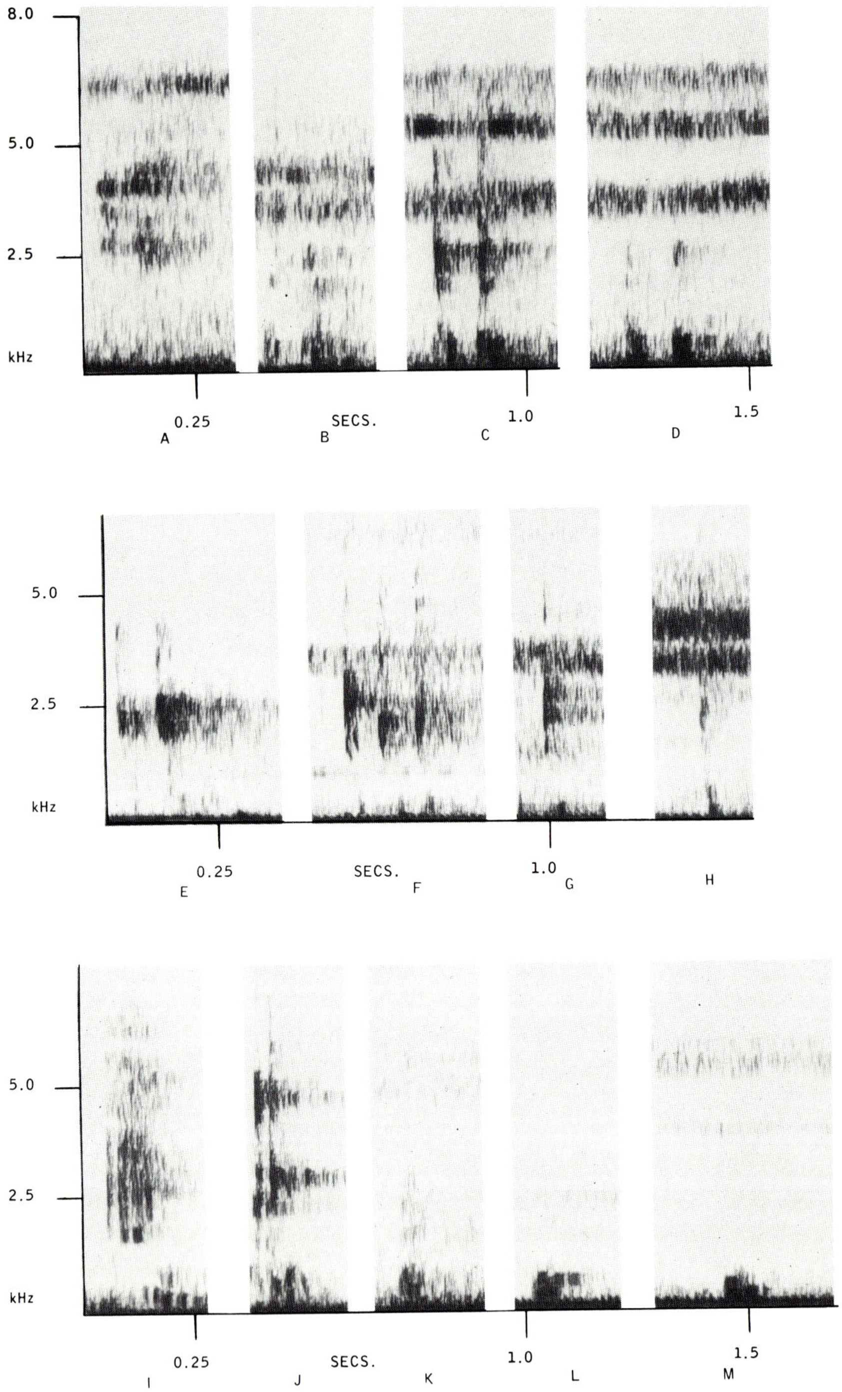
8.0
5.0
2.5
kHz
0.25
SECS.
1.0
1.5
A
B
C
D
5.0
2.5
kHz
0.25
SECS.
1.0
E
F
G
H
5.0
2.5
kHz
0.25
SECS.
1.0
1.5
I
J
K
L
M

Fig. 9: A) *p. pogonias* 2-unit staccato-chirp, Nsepe; B) *p. pogonias* 2-unit staccato-chirp (low pitched version), Idenau; C) *p. grayi* 2-unit staccato-chirp, Bifos; D) *p. grayi* 2-unit staccato-chirp (low pitched version), Bifos; E) *p. grayi* 2-unit staccato-chirp, Meyebot; F) *p. grayi* 3-unit staccato-chirp, Ongat; G) *p. grayi* single-unit staccato-chirp, Ongat; H) *p. pogonias* single-unit staccato-chirp, Idenau; I) *campbelli* sneeze, 2 portions, Kade; J) *campbelli* sneeze, 1 portion, Kade; K) *mona* sneeze, Volta Dam; L) *mona* sneeze, Nsépé; M) *mona* sneeze, Kindonge. Ignore background noise at 6.5 kHz in A); at 3.5-5 kHz in B); at 3.5, 5.5, & 6.5 in C) and D); at 3.5 in F) & G); at 3.5-4.5 in H); and at 5 kHz in M).

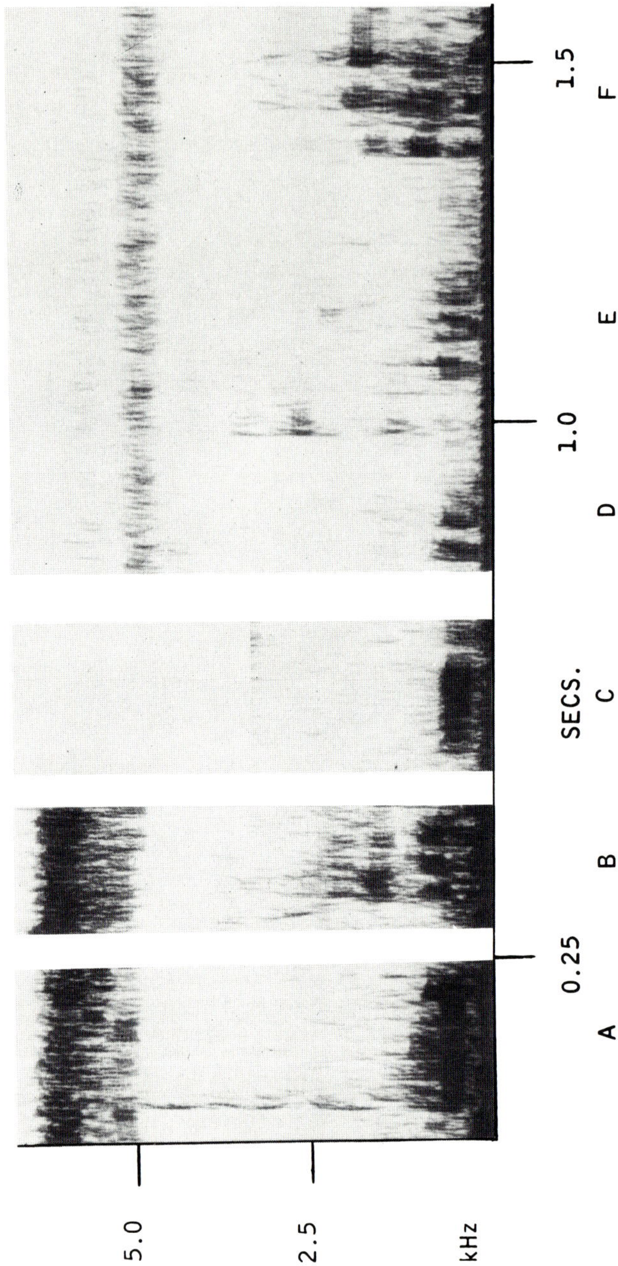
5.0
2.5
kHz
0.25
SECS.
1.0
1.5
A
B
C
D
E
F

Fig. 10: A) *erythrotis* single-unit grunt, Kindonge; B) *erythrotis* 3-unit grunt, Kindonge; C) *n. nictitans* grunt, W. Dja; *n. nictitans* chutters, Kindonge; D) 2-unit, E) & F) 3-unit. Ignore background noise at and above 5 kHz.

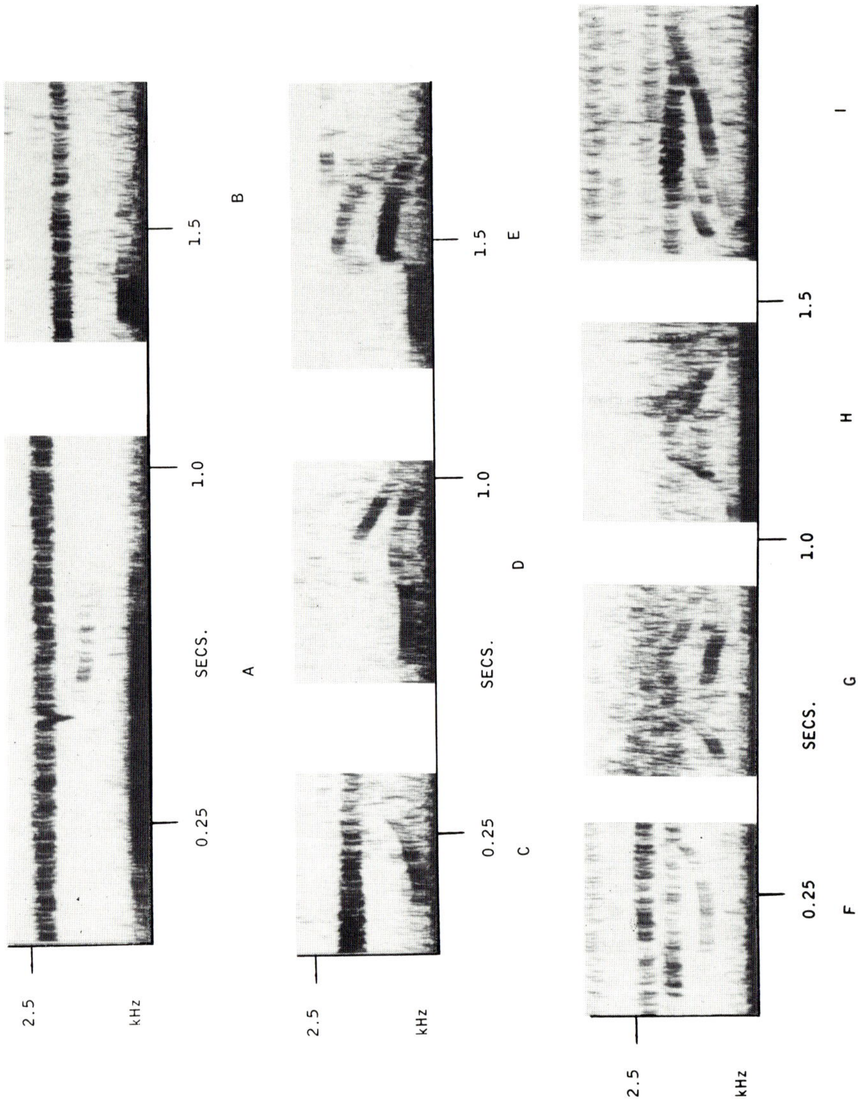
2.5
kHz
0.25
SECS.
1.0
1.5
A
B
C
D
E
F
G
H
I

Fig. 11: 11, OOO calls: A) pogonias grayi type A, W. Dja; B) mona type A, Kindonge; C) mona type B, Kindonge; D) p. pogonias type B, Nsépé; E) p. grayi type B, W. Dja; F) mona type C, Kindonge; G) p. pogonias type C, Makweli; H) p. grayi type C, E. Dja; I) mona type D, Kindonge. Ignore the background noise at 1.25 and 2.4 kHz in A); 2.4 kHz in B), C), F) & G); at 2.4 kHz in E) after the descent in the call; at 2.3 kHz in F); and at 2.3 kHz in I) where it occurs as a continuous and unmodulating band.

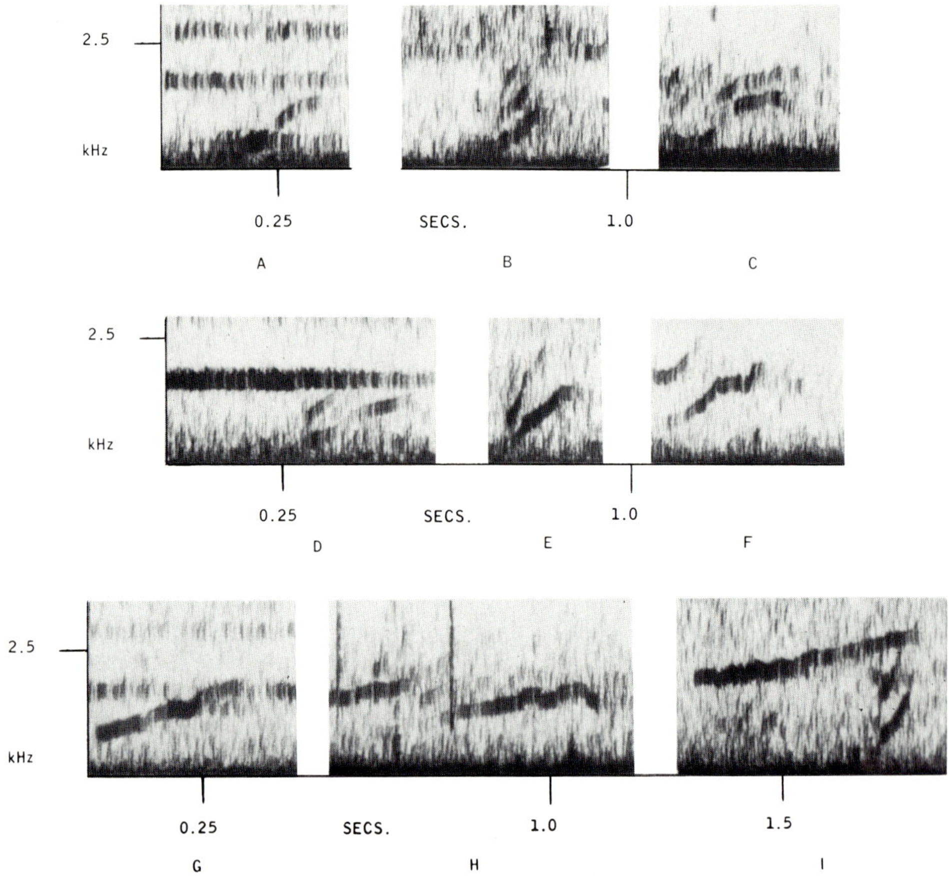

Fig. 12: 12, mona OOO calls: A) type B OOO call, Kindonge; type B, Nsépé; C) type B, Volta Dam; D) type C, Kindonge; E) type C, Nsépé; F) type C, Volta Dam; G) type D, Kindonge; H) type D, Volta Dam; I) type D is the one of longest duration and type C at the end is of shorter duration and was given by another monkey, Nsepe. Ignore the background noise at 2.3 and 2.6 kHz in A), at 2.5 and above in B), at 2.3 kHz and occurring as a continuous band in D) & G), and occurring as short-duration spikes at the beginning and near the middle of sonagram H) and ranging in frequency from about one to more than 3 kHz.

PRELIMINARY NOTES ON THE ECOLOGY AND BEHAVIOR OF THE DRILL, *MANDRILLUS LEUCOPHAEUS* RITGEN, 1824

J. S. Gartlan

INTRODUCTION

Recent studies of primate social structure, and in particular those of Kummer (1968a, 1968b) provide a paradigm for the analysis of evolutionary changes in the social organization of primates. These examples are particularly relevant to this discussion as the studies were concerned with the divergence of the social structure of one baboon, *Papio hamadryas* from a more generalized *Papio cynocephalus* type of social structure. The relationship of *Mandrillus* and *Papio* is extremely close, and indeed the drills and mandrills have frequently been classified with *Papio*, although latest taxonomy tends to regard them as generically distinct (Jolly, 1967 and this volume). Kummer's method might therefore be particularly valuable in the case of *Mandrillus*.

At the outset, the limitations of this study must be stated. These are essentially a poverty of obtained data, which precludes a precise, quantitative comparison. In 15 months of field work carried out between October '67 and December '68, a total number of 29 encounters with drill groups was recorded. These varied in duration from 27 to 426 minutes. This sample does not include two contacts with solitary animals, or one contact with a group disturbed and being shot at by poachers. The total number of hours in contact with drill groups was 77.8, for about half of which (36.8 hours) they were in good view.

Drills, *Mandrillus leucophaeus* Ritgen 1824, have a limited distrubution within the block of tropical rain forest between the Congo and Niger fivers. The approximate distribution range in West Africa, together with the location of the study area, the Southern Bakundu Forest Reserve of West Cameroon (4°.30'N., 9°.30' E.) are indicated in Fig. 1. The location of study sites within the reserve is shown in Fig. 2.

The first study area (area A) was prepared at the start of the investigation in October and November '67. The second (area B) was surveyed and prepared in April and May '68. Both were then used to different extents until the end of the study in December '68.

The study areas were bounded on three sides by tracks cut along compass lines running approximately north to south and east to west; the fourth side of both areas was a timber extraction path which ran from the main Kumba-Victoria road to Kindonge Village, within the reserve. Within these two areas a grid network of intersecting tracks was cut. These were planned to divide the areas into approximately 200 yard squares, but this was not rightly adhered to, and when appropriate old tracks or timber survey lines were encountered, they were utilized. The boundaries and tracks were numbered with white paint at intervals of 10 yards to permit the precise location of groups on duplicated maps.

These tracks permitted silent and rapid movement through the study areas. This was important in following drill groups as the noise of a person moving through undergrowth would put them to flight. The most accurate counts of drill groups were usually obtained when they crossed the cleared tracks.

The total area of the two study sites was 154.6 hectares; area A was 74.8 hectares and area B 79.8 hectares.

Groups of drills were located on 12 occasions in area A and on 17 occasions in area B. The mean duration of contact with groups south of the road (A) was 112.75 minutes (median 57.5, range 30-354). The corresponding figure for area B was 195.00 minutes. Records for solitaries and the group disturbed by poachers are excluded. The median in the latter case was 173, and the range 27-426 minutes. The mean contact duraction with groups in area B therefore approaches twice that of the figure for area A, indicating a greater degree of tolerance of the observer's presence north of the road. The figures for mean visual contact with groups show this trend even more clearly. Mean visual contact in area A was 34.33 minutes, (median 21.5, range 2-100), whereas the corresponding figure in area B was 105.64 minutes (median 81, range 11-269). This indicates that animals in area B, having been located, were less shy of being observed than were groups south of the road.

GROUP DYNAMICS

GROUP SIZE

Two main variations in social structure have been described within the Papio genus (Kummer, 1968a). The first is that of the savanna or P. cynocephalus types. Kummer (loc.cit.) argues that -- "the wide distribution of the cynocephalus type of organization in the genera related to Papio, as well as among the New World monkeys, (suggests) that this type of organization is the older pattern from which the hamadryas type branched off as a specialization to the dryest part of the area inhabited by baboons." In the same work, Kummer (loc.cit.) cites ecological factors of probable significance in the social divergence of P. hamadryas from the more common Papio social structure.

DeVore and Hall (1965) give details of group size in Papio (excluding P. hamadryas). Repeated counts of more than 2000 baboons in Kenya showed a variation in group size from 9 to 185. In southern Africa, a survey showed a range from 8 to 109. These authors state that their data indicate a central tendency (mean and median) of 30 to 50. However, wide variations both within and between areas were observed.

In P. cynocephalus the troop is the social unit and exhibits temporal continuity. The social boundaries are also defined by both the frequency and the type of communication with conspecifics. The social unit is discrete, and much of the life of an individual baboon occurs exclusively within this unit. In P. hamadryas on the other hand, the size of troops is greater and varies between 12 and 750 (Kummer and Kurt, 1963). However, in this species the troops seemed to be ecologically determined aggregations and not integrated behavioral units. The basic and persistent social units are one-male groups. These, as in the case of the larger P. cynocephalus groups, are behaviorally discrete units. Subtroops of a fairly constant size and composition are occasionally observed, however, and Kummer considers that these may correspond to the P. cynocephalus group, and gives behavioral evidence for this point of view.

The fragmentary evidence available from the literature suggests that the social structure of M. leucophaeus may not be similar to that of P. cynocephalus. Thus Jeannin (1936) gives a figure of between 15-20 individuals per troop, and Bourgoin (1955) gives a marginally larger range of from 10 to 30. Sanderson (1940) says that groups may range up to 40 individuals. However, Malbrant and Maclatchy (1949) say that drills are to be encountered in "grandes hordes bruyantes". These figures, it should be noted, are rough estimates and are unsupported by tabular data.

The distinction between relatively small groups and much larger ones (grandes hordes) is reflected in the data collected in this study. Accurate counts were difficult to obtain in the observation conditions prevailing at Southern Bakundu. Counts of groups were therefore classified as "reliable" if I was fairly confident that the number counted was approximately the total number of individuals present, or "poor" if this criterion was not reached. The classifications were generally based on the amount of noise, (both of vocalization and incidental noise), signs of activity such as shaking of foliage, areas of foraging, and (especially in the case of small groups) of systematic searches for other animals in the immediate vicinity. In general the most complete counts were obtained when groups were slightly alarmed and ran off across one of the prepared tracks. In other cases contact was usually maintained long enough for the estimate of animals not seen to be considered as reliable.

The smallest group which I was confident represented a reliable count was of 14 animals (17 June '68). This included an adult, a subadult and a juvenile male, and three other juvenile animals, with eight other unidentified animals. When the group had left the study area, a systematic search confirmed that no other drills were, or had been, foraging in the vicinity of this small group. The largest reliable count was of 179 animals and was obtained two months later, at a point only 120 meters from the smallest group count. This large group included at least three adult males, two subadult males, twenty-one adult females, a subadult female, a juvenile male and three other juveniles, twenty infants and 128 other animals. The complete list of group counts in the two study areas, together with an age-sex analysis and an indication of reliability are given in Tables 1 and 2. Figs. 3 and 4 also indicate for all counts the last position of the group, together with the number of animals counted and an indication of reliability (r=reliable or p=poor). The direction of movement of groups leaving the study areas is also shown by means of an arrow.

The fact that these directions of movement were relatively constant within each study area, and different between the two areas; the fact that no habituation of animals occurred south of the road, and the fact that an identifiable individual was not observed in area A, are taken as indicative that observations made in the two areas A and B are not of the same population. No evidence of drills crossing the road was ever obtained, although such movements cannot be theoretically excluded. In the following section, data from the two areas will therefore be considered as separate.

DeVore and Hall (1965) provide data on the home range size of P. cynocephalus groups in savanna, and Rowell (1966) gives data for the same species in a forested habitat. In Nairobi Park the former authors record ranges of from 2.0+ sq. miles to 15.5 sq. miles. The mean was 8.97 sq. miles (standard deviation 4.81, n=8). In the forest, Rowell (loc. cit.) gives figures for two groups of 1.5 and 2.0 sq. miles. In the Cape Reserve for a sample of three groups the ranges are from 3.5 to 13.0 sq. miles (mean 7.4, standard deviation 4.97). The mean home range size of both savanna samples combined is 8.54 sq. miles (standard deviation 4.65). DeVore and Hall stress that these are home ranges and not territories. They state, however, that this does not imply that groups moved about without reference to any fixed boundaries, and noted that, on the contrary, baboon groups spend most of their time in circumscribed areas from which it was very difficult, if not impossible, for a human observer to drive them out.

Reference to Fig. 3 indicates clearly the relative positions where different counts of animals within area B were made. It should be emphasised that these groups were all encountered initially either inside the study area or immediately adjacent to it. If they moved out of the area, they were followed; the positions

TABLE 1.

GROUP COUNTS, RELIABILITY AND AGE-SEX ANALYSIS OF GROUPS IN AREA A.

DATE	No.	Rel.	A♂	S♂	A♀	S♀	J♂	J♀	J	I	u/d
21 Oct. 67	4	P	1	-	-	-	-	-	2	-	1
23 Oct. 67	16	P	2	1	4	-	1	-	6	2	-
27 Nov. 67	39	R	2	-	2	-	2	-	4	2	27
28 Dec. 67	2	P	1	-	-	-	-	-	1	-	-
4 Jan. 68	6	P	2	-	1	-	-	-	3	-	-
8 Feb. 68	7	P	3	-	2	-	-	-	1	1	-
5 Mar. 68	13	R	1	1	6	-	-	1	2	1	1
18 Mar. 68	6	P	2	-	-	-	1	-	1	-	2
25 Apl. 68	3	P	1	-	-	-	-	-	2	-	-
12 Aug. 68	19	R	1	1	2	-	1	-	1	1	12
4 Oct. 68	1	P	1	-	-	-	-	-	-	-	-
7 Oct. 68	6	P	2	-	-	1	-	-	2	-	1

Reliability: P = poor; R = reliable.

shown are the last positions recorded for that particular day. It can be seen that most of these are actually within the boundaries of the study area, and mostly within one half of it. Thus this extremely large variation in group size was observed within an area of 39.9 hectares or 0.018 the mean home range size calculated from DeVore and Hall's data, or 0.087 the mean home range size of Rowell's forest population. Similar results are obtained for area A, south of the road. Even if the data for both areas are considered

TABLE 2.

GROUP COUNTS, RELIABILITY AND AGE-SEX ANALYSIS OF GROUPS IN AREA B.

DATE	No.	Rel.	A♂	S♂	A♀	S♀	J♂	J♂	J	I	u/d
25 Oct. 67	11	P	1	1	2	1	-	-	5	1	-
17 Nov. 67	2	P	-	-	-	-	-	-	2	-	-
13 Jun. 68	14	R	1	1	1	1	-	-	3	1	6
17 Jun. 68	14	R	1	1	-	-	1	-	3	-	8
18 Jul. 68	57	R	3	2	2	-	-	-	8	2	40
26 Jul. 68	88	P	2	2	21	-	2	-	5	21	35
5 Aug. 68	179	R	3	2	21	1	1	-	3	20	128
15 Aug. 68	13	P	1	1	2	-	-	-	8	1	-
5 Sep. 68	47	P	1	-	5	-	-	-	5	5	31
18 Sep. 68	1	R	1	(single adult male)							
27 Sep. 68	44	R	3	2	7	-	-	-	3	1	28
9 Oct. 68	68	R	4	1	3	-	1	-	4	3	52
21 Oct. 68	10	P	1	2	3	1	1	-	-	2	-
29 Oct. 68	59	R	3	2	1	-	-	-	2	-	51
31 Oct. 68	130	R	5	2	3	-	2	1	1	3	113
12 Nov. 68	14	P	2	1	3	-	-	-	5	3	-
15 Nov. 68	-	-	(shooting)								
29 Nov. 68	126	R	3	2	2	-	-	-	2	1	116
9 Dec. 68	1	R	1	(single adult male)							
13 Dec. 68	8	P	2	1	5	-	-	-	-	-	-

Reliability: P = poor; R = reliable.

together (and reasons for not doing so have already been given) the general principle remains clear. It seems reasonable to conclude, therefore, that observations of drill groups made in the study area indicate a different pattern of distribution than would normally be observed if the population were dispersed in typical P. cynocephalus pattern in discrete home ranges with mutual avoidance at territorial boundaries. Similar observations to those described above for M. leucophaeus have occasionally been made on Papio near water holes, or in sleeping groves at Amboseli. These are clearly ecologically determined modifications of a social structure which would not normally permit behavior of this type. There was no evidence of simple ecological causation in the Southern Bakundu population of M. leucophaeus.

A further possibility must therefore be discussed. This is that groups of different size, ranging from about 14 to almost 200 animals wander at random through the forest, and were merely located by chance in the same small area of forest. This hypothesis is compatible with Malbrant and Maclatchy's (1949) conclusion that drill groups wander through the forest "sans points d'attache bien précis".

There is observational evidence from my study which would tend to contradict this view. One individual, a subadult male became recognisable to be during the course of the study. This animal was seen in area B on several occasions, which tends to argue against the random wandering theory (especially in view of the small size of the study area). Examination of Figs. 3 and 4 where the final movement (often the flight path) of groups is indicated shows that this tends to be consistent. This would imply some knowledge of the habitat by the animals, which is more likely to occur in animals living in a restricted area than it is in animals wandering at random throughout the forest. Habituation in the context of this study would also tend to indicate attachment to a particular area.

The data on habituation of animals north of the road may be summarised as follows. Data on total contact with groups and data on visual contact with groups in both study areas were compared. Twelve contacts in area A gives an n of 12 in both categories; total contacts in area B were 17. Total contact with groups in area A for the first six contacts was 884 minutes, and in the last six contacts was 469 minutes. This is a percentage decrease of 46.9 percent. The figure for visual contact in the first six contacts was 262 minutes, and in the last six contacts was 150 minutes, a percentage decrease of 42.7 percent. This study area was the nearer of the two to human habituation, and adjoined the "chop" farms of Kindonge Village. Harassment of drills could therefore be expected to be more prevalent in area A.

Figures for area B, (more distant from Kindonge) on the other hand, show a percentage increase. The number of contacts in this area was 17, and the first eight contacts were therefore compared with the last eight in both categories. Comparison of total contact time for the first eight contacts in this area, compared with the last eight (1397:1753 minutes) shows a percentage increase of 25.4 percent. Data for visual contact (807:908 minutes) similarly shows a percentage increase, this time of 12.5 percent.

One subadult male drill (Paleface) was of lighter facial and coat color than other members of the population. It was observed on five occasions between 9 October and 13 December, 1968. All observations of this animal were in area B. The movements of the group in which this animal was observed are indicated in Fig. 5. Of the five sightings of Paleface, three reliable counts of 130, 68 and 126 were obtained. One may draw the tentative conclusion therefore that individual animals remain in the same general area for periods of several months. Also that groups observed in similar locations on successive occasions, and including identifiable individuals, may show fluctuations in total group size.

Data given by Rowell (1966) on a forest-living population of *Papio* provides a baseline comparison for characteristics of the *Mandrillus* population. In her study the number of adult males was approximately the same as the number of adult females, and therefore markedly different from the extreme reduction in adult male numbers observed in *M. leucophaeus* groups. As has already been noted, the distribution of that *Papio* population indicated by her counts is different from the spatial distribution of the *M. leuco-*

phaeus population. She does, however, record the occasional breaking up of a large group into smaller parties which followed different routes during part of the day. When groups became separated in the forest, the alarm-bark was used antiphonally to facilitate re-location by the separated sub-units, in much the same way as *M. leucophaeus* groups use the distinctive crowing call* in the same situation.

These fragments of evidence, although not conclusive in themselves, may perhaps be taken to indicate that although the distribution patterns of *M. leucophaeus* are generally unlike that which would be expected from a savanna *P. cynocephalus* pattern, neither is it entirely compatible with the theory of groups moving at random through the forest, and not is it similar to the social structure of *Papio* living in a forested habitat.

SOCIAL UNITS

Kummer (1968a) has demonstrated that the large troops in which *P. hamadryas* habitually sleeps are not social units as much as ecologically determined aggregations. These split up during daily foraging into sub-troops (which retain some features of *P. cynocephalus* groups) and into one-male units.

There is evidence that the large groups of drills observed at Southern Bakundu may in fact be aggregations of smaller units, and which are therefore fairly flexible in composition. Such a finding would be compatible with the distribution data discussed previously, and also with the data on the identifiable subadult male being observed in different sized groups and with the consistency of flight paths, etc. Field evidence on this point centers on observations of two phenomena, a) the coalescence of small groups of drills to form larger aggregations, and b) the splitting of large units to form smaller groups.

*Crowing call (Refs. G, H). Typically this call begins with discrete units of .04 - .06 sec. duration, separated by intervals of .25 sec. The intervals shorten (.06 sec.) and merge into a continuous call. This may persist for 1.8 sec. or longer; the main energy (in a juvenile sample) was from 750 - 1350 Hz.

Clear evidence of the latter behavior was observed in area B on 31 October '68. The large group (130 animals) having spent several hours foraging, began to move off in a north-easterly direction. The split took place at about 1449 hours, with a subgroup leaving the large group and moving directly north. The subgroup comprised 37 animals, including two adult males and an adult female with a ventral infant. At 1558 hours the subgroup was 300 meters north of the large group - well out of visual contact. It was also silent and was not maintaining vocal contact with the rest of the group. At this time (1558) the subgroup then crossed track 4 and moved off in a south-easterly direction towards the rest of the group. The subgroup crossed the track in 15 seconds, giving some idea of their compactness.

Struhsaker (pers.comm.) records splits and reunions of large (n=200) and medium (n=40) savanna baboon groups at Amboseli. Rowell (1966) records them in her forest population of Papio to a limited extent. In the savanna such a split is potentially not serious as long as the animals remain in visual contact with one another. (A study of the spacing behavior of savanna Papio groups would be extremely valuable in providing baseline comparative data for the evaluation of social changes). A distance of 50 meters in the savanna is of little consequence, whereas the same distance, or even less in the forest means that all visual contact is lost.

The clearest example of the joining of small groups to form larger units was observed in area A on 27 November '67. Drills were heard at 1152 hours and the volume of noise indicated a small, fairly compact group. The first animal was seen at 1241. The group was moving in a southerly direction, running off 100 - 200 meters when hearing or seeing the human observers. At 1256 hours the group descended into a ravine and climbed up the other side. They then became less shy of the observers and permitted them to sit and watch without apparent alarm behavior. The southerly direction of movement ceased and the animals sat and moved around, foraging and feeding. This group comprised about 20 animals, and included a single adult male, a juvenile male and an adult female with ventral infant. These animals remained in the immediate vicinity of the ravine until 1326 hours. At this point I heard a noise behind me coming from a northerly direction. This was another small, compact group of drills (a circle of 35 meters diameter would have contained all 20 or so). They were moving rapidly (3-5 k.p.h.) and silently towards the first group, with no sign of foraging or feeding behavior. There was a single, very large adult male in the center of the group. These animals came to within 20 meters of us, hidden on the northern edge of the ravine, and did not see us until a juvenile male, moving through the undergrowth at 4.5 meters gave alarm barks. The whole group then fled to the ravine, down into it to the place where the first group was sitting; both then joined and ran off together in a southerly direction. Apart from two or three crowing calls at 1332 hours, the two groups had certainly been out of visual and auditory contact for an hour and a half at least.

A further example of joining of small units was observed on 26 July '68. Sounds of drills were first heard in area B at 0952 hours. The first animal of the first group, a subadult male, was seen at 0955 hours. This animal also saw us and gave the two-syllable alarm bark. At 1001 crowing contact noises were heard coming from another group which was between 70 and 100 meters away from the one being observed. At 1006 the first group replied with crowing. The two sources of calls gradually drew closer to one another until at 1025 hours animals of the second group were observed crossing a track to join the first. The second group comprised 22 animals and included an adult and a subadult male, an adult female and 19 other animals. The two groups joined at approximately 1044 hours, and the contact noises then ceased. The total group comprised at least 88 animals, including two adult males, a subadult male, 21 adult females with ventral infants, five juveniles and at least 35 other animals. This large group continued moving and foraging as an entire unit until 1400 hours when contact with them was lost in heavy rain.

It is pertinent here to note that all the instances detailed above include instances of the call labelled "crowing". This call was heard on 14 contacts with drills out of a possible maximum of 30 (contacts with single adult males being excluded). It was heard at the highest frequency and intensity on 15 November '68, when a large group had been dispersed by shooting. It was also heard in area A on 7 October '68, when the observer inadvertently split a group which was foraging on both sides of the track. Some members ran off in a south-easterly direction, others in a south-westerly. An extremely high rate of these calls (one every one or two minutes) from both groups, gradually converging, was then heard for ten minutes. All calling then ceased when the groups were presumed to have joined again.

It may be relevant here to note that Malbrant and Maclatchy (1949) record the most common vocalization of M. leucophaeus as a 'trrrii' "perçant et aigu". In their collecting of species they may well have shot at and dispersed groups which would then have given the crowing call.

That this call is associated with the re-joining of small groups is emphasized by the fact that when the largest group count was made (179 animals) the group was contacted early in the morning, and seemed then to be a large unit. No evidence of joining or leaving of groups was obtained during the entire contact of over seven house. Much of this time was spent within an area some 200 x 200 meters, feeding, resting and idly foraging. When the group finally moved off, quite compactly, the mobilizing two phase grunts of adult males were heard, but no crowing call had been heard throughout the entire day.

A similar call, very loud and piercing was heard on two occasions (15 August '68 and 5 September '68) when groups which were foraging and fairly dispersed were suddenly caught in torrential rain, which acting as an acoustic blanket, drowned out all foraging and climbing sounds. Groups of juvenile animals were then both heard and seen giving the loud crow whilst engaged in fairly agitated behavior. On both occasions the calls were answered by other animals, and the juveniles then moved off in the direction of the callers. The animals involved in this behavior, juveniles, are those most commonly involved in giving the less intense crowing calls. This may suggest that the calls may be derived from early infant contact calls, which may have been retained as functional in this context of promoting the joining of groups. It is a call which is not heard as a normal part of the social repertoire of either P. cynocephalus or P. hamadryas.

These data would seem to suggest that the large groups in which drills are commonly encountered are not merely ecologically determined aggregations in the manner in which Kummer (1968a) has described for P. hamadryas, but rather that they serve some social function, and are maintained in part by a call not appearing in Papio species, which facilitates and promotes the formation of these large units.

Struhsaker (1969) in a simultaneous study of West African cercopithecines, on the basis of contact with groups of M. leucophaeus in observation conditions often better than those prevailing at Southern Bakundu, noted that groups of 20 or less had only one adult male, whilst those greater than 20 usually had more than one adult male. This may suggest therefore that the smaller units are primarily one-male-groups, which may aggregate into larger units. If this is so, then one would expect the ratio of adult males in large groups to approximate to the ratios 1:20.

The total numbers of animals reliably counted was therefore used to estimate the total number of adult males which would be expected on this hypothesis. (Adult males are particularly conspicuous animals, and most easily identifiable in the field). For counts of 20 and under, it was assumed that one adult male would be expected in the group. The expected number of males was then compared with the numbers actually observed in groups. These data are shown in Table 3.

A linear correlation run between these two sets of data yields the value of $r=0.69$. This is significant at the 0.05 level. The estimating equation is $y=1.39 - 0.69x$. A Kolmogorov-Smirnov one-sample test between the two sets of data fails to reach significance. The null hypothesis of no difference between the two sets of data must therefore be retained, indicating that the two sets of data are drawn from the same population.

These data are therefore consistent with the view that the large groups of drills (over 20) seen at Southern Bakundu, are in fact aggregations of one-male-units. There exists some field evidence for this view. First, all reliable counts of groups under 20 included only a single adult male. Second, in the example mentioned previously of the joining of two small groups, (27 November '67) both, one of 19 and one of 20, each had only a single adult male. Third, the small group which was mentioned as joining another group on 26 July '68, comprised 22 animals, including a single adult male, a subadult male and an adult female. This was a complete count of the small group, but no reliable count was made of the large group, although it included at least 88 animals.

TABLE 3.

TOTAL NUMBERS OF ADULT MALES OBSERVED (O) COMPARED WITH NUMBERS EXPECTED (E) ON THE BASIS OF 1/20 GROUP SIZE.

O	E
1	1
1	1
3	3
3	9
3	2
4	3
3	3
5	7
3	6
2	2
1	1
1	1

With a ratio of adult males to total group size of approximately 1:20, one would expect some exclusion of adult and perhaps subadult males from groups. This is upheld by field observation. Only two records of solitary animals were obtained throughout the entire study (18 September and 9 December '68) and both were in area B. Both were fully adult males. This also holds true for

Struhsaker's data. He observed six solitary animals. Of these, two were either adult or subadult males, three were adult males, and one was a subadult male. No evidence for the formation of all-male groups was collected during the study, although their presence cannot be discounted.

The existence of one-male groups with a maximum size of approximately 20, and the aggregation of these into larger, multi-male units makes the apparent ambiguity of earlier reports explicable. It seems likely that the small one-male-groups may congregate into the "grandes hordes bruyantes".

This ratio of one adult male per 20 individuals may be compared with data for other cercopithecine social groups. Hall (1963) gives the constitution of 7 groups of *Papio ursinus*. The mean of the ratios calculated from his data gives one male per 12.6 of other group members (standard deviation = 4.52). Rowell, (1967) gives data on the composition of two wild groups of *Papio cynocephalus* which indicates a ratio of one adult male per 4.09 members of the population. Kummer and Kurt (1965) give the percentage of adult males in their free-ranging, wild population of *P. hamadryas* as 22.9 percent or almost one male for every four other members of the population. Hall (1965) also gives data for groups of *Erythrocebus patas*. One group, of five animals, lacked an adult male. Eleven other groups, ranging in size from 9 to 31 (Mean: 20.27, S.D.: 7.4) had only a single adult male. In the forests of N. Honshu, unprovisioned troops of *Macaca fuscata* are often found in groups of about 20, each having a single adult male, Struhsaker (pers. comm.). In these cases, solitary males are common and are often found moving on the periphery of groups. In the actual proportion of adult males in groups, therefore, *M. leucophaeus* is different from both *Papio hamadryas* and *P. cynocephalus* and approximates to the ratio of *E. patas* and to wild, forest-living populations of *M. fuscata*.

Sabater Pi (pers. comm.) on the basis of his studies in Rio Muni considers that there is "a marked parallelism" between the drill and the mandrill. In the latter species also small groups apparently unite to form larger ones (grandes hordes). He has also observed many solitary adult males.

It would be valuable to know the extent to which drill one-male groups are consistent units. Thus in P. hamadryas the adult male maintains a group which consists of the same individuals and actually controls the integrity of the group by means of herding behavior. It is possible that an adult male drill leaving a large group might take with it a random collection of about 20 other individuals, the constitution of the group therefore varying on different occasions. Data currently available do not permit conclusions on this possibility.

DIFFERENTIATION OF FUNCTION WITHIN SOCIAL GROUPS

Until recently there has been a tendency in primatology to think of social groups as determined and structured by variations in the strength of unitary mechanisms. Recent work has demonstrated the value of a more overtly evolutionary approach. If the social group is considered as an adaptive unit, it follows that the organization of the group should reflect certain aspects of the ecological pressures functional in a particular habitat. It is to be expected therefore that an analysis of function within the social group may serve to identify some of these ecological pressures.

This approach complements that of Rowell (1967) who says that variations in any of the factors of social organization, defined as the distribution of communicative motor patterns and responses, will be correlated with differences in the resultant organization. She predicts that a group in which 5 threats are seen in an hour will have a different type of organization form one in which 55 threats are seen, even if the contexts in which they occur and the classes of animals involved are the same. It is therefore here proposed to examine certain behavior patterns of M. leucophaeus to establish the extent of functional differentiation within the group.

Two of the most common vocalizations heard in encounters with drills are, firstly, a deep two-phase grunt or roar (Type A)*

Typical Type A. (Ref. 773). Two deep grunts or roars of .4 - .5 sec. duration, separated by an interval of .5 - .6 sec. Main energy and secondly a low-intensity two-phase huffing grunt (Type B). It should be emphasised that at the extremes indicated these calls are distinct, but that in practice they grade into one another, and that the majority of intermediate forms cannot be classified one way or the other with any degree of confidence. It is probably Type A to which Sanderson (1957) refers when he writes that drills "growl and roar with a kind of pumping action." There was evidence in this study that Type A grunts were made exclusively by adult males, with Type B being made by both adult, subadult and possibly older juvenile males. The Type A grunts were invariably associated with movement of the whole group, and were heard either when the group was involved in a progression (as distinct from random foraging movements) or immediately before it. It was not heard when the group was sitting, resting or feeding, except when acting to terminate these behaviors. The probable function of Type A grunts is illustrated by observations made on 5 August '68. On this day a group was kept in close contact for over seven hours and remained for most of that time within an area 200 x 200 meters. No deep Type A grunts were heard between 0936 hours and 1542 hours. At this latter time the group was fairly dispersed, foraging, feeding, playing and chasing. At 1542 hours a series of deep Type A grunts was heard. Immediately 15 animals which were foraging away from the main body of the group ceased all other activity and started to move back fairly rapidly towards the source of the grunts. By 1545 hours the group was heading north in a closed group progression at a fairly rapid rate; this movement was accompanied by Type A grunts. All observations of these calls are consistent with the view that it initiates and promotes the movement of dispersed group members towards the source of the calls and so facilitates group cohesion at the particularly critical time when group progression is taking place.

There is evidence that in _M. leucophaeus_ a second means of

---

from 0 - 400 Hz.

*_Typical Type B._ (Ref. C). Series of rapid grunts or pants of .08 - .20 sec. duration, separated by intervals of .07 - .18 sec. with main energy between 50 - 950 Hz.

promoting group cohesion has been evolved, and which supplements the essentially passive or unenforced method described above. During the course of the stüdy, in 77.8 hours of contact with drills, 90 loud squeals or series of squeals were heard. Of these 49, or over half, were immediately associated with male grunts of all intensities between Types A and B, but in particular the less intense Type B ones. During these 77.8 hours, 44 chases of one or more animals by another were heard or seen. Twelve of these were silent without calls by either the chaser or the victim. Of these 12, subadult males were seen to be the chasers in seven instances, involving "herding" or "rounding-up" behavior. In this behavior a subadult male which has been characteristically sitting relatively close to the observer will, when a juvenile or group of juveniles ventures too close to the human observer, rush towards them, head them off, and chase them back towards the main body of the group. In a further five chases by subadult males, the victims gave squeals or squeal series. Subadult chases are only rarely accompanied by Type B grunts, and in about half the victim squeals, and in the other half remains silent.

In 16 other chases an adult male was involved to judge from the fact that deep two-phase grunts were heard during the chase. Squeals were heard in all these 16 chases. Nine other cases occurred in which squeals or squeal series were heard, but in which two-phase grunts were absent (as in most subadult male chases, which they may well have been). In only two cases in which two-phase grunts were heard, were squeals not heard. Thelma Rowell (pers. comm.) notes that in forest Cynocephalus the spectator to a chase within the group may make deep grunting noises. Some herding was also seen; after contact between the herder and victim had been established, the herder would lead the other with a prancing gait.

If a squeal or squeal series can be considered as evidence of an agonistic or aggressive interaction, the mean number of aggressive incidents per hour in M. leucophaeus is 1.15. Hall (1965) noted that in Erythrocebus patas only 49 aggressive encounters were seen in 627 observation hours, or 0.08 per hour. For Papio ursinus Hall (1962) gives a figure of 29 incidents in 190 observation hours, or 0.15 per hour.

Group sizes in these samples were different, so approximate comparisons may be made on the basis of incidents per thousand mean monkey/hour, calculated as the mean number of incidents per hour, divided by the mean group size, multiplied by 1000. The relevant figures for the three species are then comparable. *Papio ursinus* shows three incidents per 1000 mean monkey hour, *E. patas* shows four and *M. leucophaeus* 18; different population densities may be partly responsible for these differences. This calculation does not, of course, take account of the age-sex composition of groups, as precise data for *M. leucophaeus* are not available. However, in drills it is known that the ratio of adult males is much lower than in *Papio ursinus* and one might, therefore, on *a priori* grounds have expected relatively less aggression in the drill population. The observed rates are therefore particularly striking.

It seems that in *M. leucophaeus* therefore, for whatever reason, one observes relatively high rates of aggression. An analysis of the number of chases shows that males were seen or heard to be involved in 30 (68 percent), this representing the minimum involvement of adult males in chases. On all the occasions in which an adult male was seen to be involved in a chase, it was the chaser, and there is no reason to suppose (either on statistical grounds or from the known behavior of other terrestrial Cercopithecines) that there would be any significant reversal of this tendency in incidents in which chasers were not identified. The increased aggressiveness in drill groups may therefore largely be explained as an increased participation in chases by adult and subadult males. That these chases may terminate in a fairly ritualized behavior is demonstrated by field observation.

On 9 October '68, a subadult male was seen to chase a juvenile animal which was walking towards the observers. Instead of running off immediately, the juvenile crouched and squealed loudly. There was no physical contact between the animals and the juvenile, after squealing for about 4 seconds, stood up and dashed back towards the main body of the group. Also, on 31 October '68, an adult male was seen sitting on a cleared track surrounded by three juvenile animals. The male suddenly stood up quadrupedally and gave four rapid type B grunts. The juvenile directly in front of him then crouched and squealed loudly, facing him. Again there was no physical contact between the two animals, nor was any stimulus for the interaction apparent to the observer. The other two juveniles did not show any sign of interest in the interaction and did not run off. When the juvenile stopped squealing, the male sat and yawned widely three times in succession.

It seems probably, therefore, that the high level of chasing and squealing may in fact represent herding or enforced cohesive behavior on the part of adult and subadult males. The drill groups, like those of _P. hamadryas_ may to some extent be maintained by herding behavior. If so, then _M. leucophaeus_ differs in the important respect that juvenile males are herded as well as juvenile, subadult and adult females which are the animals herded in _P. hamadryas_.

The clearest evidence for the above hypothesis comes from an analysis of the chases in which subadult males were involved. Subadult males were, it was noted, observed chasing animals or groups of animals on 12 occasions. Seven of these were silent chases with no calls from either the chaser or the victim. In most cases the start, progress and occasionally the end of these chases were observed. Subadult males perform a social vigilance function in drill groups, particularly when groups have been contacted and continue foraging and feeding, rather than fleeing at the first contact. A behavior pattern which, in other words, is seen more frequently the more habituated a group becomes. In these cases the subadult male tended to come and sit relatively close to the observer. It is no accident that the one animal which I was able to recognize consistently should have been a subadult male: they tended to sit close and still for relatively long periods. The inquisitiveness of juvenile and infant primates is well-known and well-documented, and drills are no exception to this rule. It was common for a play-group of four or five juveniles, having seen me, to approach gradually closer and closer, glancing frequently and rapidly at one another as if for mutual reassurance. Finally, one would panic and dash back towards the main body of the group, followed, a split-second later by the rest. Occasionally, however a group of juveniles would venture too close. The "sentinel" subadult male would then round up and chase the juveniles back towards the main body of the group. Of the 12 observed cases, 10 involved more than one chased animal, and the chased animals were all juveniles. Two others were chases of single animals, but suggestive in both cases of herding behavior.

The first of these was observed on 5 March '68. A group of drills was heard at 1113 hours and contacted at 1126 hours. Contact was maintained until 1407 hours when the group started to move off. However at 1414 hours a single juvenile remained behind - for some reason no adult male two-phase grunts initiated this movement. The juvenile was foraging and feeding on the slope of a ravine. Suddenly a subadult male appeared, running, at the top of the ravine, and ran down the slope towards the juvenile giving two Type B grunts. It ran straight to the juvenile, to the side of it furthest from the main body of the group, circled it, still running and dashed off back towards the main body of the group, closely followed by the lagging juvenile. No squeals were heard during this incident and there was no physical contact. However, in a second incident of a subadult male chasing a single animal, a series of squeals was given by the victim.

It seems possible, therefore, that "rounding-up" or herding behavior of subadult males may indicate that the pattern is learned on the periphery of groups before becoming a part of the adult male social repertoire. If these hypotheses are upheld, it would seem that cohesion in drill groups is facilitated through two-phase type A grunts of adult males, permitting their location by other group members when group progression starts. This behavior is supplemented by actual herding behavior on the part of both adult and subadult males. These behavior patterns account to a large extent for the high levels of aggression noted in drill groups when compared with either *Papio ursinus* or *Erythrocebus patas*. These behaviors illustrate clearly differentiation of function in males.

Other evidence of social differentiation in adult and subadult males involves threat displays and aggression towards the human observer in habituating groups. Two distinct types of threat orientated towards the observer were observed being performed by drills. The first of these was a threat jerk, with the head suddenly thrust forward, eyelids retracted and medial crest raised. The lips are generally compressed and pushed slightly forward. The posture may occur when the animal is sitting, or when it is standing quadrupedally. In the former case, the body, and in particular the head, shoulders and upper part of the trunk are often involved in the jerk. A total of nine incidents of this type of threat were observed. One was performed by a juvenile animal, and one by an adult female. The remaining seven were performed by subadult and adult males in the ratio 4:3.

A second type of threat or intimidation display was seen comparatively rarely. This was branch-shaking or rattling. The displaying animal characteristically jumped on to a branch and rattled it up and down sharply, often several times, with the whole body weight and gripping the branch with both hands and sometimes with the feet also. Seven incidents of this behavior orientated towards the observer were seen. Of these seven, one was performed at a relatively low intensity by a juvenile animal. The remaining sex were high-intensity behaviors and were performed by subadult (4) and adult (2) males.

Struhsaker (1969) records threat behavior towards the human observer by subadult males only rarely, and by adult males never. He noted that when drills detected him they either fled immediately, or adult females and immatures approached the observers and gave pant-threats (alarm barks) towards them before eventually fleeing. In my study, when groups fled there was no evidence of any defensive role by adult or subadult males, and indeed the former were often the first in the onward rush. This is in agreement with Struhsaker's data. The evidence on adult males and subadult males threatening me therefore requires explanation. Of the adult and subadult threat jerks towards me, five, or 71 percent occurred in the last four, 14 percent, of contacts. A similar, though less well-marked trend is shown by branch-shaking behavior, where all adult and subadult performances occurred in the last 50 percent of contacts. These data would indicate that these roles are performed only in habituating groups, and are reactions to a more or less familiar person, and cannot therefore be considered as evidence of defensive roles in the normal context of contact with predators.

The social group is often considered to have adaptive significance in respect of increased protection from predation. Investigative and alarm behaviors are therefore an important aspect of group social behavior. Alarm barks, which are very similar to those of P. cynocephalus, may be given by any age-sex class above that of infant. Characteristically they consist of short staccato barks, occurring either singly or in pairs, given either once or repeated at brief intervals for lengthy periods of time. An analysis of obtained data showed 59 instances of barks or series of barks. The age and sex of the calling animal was only discerned in 32 instances. Of these, five were given by adult males, five by subadult males and five by adult females. The remaining 17 were all performed by juveniles. Juveniles in other words give alarm barks more frequently than all other age-sex classes combined, possibly because there are more juveniles in groups.

It is instructive to compare this pattern of distribution with that for investigative behavior. Ninety-one instances of this behavior were observed. One instance, it should be noted, may include several animals, usually a play-group, but theoretically up to the total numbers of animals present. The behavior involved coming to look at the human observer and is not mutually exclusive with behavior such as alarm-barking, threatening, etc. To be included in this category, behavior must show evidence of movement towards the observer, and evidence of "investigation" such as a lateral side-to-side jerking of the head and shoulders, which was observed in 12 instances, or standing bipedally on the hind legs to obtain a clearer view, which was seen in 15 instances, or merely persistent looking or staring at the observer. It should be noted, incidentally, that investigative behavior differed in form in different age-sex groups. Thus, in juveniles it often involved sitting and staring from close quarters for relatively long periods. At the other end of the scale, adult males showed usually a very brief, passing inspection persisting usually for no more than a few seconds and from a greater distance than was typical of juveniles. In the 91 instances of this behavior, adult males were investigators on 26 occasions, subadult males on 23 occasions, and adult females on 11 occasions. Juveniles performed it on 72 occasions and unidentified animals on 28 occasions. It should be noted that adult and subadult males are particularly conspicuous in the field, and most easily classifiable. The majority of the 28 unidentified instances are therefore likely to be juvenile or adult female animals. Once again, therefore, where the age-sex class of investigators is known, juveniles outnumber all other age-sex classes combined. This may again be merely a reflection of the structure of the population.

A mere summarizing of the frequencies and types of behavior observed in *Mandrillus leucophaeus* groups indicates a social organization that has some differences from that of the typical *Papio* structure and which is different also from the more specialized *P. hamadryas*. Adult males initiate group movement, as in *P. hamadryas*. Subadult and adult males perform herding or rounding-up behavior, again as in *P. hamadryas*. Subadult males act as sentinels, sometimes a feature of *P. cynocephalus* organization. Adult and subadult males perform most threat behavior towards the human observer, but only in habituating groups and show no evidence of playing a major defensive role in normal interactions with potential predators, as has been described for *P. cynocephalus* by DeVore (1963). Juvenile drills are the most frequent investigators and alarm callers.

CONCLUSIONS

This account of the differentiation of social roles within drill groups emphasises the functional coherence of the social group. Some of the selective pressures causal in molding such a system require further discussion however. The adaptive significance of the fragmentation of the population into small units which aggregate frequently into large groups in a predictable manner in terms of size and composition is not immediately obvious, but may be related to ecological factors. It is possible that an analysis of obtained data might throw light on this possibility. However, that these large groups are not merely ecologically determined aggregations as is the case with *Papio* at waterholes, sleeping groves or rocks, is indicated by the fact that distinct calls have been evolved facilitating the formation of these large groups.

The reduction of adult males in heterosexual groups also requires comment. It is not conceivable that a reduction of the order described could arise merely through differential maturation rates, nor does differential mortality seem a plausible alternative. It has often been considered that the one-male-group system evolved as an accommodation to the selective pressures of arid environments, reducing the numbers of biologically surplus males competing for scarce food resources. This system does not however eliminate food-competition unless the surplus males are excluded from the home ranges of the heterosexual groups, or die earlier, or are born less frequently than females. There is almost certainly no increased susceptibility of isolated males in the forest to predation and they may to some extent be protected. The main (and in this area probably only) predation on drills is by man. He locates groups by noise and shoots them in their sleeping trees - dogs often being used to keep them from descending to the ground. It is unlikely that the unvocal solitaries will be as vulnerable as the very noisy large groups which were actually shot at and reduced during the course of this study. An alternative suggestion must therefore be sought.

Mandrillus leucophaeus is strongly sexually dimorphic, more so than the savanna and arid country baboons. From the evidence quoted, it seems likely that this dimorphism does not represent primarily a defensive adaptation on the part of males. It seems more probable that it has arisen through either inter- or intrasexual selection. This would imply differential reproductive success by males. There is no evidence that the willingness of female baboons to mate is affected by the secondary sexual characteristics or visual stimuli offered by the males; there is indeed evidence to the contrary. Kummer (1968b) records adult female *P. hamadryas* in one-male groups occasionally "stealing" a copulation with subadult, but sexually capable, males when the adult male's back is turned. The sexual dimorphism is drills is therefore most probably a result of intrasexual selection; this hypothesis is compatible with the reduced numbers of males observed in groups and would imply the exclusion of some males from breeding. In the savanna where visibility is no problem, the mere presence of a larger male might act as a deterrent to copulation, but there can be no such sanction or control in the conditions of poor visibility common in the forest. The only means by which a male could effectively prevent clandestine copulation would be to exclude other males from the group. There would be selective advantage in being able to do this, whilst at the same time maintaining the coherence of the females in the unit. This latter seems to be effected primarily through herding behavior and the evolution of Type A mobilizing grunts.

All these adaptations bear clear signs of their origin in *Papio*-like behaviors, and it would be easy to derive either social system from the other by very few changes. It would seem that the primary factor responsible for the evolution of the *M. leucophaeus* social system has been the problem of visibility and communication in the tropical rain forest.

REFERENCES

Bourgoin, P. (1955). Animaux de Chasse d'Afrique Paris: La Toison d'Or.

DeVore, I. (1963). A comparison of the ecology and behavior of monkeys and apes. In Classification and Human Evolution (Ed. Washburn, S. L.). Chicago: Aldine.

DeVore, I., and Hall, K. R. L. (1965). Baboon ecology. In Primate Behavior: Field Studies of Monkeys and Apes (Ed. DeVore, I.). New York: Holt, Rinehart and Winston.

Hall, K. R. L. (1963). Variations in the ecology of the Chacma Baboon, *Papio ursinus*. Symp. zool. Soc. Lond. **10**: 1-28.

Hall, K. R. L. (1966). Behavior and ecology of the wild patas monkey, *Erythrocebus patas*, in Uganda. J. zool. Soc. Lond. **148**: 15-87.

Jeannin, A. (1936). Les Mammifères Sauvages du Cameroun. Paris: Paul Lechevalier.

Jolly, C. J. (1967). The evolution of the baboons. In The Baboon in Medical Research Vol. 2 (Ed. Vagtborg, H.) Austin, Texas: University of Texas Press.

Kummer, H. (1968a). Two variations in the social organization of baboons. In Primates: Studies in Adaptation and Variability. (Ed. Jay, P. C.) New York: Holt, Rinehart and Winston.

Kummer, H. (1968b). Social Organization of Hamadryas Baboons: A Field Study. Chicago: University of Chicago Press.

Kummer, H., and Kurt, F. (1963). Social units of a free-living population of hamadryas baboons. Folia primat. **1**: 4-19.

Kummer, H., and Kurt, F. (1965). A comparison of social behavior in captive and wild hamadryas baboons. In The Baboon in Medical Research (Ed. Vagtborg, H.). Austin, Texas: University of Texas Press.

Malbrant, R., and Maclatchy, A. (1949). Faune de l'Equateur Africain Français. Paris: Paul Lechevalier.

Rowell, T.E. (1966). Forest living baboons in Uganda. J. Zool. 149: 344-364.

Rowell, T.E. (1967). A quantitative comparison of the behavior of a wild and a caged baboon group. Anim. Behav. 15: 499-509.

Rowell, T.E. (1968). Variability in the social organization of primates. In Primate Ethology (Ed. Morris, D.). Chicago: Aldine.

Rosevear, D.R. (1953). Checklist and Atlas of Nigerian Mammals. Lagos: Government Printer.

Sanderson, I.T. (1940). The mammals of the north Cameroons forest area. Trans. zool. Soc. Lond. 24: 623-641.

Sanderson, I.T. (1957). The Monkey Kingdom. London: Hamish Hamilton.

Struhsaker, T.T. (1969). Correlates of ecology and social organization among African Cercopithecines. Folia primat. 11: 80-118.

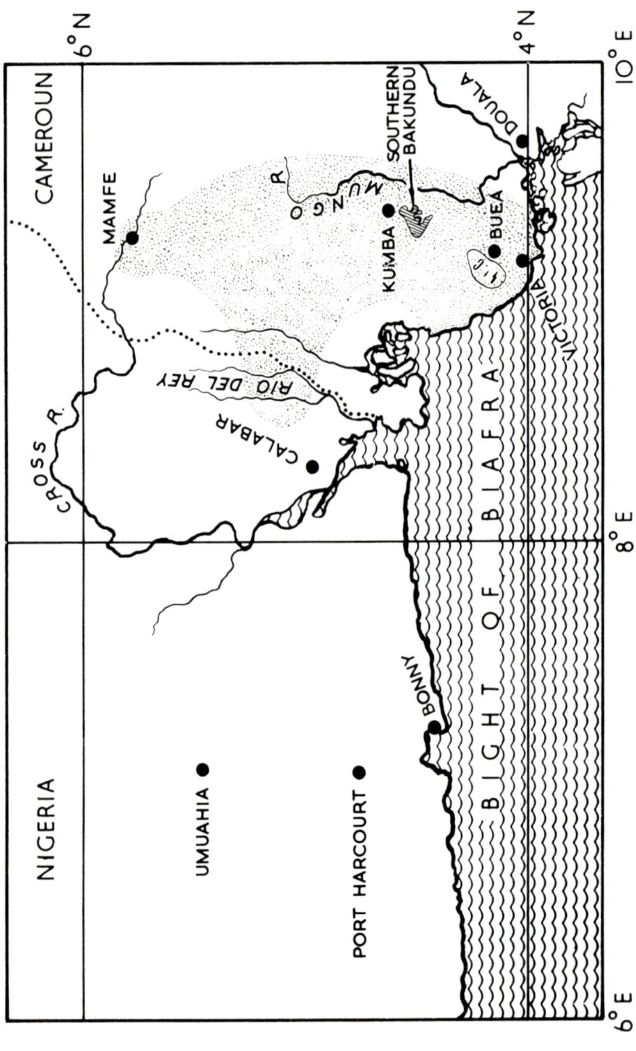

Fig. 1. Approximate distribution of Mandrillus leucophaeus in W. African tropical rain forest (after Rosevear, 1953). This figure also shows location of study site.

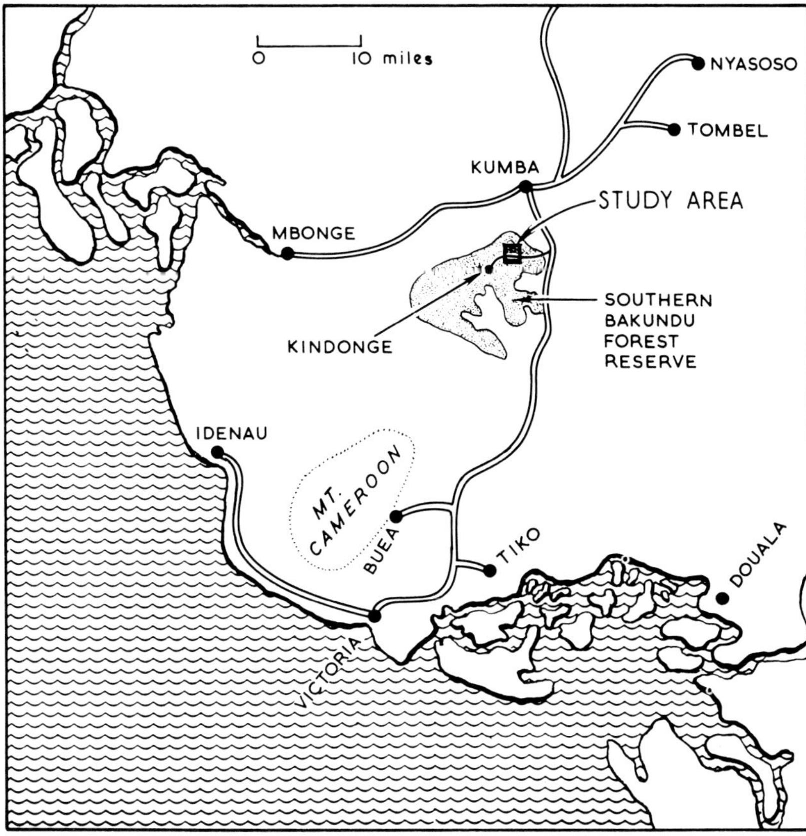

Fig. 2. West Cameroon, showing main towns, coast, Mount Cameroon and Southern Bakundu Forest Reserve and study area.

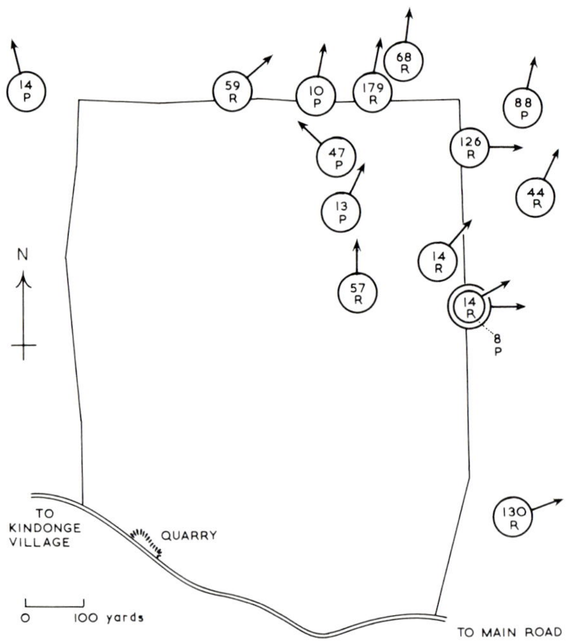

Fig. 3. Location of drill groups in area B; last recorded position (circle), final direction of movement (arrow), count and reliability (R=reliable, P=poor).

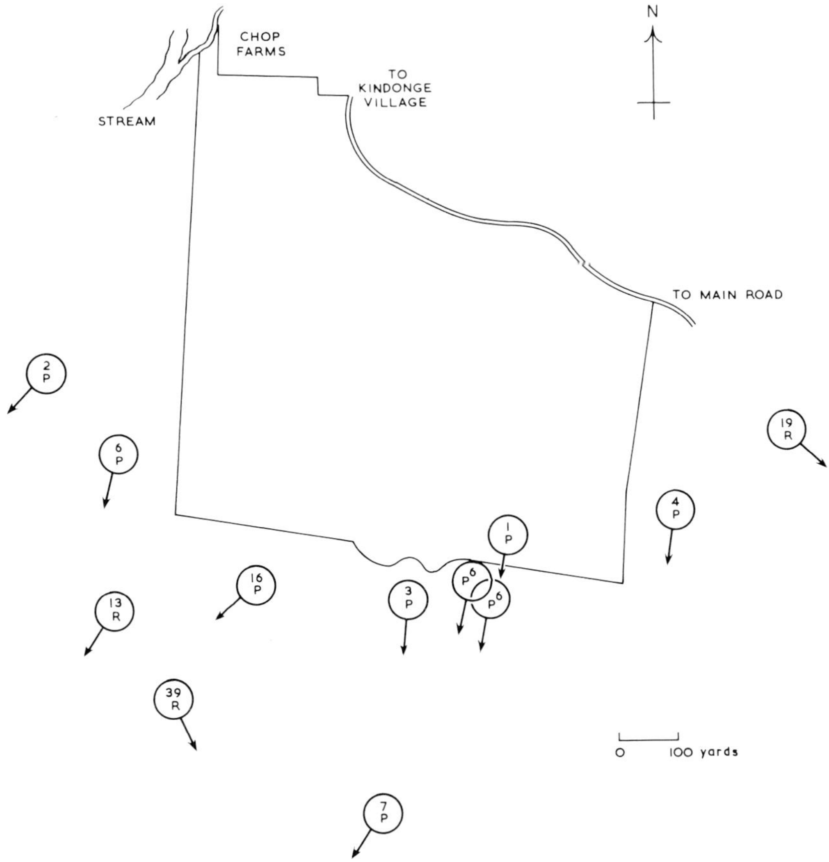

Fig. 4. Location of groups in area A; last recorded position and final direction of movement, count and reliability.

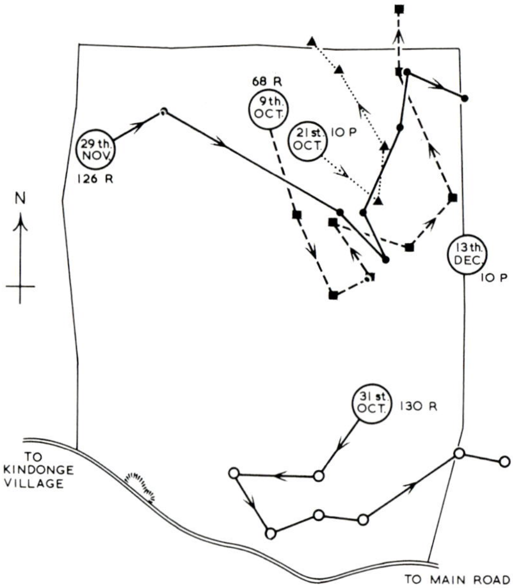

Fig. 5. Dates, sightings and travel paths of the group in which the identifiable subadult male was observed. Counts and reliability also indicated.

LEAVES AND LEAF-MONKEYS

The Social Organization of Foraging in Gray Langurs
Presbytis entellus thersites

Suzanne Ripley

INTRODUCTION

In the past few years there has been a major shift in emphasis in field studies of primate behavior. Earlier efforts were directed towards sampling the social behavior of various species in order to explore the degree of variability within the order as a whole. Recently, great interest has been shown in exploring the variability of social organization within a single species (Eisenberg, 1966; Jay, 1968); and investigating its determinants as a necessary preliminary to the understanding of adaptive radiations within the primate order. In particular, there is currently an effort to sort out phylogenetic and ecological influences on the forms of morphology, behavior and social structure.

The initial hypotheses relating ecological factors to social organization in primates were made on savanna baboons (DeVore, 1963; Hall, 1965). These hypotheses relate ground feeding in the presence of predators to large group size, multi-male troops, sexual dimorphism, male dominance hierarchies, and to the special dispersal patterns during troop movement. In gelada and hamadryas baboons the social pattern of one-male sub-units within a large loosely structured herd has been related to the dispersion and scarcity of sleeping sites and food sources (Crook, 1966; Kummer, 1968). In these species the functional complexes mentioned above appear to be such clearcut adaptations to marginal environments that it is difficult to contest the hypotheses advanced thus far. Nevertheless, Struhsaker (1969) has been able to challenge some of the hypotheses regarding the ecological determinants of group size and composition on the basis of his, and of Rowell's (1966) observations on arboreal and terrestrial species of forest-dwelling cercopithecines. As Struhsaker points out, however, comparable correlations are very much more difficult to draw for arboreal species in the forested habitats.

By far the most challenging formulation is that of Crook and Gartlan (1966) who have submitted the patterns of adaptive behavioral features among the species that have been investigated so far. They draw attention, however, to certain puzzling aspects particularly the questionable applicability of their proposed explanation for the small size of group and defence of its territory in rain forest frugivores to the leaf-eaters where comparable limits to the availability of food sources are not obvious.* Another puzzle is the presence of a leaf-eater, *Presbytis entellus*, among such accomplished omnivores as *Macaca mulatta*, *Papio cynocephalus* and *Cercopithecus aethiops* in the forest fringe-tree savanna class.

One impediment to a clearer understanding of the nature of the transitions between the adaptations found in forest, forest fringe and savanna habitats (grades III, IV and V in their schema) appears to be an undifferentiated use of "forest" and "leaf-eater"; they generalize as to the types of forest, to climatic seasonality, and to the modes of food collection or other forms of utilization in the habitat. The mode of food collection, as well as the type of food consumed and its distribution within the home range, loom large in the set of ecological variables. This is so because feeding involves so much time and may require complex supporting social behaviors. Feeding behavior in the gray langurs of Ceylon occupies some three to six hours daily, or a quarter to half of the length of the day. In howlers feeding occupies about a quarter of the day (Altmann, 1959) as it does in *Ateles geoffroyi* (Eisenberg and Kuehn, 1966). In forest-dwelling species, patterns of foraging behavior may well be more important as a formative agent for bio-social

---

* "The small size of social units in many forest frugivores is probably related to limiting conditions of food supply occasioned by the relatively stable conditions of tropical rain forest. A non-seasonal climate with a moderately constant availability of various fruits presumably allows increase in numbers to a ceiling imposed by periodic food shortages due to local food crop failures. ... The "territorial" behavior of forest groups may be interpreted as ensuring an adequate provisioning area for the individuals comprising them. It remains difficult, however, to apply this argument to all leaf-eaters."

(p.1201) Crook and Gartland *op. cit.*

organization than behaviors directed towards ensuring safe sleeping sites and predator avoidance. Foraging behavior and its social organization are not well-described for any herbivorous primate. As a result, the connections between this important variable and the social systems within which it occurs cannot as yet be clearly drawn.

This paper presents a preliminary account of the social context of foraging behavior in a small population of gray langurs in Ceylon. A complete analysis of the utilization of vegetation within the home range will follow in a later publication. In this present paper, therefore, I will sketch out the general patterns of tree-feeding and the social implications of seasonal variation in phenological phases, then compare this general pattern with that of ground-feeding and its social implications. In particular I want to suggest some of the ecological factors which may favor the following features of gray langur organization; large troop size, a relative lack of emphasis upon dominance behaviors in the multi-male troop and the strong development of territorial behavior which has proved so puzzling in leaf-eating langurs in comparison with its lack among ground-dwelling omnivores and its presence in arboreal fruit-eaters, like the gibbons.

GRAY LANGUR HABITAT

Gray langurs in Ceylon occupy the lowland mixed deciduous and evergreen forest which experiences a climate that can be characterized roughly as winter-wet, summer-dry. The distribution of langurs overlaps that of the dry zone macaque, *Macaca sinica sinica* and with one variety of the purple-faced langur, *Presbytis senex senex*, in the wetter areas of the dry zone and in gallery forest in the drier areas of the dry zone. Gray langurs inhabit alluvial and high stature forest (15 m or more); medium and low stature forests (5-15 m), low scrub with scattered emergents, and forest recently disturbed by slash-and-burn cultivation. Gray langurs apparently favor a mixture of physiognomic types such as a combination of closed and open forest, herbaceous cover (crops or grass), and open surfaces (water or rocky outcrops).

The primary study site at Polonnaruwa is located within an archaeological reserve of some two square miles. The cover is high stature closed forest, or low to medium stature discontinuous forest with scattered emergents and intervening herbaceous cover. There are two tanks with permanent water and several seasonal waterholes and watercourses. The shrub layer has been cleared from about one half of the total area leaving the lower canopy and the deciduous emergents. Despite its other drawbacks, the partially cleared area facilitates observation of the selective use of the taller woody vegetation which appears to be such a critical factor in gray langur distribution. All of the subject troops have some uncleared forest included in their home ranges so that comparisons of usage of cleared and uncleared forest can be made in each case.

The climate over the range of distribution of gray langurs in Ceylon can be characterized more precisely as fall-and-winter wet, spring humid or subhumid and summer drought with two major variations in the late winter - the presence or absence of a short dry period in February. The annual mean temperatures vary between 79° and 82°F (Mueller-Dombois, 1968). Polonnaruwa lies within the area of dry deciduous forest, a vegetation type of limited distribution in Ceylon (Mueller-Dombois, op cit). Locally in Polonnaruwa the wettest period of the year runs from mid-October through mid-January and includes the convectional rains of the Fall, which merge in the northeast winter monsoon. A short dry period follows in February and early March during which there is a great deal of plant activity, particularly flushing and flowering. Thereafter, the convectional rains in late March and April are followed by the southwest summer monsoon which blows as a dessicating wind over the dry zone throughout the months of May to September.

FEEDING PATTERNS

1. General. The food of the gray langurs consists primarily of termite clay (see Table I). These monkeys are eclectic feeders and never spend the entire feeding periods of a single day on one food source. Typically there are from one to four items which can be considered staple foods at any one point in the seasonal cycle,

and about three to eight other items that are taken as complements in lesser quantity. Often, for several days in a row, the troop may show a similar pattern of feeding insofar as they tend to revisit the same trees and areas that are active at that time. As one element of the staple diet becomes exhausted or passes out of season it is replaced by other food sources and the diet gradually assumes an entirely new aspect.

Feeding usually occupies the period from 0600-0930 and from 1500-1800. It also occurs during the middle of the day interspersed with well-defined periods of rest and mixed social behavior. Intensive feeding usually occupies about three hours per day or about half the total feeding time; however, this may vary seasonally. The most important feeding peaks are usually in the early morning at about 0600-0730 and in the late evening between 1530-1800; at these times the langurs concentrate upon their primary staple foods for the day. Apart from these peaks, the troop moves gradually through its daily feeding sequence taking complemental items *en route*.

During intensive feeding the troop is often concentrated at a particular food source. Such a concentration may be considered indicative of a preferred source, usually a tree that is undergoing a phenological change at that time. There may be several of these preferred food sources in season at once, but at no one time is there a great abundance of these preferred sources. Even a tree species that is abundantly distributed within the home range may have a long overall duration for any one phenological phase. Thus, on any one day, only a very few individual specimens of a particular species of tree are just right for harvesting and so are particularly attractive to the whole group. In contrast to these intensive periods of feeding, the casual midday feeding is likely to be done in small subgroups of four to six dispersed over a variety of sites. Here we will examine only feeding in the intensive periods when the troop is concentrated at a single source. Particularly, I will consider the behavior of the troop in individual trees.

For our purpose a tree may be thought of as divisible into some or all of the following component parts: (1) the lower trunk, the bole or unbranched major support, (2) the upper or branched trunk, (3) the primary ramages or thickest branches leading off the trunk, (4) the secondary ramages or further subdivisions usually connecting the primaries to the leaf-bearing terminals, (5) the minor ramages or those supporting the leaf-bearing twigs, (6) a terminal, or tertiary branches and foliage-bearing twigs taken together, and sometimes growing as suckers along the primary branches or trunk, (7) the "core" or the generally leafless region of the primary branches and the area of their attachment to the trunk and, (8) the "foliage zone" which includes any region of the tree where leaves principally grow.

The behavioral pattern seen during an intensive feeding period has the following structure. After an initial progression from a resting tree or area, the troop converges on the selected feeding tree. The first animals to arrive climb up directly, without hesitation, to the terminals and begin to feed. The other animals soon follow after and gradually fill up the available sites in the foliage zone. By the time three or four langurs are in the tree, the later arrivals pause at the crotch of the first primary branch before moving up and out. They may stop again at several points along the secondary branches before moving out to a terminal. These pauses apparently allow time for the selection of an unoccupied feeding site.

Once a langur arrives at the feeding site it takes up a sitting position which is presumably selected for reasons of comfort. Very frequently it rests on its callosities on the firmest, thickest portion of the terminal and grasps footholds further out on the same tertiary branch or side twig. There is some tendency to grasp footholds at a lower level than that of callosities during feeding; this appears to be the most secure sitting position for the springy ends of branches. In this preferred position the langur is generally facing away from the central core of the tree, or may be oriented laterally to it. The langur usually keeps a minimum of four to six feet between itself and its nearest neighbors when feeding in those trees in which the distance between adjacent terminals does not in itself create a wider separation. Therefore the number of langurs that the tree will accommodate at one time is limited only by the availability in the terminals of suitably spaced sitting places where the food is located. The only exception to the one-langur-one-terminal rule is in the case of consort pairs and mothers with young infants.

Most of the foods that attract the langurs are located within two to three feet of the extreme ends of the terminals (*Ficus racemosa*, *Ficus bengalensis* and *Lepisanthes trichocarpa* being exceptions). A langur places itself in a suitable terminal and eats systematically in an arc by means of a series of changes in body position and by bending the terminal twigs towards itself. Langurs eat either directly from the attached branch, plucking off the food with the fingers or with the mouth, or by breaking off a terminal twig and eating from it. One such continuous feeding sequence is referred to as a "bout".

A langur tends to move from one minimal ramage to another within the chosen terminal, in a series of minor moves. Such a spate of feeding bouts within any one terminal are referred to as an "interval". A single langur seldom if ever exhausts all the food within reach at any one feeding site in the process of harvesting the produce of the terminal as a whole.

After such an "interval" of some five to twenty minutes or so, the langur makes a major move to an entirely different terminal and repeats the same procedure. It may simply move to a neighboring terminal or it may move from one side of the tree to the other; this behavior pattern appears to depend on the distribution of other langurs within the tree at that moment. In the course of such a major move, the langur must pick its way carefully with respect to other langurs. If it approaches to within four to six feet of a fellow troop member, an encounter ensues and the moving langur may either be threatened, slapped or displaced, though it is unlikely to be chased or attacked.

There are two methods of changing terminals. In deciduous trees in leaf-fall, where visibility is good and terminals are placed near enough to hand, the langur may be able to chart a course to a new terminal through the foliage zone via the tertiary branches without returning to the central core. However, the commonest method of changing terminals, especially in the larger trees, appears to be via a secondary branch fork, if not the core itself, followed by an outward move to a new terminal. Either method may involve passing near other langurs and risking an encounter. However, since not more than one langur at a time is usually on the move, except during the initial phase of filling up the available feeding spaces, the langur that is moving has the controlling advantage; it can avoid contacting other langurs either by keeping away from them or by passing to their rear. A series of such feeding intervals separated by major moves between terminals constitutes a "feeding period". Two examples of this movement pattern may be quoted.

A time sample of flower and fruit feeding of an adult female on Holopteles integrifolia is provided (March 23, 1969) to indicate the amount of time actually spent within one terminal and the distance moved between terminals. The animal was feeding at the rate of one pluck every 3 seconds, or 20 per minute.

Time	Interval	Bout	Bout Duration (mins. & secs.)	Interval Duration (mins. & secs.)	Estimate Distance Moved (feet)	Nearest Langur (feet)	Total Langurs in Tree
16.55	1	1	0.58	.58	35	*	15
	2	2	4.35	4.35	35	15	15
	3	3	3.15	3.15	40	15	15
	4	4	.15	.15	20	20	15
	5	5	5.20	5.20	15	10	15
		6	5.00		5	10	15
	6	7	3.17	8.17	35	30	15
17.20		8	1.08		2	30	15
		9	7.20		1	20	13
		10	5.44		1	20	13
		11	0.53		3	20	13
		12	3.52		3	20	13
	7	13	2.15	20.54	12	20	13
		14	1.53		5	20	12
17.44	8	15	0.42	2.35	leaves tree		

* No information.

The timing of an adult female feeding on Manilkara hexandra flowers is given below (February 2, 1969, 09.20). It should be noted that this female ended up within 10 feet of her starting position after over 40 minutes of feeding on flowers. Her nearest neighbor was an adult male who fed continuously in the same spot for 50 minutes and then remained there, resting.

Time	Interval	Bout	Bout Duration mins. & secs.)	Interval Duration (mins. & secs.)	Distance (feet)	Nearest Langur (feet)
09.20	1	1	8.30	8.30	5	35
		2	3.07		turn	
	2	3	1.53	4.58	6	35
	3	4	5.40	5.40	2	20
		5	.45		2	
	4	6	3.42	10.07		
	5	7	1.40	1.40	2	
		8	2.10			
		9	5.35			
	6	10	3.40	11.25	rest	

After the end of a feeding period the langur usually rests. There are two varieties of rest. One is simply a short breather taken between feeding intervals or even between bouts during which the langur remains where it is, or perhaps it may make a 180° turn to face the center of the tree, thus reversing the relative heights of feet and callosities; during this phase the langur often leans its body forwards, resting on its outstretched legs. The second type of rest is longer and usually involves movement towards the center of the tree or to the horizontal primary branches in the core. This frees a potential feeding site for another langur.

The contrast between resting and feeding bouts in respect to sitting position, spatial location and social orientation is quite striking. This undoubtedly is a result of the interplay of the following factors: location of food, stability of the substrate, body weight, and hand mobility. Since langurs feed with the hands quite regularly (apparently much more frequently than do howlers: Carpenter, 1934), they need to balance themselves on an unstable substrate using only the feet and callosities in order to keep facing the food source. The larger the langur, the fewer positions are feasible at the feeding locus. On the other hand, when resting, the langur is concerned only with finding an optimal position (stable, shaded or sunny and with proper resting points including support for the back); the hands are free either for supporting the body in the foliage zone or for grooming in the central core.

The social implications of this contrast between resting and feeding bouts are clear. The standard means of expressing either disengagement from a social encounter or unwillingness to enter into such is to turn the back. In the normal sitting position for intensive feeding on flowers, fruits, seeds or flush at the ends of the terminals, the langur perforce has its back already turned to any other langur on the same level within the tree. In fact, most of the other langurs are <u>not</u> on the same level, but are either above or below; in one or another way the langur both expresses and ensures social disengagement during feeding. In effect the substrate itself fosters such disengagement by influencing neighboring langurs to sit parallel to, and spaced out from, one another during feeding bouts. On the other hand, during short rests when the langur turns and faces the center of the tree, for reasons of comfort or support, it exposes itself to potential interactions by being clearly unengaged and by facing on-comers. It would appear that resting is actually or potentially sociable whereas feeding is not.

In addition to the influence of the substrate in fostering social separation while feeding, there are a number of body signals expressing location and sexual and individual identity that assist in the process of spacing during long feeding intervals. These are effective, even in trees with particularly dense foliage zones. The most obvious signal is the pendant tail which is often visible and identifiable when the owner may, itself, be invisible, in thick foliage at the end of a terminal with back turned. The shape of the tail tuft is one of the most distinctive of individual characters. The location of the adult males of the troop can be determined from below by the vivid design of the rump which is placed nearer to the core than are the feet and thus is not hidden in the foliage zone. When the male sits on one callosity on a thin tertiary branch, the pink, gray and black pattern of its rump is easily distinguishable at a distance from that of females or subadult males.

At the end of a feeding period the langurs are likely to leave the tree as they came, one by one, by the various exit routes -- into a neighboring tree or straight to the ground. Traffic jams at egress points seldom occur except when the animals are being threatened or frightened (particularly by humans), when they all attempt to leave the tree at one time.

The only break in the orderly feeding system just outlined, occurs when some auditory stimulus triggers off a display-jumping sequence by a number of the feeding langurs. During these displays all feeding stops and the entire tree reverberates with the crash of leaping, bounding, whooping langurs, the sound as the larger animals ricochet off them. During a display-jump sequence all parts of the tree are used with many langurs moving rapidly in all directions. The langurs not participating in the display usually stop feeding and cling tightly on to the terminal branches with all four extremities to avoid being shaken off. During such a display the feeding site is transformed into an arena and all patterns of behavior appropriate to a feeding context are altered.

In general, then, each major move of a langur between terminals for successive feeding intervals, alters the physical proximities of the individuals within the social field of a tree where intensive conjoint feeding is in progress. In effect, these changes of position tend to disrupt the others in their feeding. In view of the relatively unnutritious character of the langur diet, the poor absorption of the ingested materials and the need to spend a great deal of time feeding, it is advantageous that such disruption should be controlled as much as possible. Such control is effectively maintained by breaking up the social field into a number of exclusive sub-areas or terminals within which each langur remains feeding for a long period of time. The longer that each remains within one sub-area, the fewer major moves are necessary, and the less the social field is disrupted.

Two factors contribute to the length of time an individual

langur remains in one terminal. Since langurs will feed on flowers, fruits, seeds and leaves in more than one stage of ripeness or maturity, each single locus within a terminal provides a more abundant food source than if the monkey were selective about the stage of ripeness. Furthermore, since a langur has to masticate each mouthful of food immediately upon collecting it, speed of consumption is slow. Lacking the cheek pouches characteristic of the other subfamily of Old World monkeys, a langur cannot feed "smørgasbord style", collecting food to swallow later but must eat sitting down in one place - "banquet style".

The feeding system of the langur is so simple and so effective in permitting large numbers of monkeys to feed in rather close proximity for long periods of time with minimal conflict or competition, that the few aggressive encounters that do occur are startlingly obvious. In fact most encounters do not occur in a feeding context but during the more sociable periods of rest, grooming, progressions and social gatherings: they tend to occur in the more open zones of the tree, namely in the region of the central core and horizontal primary branches, the trunk, the base, the exposed roots and the adjacent ground. The arrangements are altered in subtle ways by the nature of the woody vegetation that serves as substrate. These variations will be examined in a future paper which will also attempt a classification of the woody vegetation of the home range from the standpoint of langur usage.

2. Seasonal and Perennial Tree Sources. The general pattern of the feeding behavior of langurs in the trees has been described; it appears to be roughly the same for any type of food whether it be flush, flower, fruit, seed or leaf (but not bark or sap). This general pattern however has two significant variants which are essentially attributable to two different kinds of food sources. These have been characterized as "fruit" and "leaf" diets in the literature on primates, but this generalization requires amplification. In the first variation, flowers, leaf-flush and seeds are very like fruit in that they are distinct, ephemeral phenological changes or phases; they are distributed on the tree, of course, in a similar manner. All are in contrast to features of leaves which are often perennial, more abundant, and more evenly distributed on the tree. In an

attempt to sort out the social effects of foraging the first group of food sources will be called "seasonal" sources, and the second group "perennial" sources.

The food list shows that the sobriquet "leaf-eater" is perhaps not quite an appropriate term for the diet of the gray langur. The large fruit-flower-seed component in the langur diet and their observed preference for these sources suggests that part of the gray langur's foraging behavior must be considered as typically "fruit-eating" or dependant on seasonal sources, rather than "leaf-eating or dependant on perennial sources. Some of the social behaviors which would seem to support the first pattern of foraging on seasonal sources will be briefly examined.

The utilization of seasonally shifting sources of food implies a large home range and regular changes in the pattern of its usage based upon an awareness of the phenological changes as they occur. Such a pattern requires for its effectiveness a traditional knowledge of the changes occurring in the home range throughout the year. The large troop size in *P. entellus* means that a pool of traditional knowledge available to any one group is increased as is the group's ability to conjointly exploit the home range. This pattern therefore implies frequent, coordinated movement by the group within the home range.

These features are reminiscent of the otherwise dissimilar pattern of ground feeding among savanna-living primates where synchronized phenological changes are also a major bio-climatic factor. It is certainly more evocative of the seasonal savanna than of the typical rain forest habitats where phenological changes may not be synchronized even among the individuals of one plant species, let alone synchronized with a particular, seasonally recurring, climatic feature, such as drought (Koelmeyer, 1959, 1960; McClure, 1964, 1966) among a whole assemblage of species. A rain forest leaf-eater, or even a fruit-eater, undergoes less distinct peaks of seasonal abundance or scarcity in its plant resources, or none at all, and therefore may not undergo the implied annual variation in the value of food resources. (It should be emphasized, however, that there is very little data as yet on phenological changes in rain forest habitats.)

Also included in the repertoire of social behaviors which can support this seasonal foraging pattern is a particularly aggressive form of territorial defence (Ripley, 1967). Due, in part, to their characteristic daily progressions over almost all parts of their moderately sized home ranges, the langur troops are quite familiar with their boundaries and with the neighboring troops beyond. While there may be no direct need, in terms of current food shortages, to expand the home range, there may be a continual exploration beyone its boundaries for preferred seasonally abundant foods; this behavior will thus bring neighboring troops into contact, and may result in fights. This may be one explanation for the apparent lack of change in "boundaries" following these fights.

This was quite clearly demonstrated during the spring of 1969 by the fights occurring between one troop and its neighbor; usually two fights or more occurred during each day. These occurred in or near certain preferred trees. In one instance a particular banyan tree (*Ficus bengalensis*) in fruit grew on one side of the home range boundary and a large *Holoptelea integrifolia* in flush, flower and fruit lay on the other, as well as a grove of *Walsura piscidia* in flush and flower at another part of their common boundary. These trees were the objects of periodic raids by the neighboring troop. Many other *Walsura piscidia* trees grew within the home range of the raiding troop, but they preferred to raid the trees "belonging" to the next troop.

This preliminary account indicates the way in which part of the foraging pattern of gray langurs in their use of the forested component of their home range corresponds with that of fruit-eaters and omnivores. Some behavioral concomitants, which may support this hypothesis, have been put forward.

Gray langurs also eat a large amount of leaves and undoubtedly these are often the staples that sustain them in the long run. Indeed, some aspects of their behavior are more typical of "leaf-eaters" than fruit-eaters". These features are better known and can be listed briefly. In this regard, the smaller subgroups into which the troop breaks up for a large part of the day's activities, the long periods they spend sitting at one spot while eating, as well as the general infrequency of intra-troop dominance encounters should be noted. Additional features, perhaps, are the lack of alliances among adult males, the tenuous relationship between adult males and infants and the lack of a protective role (in the sense in which it has been described for baboons) exercised by the adult langur male toward the troop.

GROUND FEEDING

We can now turn to the third major complex in the gray langur foraging pattern: ground feeding. All gray langurs spend some part of every day foraging on the ground where their social situation differs radically from that in the trees. On the ground, individuals are all on the same level. Differences in body size are clearly apparent. The visibility is relatively unobstructed, especially in the open areas of discontinuous forest and herbaceous cover where ground feeding tends to occur. There are no interaction screens, such as foliage, which can reduce the effect that each individual immediately exerts upon all the others in the social field. Although langurs on the ground maintain the same extent of inter-individual spacing as in the trees (circa four to six feet), it _appears_ to be much less effective in terms of social distance. In short, the ground as a substrate appears to transform the parameters of social interaction occurring there quite remarkably. The control of encounters generally, and of dominance interactions in particular, is not as easy to maintain; it must be done, as it were, in spite of the substrate.

There are two forms of foraging pattern on the ground. The first comes into play when the langurs are feeding collectively on herbs. In Polonnaruwa, the preferred ground food is a small thorny herb, _Mimosa pudica_, on the leaves of which the langurs feed nearly every day. _M. pudica_ grows in clumps among rocks and in depressions. When feeding on these plants the langurs are seen in groups of as many as 16 animals or more, all within close proximity of one another. An example of this movement pattern in the case of an adult male is given below:

Time	Bout	Bout Duration (mins. & secs.)	Distance Moved (feet)
16.30	1	.40	1
	2	.30	40
	3	1.07	*
	4	.35	1
	5	1.21	5
	6	.30	10
	7	.25	*
	8	.15	1
	9	.07	7
	10	.30	progression

* No information.

The second form of foraging pattern comes into play when langurs conjointly feed on fruits and seeds, which have dropped from the trees. The area where the food is to be found is circumscribed by the area of windfall from the crown. In either form a langur may feed within an arc circumscribed by its sitting locus, but when the supply within its reach has been exhausted it moves a few feet in one direction or another. This inevitably brings it into close contact with a feeding neighbor. Unlike tree feeding, there are no natural divisions of the feeding space.

Similarly ground feeding differs from tree feeding in that there is a reduced abundance of the food at a single feeding locus by comparison with its plenitude in trees. A langur on the ground more quickly exhausts the food supply in the immediate environment and, thus, has to move sooner than it would need to when foraging in trees.

Whenever a langur moves in this way, it diminishes the inter-individual distance between it and its neighbors. As a result of this encroachment the neighbors move. The wave of displacement eventually involves the whole feeding group. The visual impression of langurs feeding on the ground is that at least one member of the group is always in the process of shifting its position.

The movement is not organized in the clumped pattern described for harvesting seasonal food sources in trees. In the two ground-feeding patterns each minor move has roughly the same social effect as has a major move between terminals in the tree-feeding pattern. Effectively there are, therefore, many more socially dislocative moves per equivalent feeding period on the ground compared with the tree-feeding pattern. In these circumstances the frequency of potential encounters appears to be extremely high, particularly in respect of displacements. In fact, it is often difficult to tell if the movements one sees on the ground are due to exhaustion of a good source at one locus or to a displacement of one animal by another.

During a period of ground feeding on herbs, the entire group maintains a direction of movement such that all minor moves contribute to a single line of travel and serve over a short period of time to displace the entire feeding group to a new area without ever breaking up the social pattern just described. The avoidance of encounters can be achieved here in the same fashion as in the trees. The langurs all face roughly in the same direction - the line of travel - so immediate neighbors usually maintain a dorso-ventral orientation to one another expressing inter-individual disengagement from encounters. If, in fact, two langurs come face-to-face during ground feeding, a short lunge, a slap, a stare-threat are likely to occur. Each member of the group is continually scanning its neighbors, and the social tonus of the entire co-feeding group is heightened.

While in such a condition of enhanced social tonus and inter-individual awareness, the safety of the entire group is strengthened. At Polonnaruwa even in the most exposed areas of the cleared portion of the home range, where langurs are often chased by men and dogs, it is impossible to catch an adult gray langur; however, infants are sometimes caught. It is interesting to note, in this connection, that deaths among adult langurs, thus far observed, apparently from attacks by dogs, principally involve *Presbytis senex*. This species only rarely crosses the ground between trees and then always at a run and usually alone. The chances of a dog catching *P. senex* during one of its rare visits to the ground would appear to be practically nil compared to the chances provided by the gray langurs which frequently feed on the ground. Yet, the fact of the matter is that gray langurs are seldom caught in this way, despite their greater exposure. My impression is that this has to do with a heightened awareness of outside danger by the social group in P. entellus compared with that of the individual P. senex. What occurs is somewhat similar to the situation in deer; at all times one or other of the group is on the qui vive; their scanning activity is, as it were, pooled. As a simple extension of this effect the following hypothesis can be developed: up to a limit the larger the group size the greater is the protection effect.

There is some further evidence which appears to support this association of large group size with a ground-feeding habit. The largest troops of gray langurs seen in Ceylon consist of 40 - 60 animals and contain as many as 8 - 10 adult males. Seven troops of this size have been found in one area of Yala National Park. They inhabit the forest surrounding one of the largest fresh water reservoirs or tanks. On one occasion they were observed feeding on water plants on the exposed tank-bed continuously for $5\frac{1}{2}$ hours. This period ran from 1000 hours, when observation started, to 1530 throughout the hottest time of the day. On another occasion troops were observed feeding for at least $2\frac{1}{2}$ hours and probably longer. During this prolonged feeding interval on the ground, they were at least half a mile away from the trees. At the particular time of year when these observations were made, there was ample food in the forest; the langurs simply preferred the water plants. On one occasion a troop of more than 40 animals, alarmed by something, began to run back to the forest edge. The distance was so great however, that their anxiety subsided before they reached the safety of the trees, so they stopped running and resumed feeding. These troops also engaged in inter-troop fights on the open tank bed when they encountered other troops feeding near the water's edge.

Unlike the conditions at Polonnaruwa there langur troops live in a wild-life sanctuary where there is a full complement of animals including potential predators such as leopards, smaller cats, jackals and crocodiles. The langurs confine their ground feeding to the midday period when the hunting carnivores are least active. At this time, moreover, they do not have the benefit of partial protection from the herds of deer that join the langurs on the ground for feeding later in the afternoon.

In this sanctuary, the combination of extremely large group sizes (40 - 60) with a closely packed social space during the long periods of ground-feeding lends support to the suggestion that in *P. entellus*, at least, ground feeding the large troop size can be correlated. As we have seen, an entirely different pattern of inter-individual relationship obtains during ground-feeding as compared with tree-feeding. This is particularly true when one considers the difficulty of maintaining social tranquillity during the long periods of slow feeding when individuals in the troop are readily accessible, and when individual movements during feeding are greatly speeded up. The extent to which other features of social organization are altered during this prolonged ground-feeding is, at present, not known. Theoretically the most significant feature might be an alteration in the role of the adult males towards the troop and towards each other, including the development of a dominance order, or inter-male alliances, in the direction of the baboon/macaque pattern. It is not known at present what other social arrangements might contribute to the safety of the troops in such exposed situations. One might anticipate that an increased tolerance for compacted group feeding, which also controls inter-male dominance interactions, is sufficient. Further field observations should serve to clarify this problem.

CONCLUSIONS

Gray langurs in Ceylon have evolved a complex mosaic of at least three distinct foraging patterns. A ground-feeding pattern is superimposed on a tree-feeding pattern with two aspects: feeding on seasonal plant sources on the one hand and on perennial sources on the other. In separating these patterns analytically, both the disposition of the food source in the substrate and the seasonal periodicity of its availability must be considered. The three patterns and their associated social behaviors, have enabled this particular species of primate to colonize much of the Indian subcontinent as well as much of Ceylon. Successful occupational zones are, particularly, the drier areas of the seasonal semi-deciduous forest. Like the terrestrial vervets, the baboons of Africa, and the macaques of Asia, gray langurs have not undergone extensive speciation, but they do show remarkable variability in patterns of social organization like the baboons.

In brief recapitulation, then, the adaptation of langurs to foraging on mature leaves in the absence of cheek pouches has imposed certain basic patterns of feeding behavior. During each minute of feeding time about 2-3 leaves are ingested and each mouthful must be completely masticated before the next is ingested. Feeding is carried out while individual animals are widely dispersed in a social field in which preferred feeding loci are not distinguished and no special body orientation with regard to the food source is advantageous. There is little reason for competition for particular places and little advantage in one tree over another. There are no peaks of abundance or concentrated sources. Small groups of 5 - 6 animals can easily remain, without conflict, in a single tree and maintain their inter-individual distances. Hidden by the foliage and sitting at different levels they may, in fact, require means of maintaining social contact, rather than mechanisms for ensuring dispersal.

The seasonal harvesting of flush, flowers, fruits and seeds concentrates the members of the social group in the particular trees which are phenologically active. The feeding rate is higher than in perennial feeding, about 24 plucks per minute as compared with the 2-3 of the latter. While feeding the animals are oriented in a particular direction which enables them to maintain a relatively secure sitting place at the outer edge of the foliage zone in individual branch terminals where this type of food is located. Feeding areas and postures usually differ from the areas and postures chosen for rest. The higher potential interaction rate that could result from the concentration of troop members into a small social field during feeding is controlled by a natural subdivision of the feeding zone into small sub-fields - the branch terminals - which provide the major sitting supports for the individual langurs. Minor movement within these branch terminals does not necessarily lead to encounters with other troop members. When making a major move to a different terminal, a langur may avoid interactions by (1) charting a course through the tertiary branches when it can see the location of the other langurs and avoid them or (2) by using the foliage as an interaction screen or, (3) by moving back into the central core before moving out to a new terminal.

Movement between trees in a large home range is usually accomplished by the progression of the entire troop on the ground, in a linear fashion, to trees "in season". Knowledge of access routes and familiarity with the wide variety of foods, some of which may appear only at long intervals, is necessary. A moderately large group size (circa 25 individuals) enlarges the storage capacity of such knowledge as is available to the troop as a whole. Large troops also contribute to the effective maintenance of priority rights of a langur troop to certain trees which lie at the border of its home range. The small number of trees in season at any one time, combined with a moderate home range area (15-30 hectares) ensure that such inter-troop encounters will occur regularly. All troop members participate in such encounters. Large group size also facilitates group travel on the ground between core areas of the home range.

The third pattern, that of feeding on the ground, brings troop members into even closer proximity and requires even more effective spacing machanisms to control the interaction rate within the troop, particularly those encounters which may lead to conflict and to displacements. Individual troop members are equally inter-accessible; they are observed to move about to new feeding loci at a higher rate within this social field that has no natural subdivisions imposed by the substrate. The frequency of interactions, is, apparently, still controlled by the device of strategic bodily orientation with respect to nearest neighbors; in this way physical proximity may be maintained without unnecessarily precipitating face-to-face encounters. Such a device is particularly necessary insofar as an increase in group size facilitates its capacity for self-protection, without at the same time exposing it to increased intra-group aggression.

The two main types of ground feeding observed in *P. entellus* are: (1) on seeds and dropped fruits within a restricted social field defined by the area of windfall beneath the tree-crown; (2) herbs in which a single direction of travel is maintained by small individual moves of the members of the feeding groups.

Finally it may be noted that the gray langur, characterized by a flexibility in foraging patterns, is particularly well adapted to a wide range of habitats, expecially those which experience a period of extreme dryness and an annual round of phenological changes. In the semi-deciduous dry tropical forests of Ceylon langurs can harvest the deciduous emergents, the evergreen canopy and the ground herbs through the medium of these three patterns of feeding behavior. This adaptive pattern appears to depend on a single critical physiological feature, namely the ability to eschew free water for drinking for periods of several weeks duration at least. Special ecological and physiological studies in the drier areas of gray langur distribution would enable us to determine the extent to which the metabolic water requirement is made up from dewfall, condensation, and the bound and free water content of the diet; they would also help us to determine the nature of special physiological adaptations, if any. The ability to eschew free water for drinking, in combination with long continuous, sedentary, peaceful feeding in large groups, which allows troops of gray langurs to exploit ecological zones that remain virtually closed to the otherwise ubiquitous macaque.

ACKNOWLEDGEMENTS

The fieldwork reported in this paper was supported by Smithsonian Foreign Currency Grant No. SFC-9-7004 and National Institutes of Health Grant No. 1 RO1 MH15673-01 in 1968-70, and National Institutes of Health research grant supplement M-5542 in 1962-3.

The assistance rendered by Mrs. Annette Hladik of our project, by Drs. Dieter Mueller-Dombois and F. Ray Fosberg of the Smithsonian Ecology Project and Flora of Ceylon Project members is gratefully acknowledged.

REFERENCES

Altmann, S.A. (1959). Field Observations on a Howling Monkey Society. J. of Mammal. 40(3): 317-330.

Carpenter, C.R. (1934). A field study of the behavior and social relations of Howling Monkey (Alouatta palliata) Comp. Psych. Monag. 10:(2). Reprinted in Naturalistic Behavior of Non Human Primates. The Pennsylvania State University Press.

Crook, J.H. (1966). Gelada baboon herd structure and movement - A comparative report. Symp. zool. Soc. Lond. 18: 237-248.

Crook, J.H. and J.S. Gartlan (1966). Evolution of Primate Societies. Nature Lond. 210: 1200-1203.

DeVore, B.I. (1963). A comparison of the ecology and behavior of monkeys and apes. In Classification and Human Evolution Chicago: Aldine.

Eisenberg, J.F. (1966). The social organizations of mammals. Handbuch der Zoologie 10(7): 1-92.

Eisenberg, J.F. and R.E. Kuehn. (1966). The behavior of Ateles geoffroyi and related species. Smithsonian Misc. Coll. 151(8); 1-63.

Hall, K.R.L. (1965). Social Organizations of Old-World monkeys and apes. Symp. zool. Soc. Lond. 14: 265-289.

Jay, P.C. (1969). Primates: Studies in Adaptation and Variability. New York: Holt, Rinehart and Winston.

Koelmeyer, K.O. (1959). The periodicity of leaf change and flowering in the principal forest communities of Ceylon. The Ceylon Forester 4(2): 157-189.

Koelmeyer, K.O. (1960). The periodicity of leaf change and flowering in the principal forest communities of Ceylon. The Ceylon Forester 4(4): 308-364.

Kummer, H. (1968). Social Organization of hamadryas baboons: a field study. Chicago: University Press.

McClure, H.E. (1964). Some observation of Primates in Climax Dipterocarp Forest near Kuala Lumpur, Malaya. Primates 5: (3-4): 39-58.

McClure, H.E. (1966). Flowering, fruiting and animals in the canopy of a tropical rain forest. The Malayan Forester 29(3): 182-203.

Mueller-Dombois, D. (MS). Ecogeographic analysis of a new climate map of Ceylon with particular reference to vegetation. Mimeo Report No. 8, Smithsonian Ecology Project, Ceylon.

Ripley, S. (1967). Intertroop encounters among Ceylon gray langurs. In Social Communication among Primates (Ed. S Altmann) Chicago: University Press.

Rowell, T.E. (1966). Forest-living baboons in Uganda. J. Zool. Lond. 149: 344-364.

Struhsaker, T.T. (1969). Correlates of Ecology and Social Organization among African cercopithecines. Folia Primat. 11: 80-118.

APPENDIX

LIST OF WOODY PLANT FOOD SPECIES AT POLONNARUWA

EMERGENTS	L	v	*	O	o	s
1. *Adina cordifolia* (Roxb.) Brandis	L	v				
2. *Bridelia retusa* (L.) Spreng.		v	*	O		
3. *Chloroxylon swietenia* D.C.	L	v	*	?		s
4. *Elaeodendron glaucum* (Rottb.) Pers.	L	v	*	O	o̲	s
5. *Ficus bengalensis* L.		v		O		
6. *Ficus racemosa* L.		v		O		
7. *Ficus religiosa* L.		v		O		
8. *Holoptelea integrifolia* (Roxb.) Planch.	L	v	*	O	o	
9. *Madhuca longifolia* (L.) Macbr.		v	?	O	o	
10. *Manilkara hexandra* (Roxb.) Dubard		v	*	O		
11. *Pterospermum canescens* Roxb.		v		O		
12. *Samanea saman* (Jacq.) Merr.	L	v	*	O	?	
13. *Schleichera oleosa* (Lour.) Oken.		v	*	O	o̲	
14. *Sterculia foetida* L.		v	*	O		s
15. *Syzygium cumini* (L.) Skeels		v	*	O	o̲	
16. *Tamarindus indica* L.	L	v	*	O	o	

CANOPY	L	v	*	O	o	s
1. *Atalantia missionis* (Wight) Oliv.	L	v	*	?	?	
2. *Aglaia roxburghiana* (Wight + Arn.) Mig.		v		O	o	
3. *Alangium salviifolium* (L.f.) Wangerin	L	v	*	O		
4. *Alseodaphne semecarpifolia* Nees			*	O	o̲	
5. *Azadirachta indica* A. Juss		v	?	?	?	
6. *Bauhinia racemosa* Lam.		v	?	?	?	
7. *Berrya cordifolia* (Willd.) Burret	L	v	?	?	?	
8. *Butea monosperma* (Lam.) Taub.		v				
9. *Cassia fistula* L		v	*		o	
10. *Cassia roxburghii* DC	L	v				

CANOPY (contd.)

11. *Cocos nucifera* L.			*			
12. *Diospyros montana* Roxb.	L	v	*	O	?	
13. *Diospyros ovalifolia* Wight				O		
14. *Drypetes sepiaria* (W A) Pax Hoffm.		v	*	O	o	
15. *Feronia limonia* (L.)Springle	L	v	?	?		
16. *Ficus amplissima* J.F. Sur		v		O		
17. *Ficus retusa* L.		v		O		
18. *Ficus mollis* Vahl.				O		
19. *Ficus parasitica* Koen.				O		
20. *Glenniea unijuga* (Thw.) Radlk.		v	*		o	
21. *Grewia polygama* Roxb.		v		O	o	
22. *Ixora arborea* Roxb. ex Sm.	L	v		?		
23. *Lannea coromandelica* (Houtt.) Merr.		v	*	O		s
24. *Lepisanthes trichocarpa* (Thw.) Radlk.		v	?	O		
25. *Mangifera indica* L.		v		O		
26. *Morinda tinctoria* Roxb.	L	v	?	?	?	
27. *Muntingia calambura* L.		v	*	O	o	
28. *Peltophorum pterocarpum* (DC) Backer ex Heyne		v	*			
29. *Premna tomentosa* Willd.		v	*	O	o	
30. *Sapindus emarginatus* Vahl.		v				
31. *Streblus asper* (Retz.) Lour.		v	*	O	o	
32. *Strychnos potatorum* L.f.	L	v	*	O	o	
33. *Thevetia peruviana* (Pers.)Merr.			*	O		
34. *Vitex pinnata* L.		v		O		
35. *Walsura piscidia* Roxb.		v	*	O	o	

SHRUBS

1. *Canthium coromandelicum* (Burm.f.) Alston		v	*
2. *Carissa spinarum* L.		v	*
3. *Carmona microphylla* (Lam.) G. Don	L	v	

SHRUBS (contd.)

4. Glycosmis pentaphylla (Retz.) Corr.	L	v	*	?
5. Phyllanthus polyphylla Willd.	L	v		
6. Randia dumetorum Lam.	L	v		
7. Ricinus communis L.	L			
8. Zizyphus oenoplia (L.) Mill	L	v		O

CLIMBERS

1. Ventilage maderas patana	L	v		

KEY TO TABLE: L = leaves; v = flush; O = fruits; o = seeds; s = sap; * = flowers.

Note: a line under the symbol for seeds indicates that they are commonly gathered from the ground as well as from trees.

PART IV

ECOLOGY AND SYSTEMATICS

THE MONKEYS OF SUNDALAND

Ecology and Systematics of the Cercopithecids of a Humid Equatorial Environment

Lord Medway

1. INTRODUCTION

At the southeastern extremity of Asia the Sunda Shelf bestrides the equator from about 10° N to 9° S, its outer limits defined by the deep waters of the Macassar Straits and the Indian Ocean. At present much of the Sunda Shelf is submerged to shallow depths not exceeding 100 fathoms (Figure 1). In past millenia the relative proportion of exposed land surface, conventionally known as Sundaland*, has varied with fluctuating sea levels and local orogenic movements. Although large areas of the principal land masses are ancient, existing coastlines represent a moderately high sea level and were determined comparatively recently in Pleistocene or post-Pleistocene times (Ho, 1964). In several places, including both shores of the Malacca Straits and much of western and southern Borneo, broad alluvial plains have been deposited in the last 5000 years. Parts of the shoreline are still extending seawards, in places at rates of 10-12 m per year.

Yet as a whole since the Mesozoic era the lands of the Sunda Shelf have had no close relations with neighbouring continents. Even at periods of greatest exposure, apparently no eastward connection with the Australasian region was established and the link with continental Asia was not greatly enlarged. In modern times climatic factors reinforce geographical isolation, for both the narrow land connection of peninsular Thailand to the north and the chain of islands to the east coincide with the transition from an equable, more or less perennially humid climate prevailing in central Sundaland to a more variable climate marked by a distinct alternation of wet and dry seasons.

Due to these three influences -- partial isolation from both Asia and Australasia, intermittent connection between islands, and a distinctive climate -- Sundaland has been an important center of plant and animal evolution. Both its flora and fauna are extremely rich and diversified, although as in most tropical regions many

---

* This term is preferred to the alternative and exact equivalent in a zoogeographical sense, 'Malaysia.'

species are represented by comparatively few individuals.

The mammals of all parts of Sundaland are sufficiently well known for valid zoogeographical deductions to be drawn (Chasen, 1940). At the family- and genus-group level the fauna is clearly related to that of continental Asia to the north. But at the species-group level a high proportion are Sundaic endemics. Some endemic species are restricted to a single large island, but several are widely distributed in Sundaland although not occurring outside its limits. Thus, despite local differences, the mammals of the component parts of Sundaland show closer affinities with each other than with the mammals of any other region. This faunistic coherence, also found among other vertebrate classes, has led zoogeographers to recognize the area as a distinct subregion of the Oriental Region.

The faunal characteristics outlined above are equally well exemplified by the monkeys of Sundaland.

The early fossil record is limited but there is no evidence that there were monkeys in Sundaland before the Pleistocene. By the Middle Pleistocene, however, both cercopithecines and colobines were present, and were represented by forms separable only as subspecies of extant species (Hooijer, 1962). It is reasonable to assume that, during the period of invasion and radiation in Pleistocene Sundaland, the monkeys have responded to general selective pressures of the environment which have affected other contemporary mammals of the region in a similar manner. In the following pages likely factors are noted, and the degree to which they have been important in cercopithecid evolution is examined.

2. THE ENVIRONMENT

1. CLIMATE

Most of Sundaland lies between the 5th parallels of latitude, within the limits of the humid equatorial environment. Mean air temperatures are high, typically about 28°C at sea level. The range of annual variation of mean temperature is narrow (at

Kuala Lumpur, a mere 1°C, Dale, 1963) and significantly lower than the daily range. Mainly because of the frequent incidence of cloud, the high maxima of summer or dry-season characteristic of the monsoon environment are not generally experienced. On small islands or coastal locations (e.g. Singapore Island), where the uniform conditions of low latitude combine with maritime effects, the climate is among the most equable in the world.

Ecologically important variations in temperature are associated with change in altitude. Freezing level in the atmosphere at these latitudes is higher than any mountain in Sundaland, and there are no areas of permanent frost. But temperatures drop progressively with increasing elevation, at about 1°C per 100 m, so that mean temperatures in the highlands of Sundaland may fall to levels associated with the temperate zone.

At all elevations rainfall is high. The average annual total exceeds 2500 mm a year over much of Sundaland, and in places is more than 4000 mm. In Java there is a progressive decrease in rainfall from west to east. A similar pattern is seen in Borneo at low latitudes, although the northeast and southeast coasts both experience high rainfall (Banks, 1935). Throughout the region, the year to year variation is typically large and in some places may be as great as 40 percent of the mean. No long-term pattern has been discerned, but in Malaya it is clear that variations in annual rainfall have a significant effect on sporadic fruiting cycles in the virgin forest (cf. McClure, 1966; Medway, in prep.).

It is important to note that comparable climatic conditions have prevailed in the region for a long time. The late Pleistocene Ice Ages were apparently represented in equatorial regions by a lowering of the mean temperature not exceeding 5°C (Moreau, 1966) and perhaps a higher and more consistent rainfall. Archeological work has shown that the later Upper Pleistocene fauna of Borneo 50,000 years ago was largely identical in composition with that of today (Medway, 1964a), implying similar environmental conditions.

2. VEGETATION

The natural climax vegetation throughout Sundaland is tall evergreen tropical rain forest, providing a continuous, closed canopy interrupted only by the watercourses of large rivers. Even the banks of estuaries and muddy seashores are vegetated to below highwater mark with mangrove (dominated by the tree family Rhizophoraceae), which merges without a break into terrestrial forest on its landward side. Although patches of scrub, bush and even open savanna-like vegetation can nowadays be seen in many places, there are with few exceptions artificial habitats made and maintained by the activity of man.

The climatic climax vegetation below 4-500 m is Lowland Dipterocarp Rain Forest, dominated by the tree family Dipterocarpaceae which has only a limited range outside Sundaland. The upper canopy is closed, usually about 40 m high with emergents rising 10-12 m further. At higher elevations this forest is replaced by successive climatic climax formations as follows: at about 1500 ft (circa 460 m) by Hill Dipterocarp Forest; at about 3000 ft (910 m) by Submontane Tropical Rain Forest in which dipterocarps are rare, and oaks (Fagaceae) and Lauraceae are prominent; at 5-9000 ft, depending on exposure and topography, by a Montane Tropical Forest, dominiated by Ericaceae (including many rhododendrons), Myrtaceae and conifers. At extreme heights (on Mt. Kinabalu) the forest is dwarfed, and sedges, grasses and herbs predominate, among them many species belonging to circumboreal genera (Poore, 1964).

Although the forest as a whole is evergreen, there are many tree species in it that exhibit cycles of leaf change, flowering and fruiting. Three main types of periodicity can be recognized: (i) annually recurrent cycles; (ii) non-annual, possibly endogenous cycles which may be shorter or longer than one year; (iii) sporadic. In an area of virgin forest at Ulu Gombak which has been studied since 1961 (McClure, 1966; Medway, in prep.), preliminary analysis indicates that 42 percent of the 41 tree species under observation exhibit annual fruiting cycles, not wholly in synchrony but concentrated in the months June-September;

10 percent exhibit cycles of less than a year (an additional 3 species may have cycles longer than one year); and 42 percent are sporadic, apparently being stimulated to flower and fruit only by prolonged periods of limited rainfall. The last group includes all 10 species of dipterocarps present which thus provide a very unreliable source of food to frugivorous animals. The group that shows most frequent fruiting contributes to the greater richness of this habitat as compared with virgin forest.

3. THE MAMMAL FAUNA

1. GENERAL

Most native Sundaic mammals are specialised forest-dwelling forms. Comparatively few species are able to exploit open habitats or cultivated land. In Malaya, trapping for small ground mammals (i.e. rats, squirrel, treeshrews, insectivores) yielded only 3 species in rice-fields, 5 in grassland and bush and 7 in secondary forest, as opposed to 13 species in undisturbed lowland forest (Harrison, 1962a). Among both birds and mammals, the natural habitat of those species which have successfully invaded cultivated areas such as suburban gardens, plantations, and other partly deforested land, appears to be the forest edge, including riparian vegetation alongside large rivers and the coastal mangrove (cf. Nisbet, 1968). The mangrove itself may also support a limited number of exclusive species of which mammals are less numerous than birds.

In the tall inland forest, available food in the form of growing shoots, flowers and fruits, is concentrated in the upper canopy. In undisturbed tracts, the carpet flora is limited, and the ground level is consequently poor in living photosynthetic plant tissue. The fauna is correspondingly rich in arboreal species, which in Malaya constitute 48 percent of all mammals other than bats (Harrison, 1962b) and in the lowlands of North Borneo 45 percent (Davis, 1962), compared with 15 percent in the temperate forest of northeastern U.S.A. (Davis, 1962). With the exception of a few large ungulates, phytophagous mammals are overwhelmingly arboreal (Harrison, 1962b). Reflecting the complex structure of the forest, the arboreal fauna can be divided into vertically stratified ecological groups, each specialised to exploit a different zone between the jungle floor and the upper canopy.

Probably as a direct response to the complexity of the habitat, the mammal fauna of the forest is highly diversified and rich in species. In Malaya, Medway (1966b) reported at least 93 species (4 more have since been found) occurring in the upper Gombak Valley, an area of circa 110 sq km of partly disturbed hill forest, and Harrison (1969) recorded at least 35 species in 14.6 ha of disturbed lowland forest. The known mammal fauna of Malaya and Singapore together (circa 149,000 sq km) comprises 198 species (Medway, 1969), a much greater total than in any equivalent area of the temperate zone. Yet population densities are low, measured either as counts of individuals or in terms of biomass. The average rate of catch in small mammal trapping programs, at about 5 individuals per 100 trap nights, is much lower than rates in temperate woodland (Medway, 1964b, 1966a). The average density of mammals in a tract of lowland forest has been estimated at only 7.9 individuals per ha (Harrison, 1969).

Many species are extremely rare. Several endemic to Sundaland are still known only from the type or, at best, a very small series. In some cases the rarity is spurious, reflecting narrow specialization (Medway, 1964d). The uneven coverage by collectors has also created problems for the systematist working on the Sunda fauna. Collection localities often inadequately cover the full range of a species, leaving gaps which may be crucial.

2. ISOLATION: THE GREATER SUNDA ISLANDS

In this area of shallow seas and scattered islands, isolation has obviously been a recurrent factor in the evolution of non-flying terrestrial mammals. It is clear firstly that, although not proving a total barrier, the deep water of the Macassar Straits has restricted contact with Celebes. Several mammalian families (including Cercopithecidae and Tarsiidae) occur in both Sundaland and Celebes, but there are few common species and no evidence of recurrent gene exchange.

An interesting comparison is provided by the chain of Mentawi Islands which lie off the western edge of the Sunda Shelf, separated from Sumatra by a deep channel 80-130 km (50-80 miles) wide. The depauperate mammal faunas of the four main islands are essentially similar and are clearly derived from the fauna of Sumatra. Excluding bats, four of the eighteen species present (among them two monkeys) have diverged sufficiently from Sumatran congenerics to be classified as distinct endemics (Chasen and Kloss, 1927).

A measure of the relative duration of isolation of each of the three large islands of Sundaland can be derived from the present distribution of mammals of continental Asian origin. Of the three large islands, despite instances of close relation with Sumatra (presumably via the islands of Bangka and Belitung), Borneo has the fewest species in common with mainland Asia (Chasen, 1940). The greater isolation of Borneo is also indicated by the higher proportion of endemic genera and species among its mammals. Conversely, Malaya contains fewer endemics and more continental Asian species than any of the islands, suggesting that immigration continued overland after present shorelines were established. Several species in Malaya are evidently limited by factors other than the sea barrier as they do not extend the full distance to the southern tip of the peninsula.

The past distribution of certain mammals may have been more extensive than it is today. The present ranges of the orang-utan (Sumatra and Borneo), siamang and tapir(both in Malaya and Sumatra) are demonstrably relict. The orang is known as a Pleistocene fossil in Java and China (Hooijer, 1948); fossil remains of the siamang have been found in Java (Hooijer, 1960); and the tapir has been found both in stratified cave deposits in Borneo dated by C^{14} through the late upper Pleistocene to 8000 years B. P. (Medway, 1960), and in deposits of Pleistocene age in Java.

Three mammals occurring in Borneo show affinities with forms from Indochina. These are the smooth-tailed treeshrew (<u>Dendrogale</u>), the ferret-badger (<u>Helictis</u>) and the banteng (<u>Bos sondaicus</u>); the last two also occur in Java. This pattern of distribution suggests a limited invasion of Sundaland along the drowned northeast shore of the Sunda shelf at some past period of emergence.

3. ISOLATION: THE SMALL ISLANDS

On the lesser islands and archipelagos of the Sunda Shelf, mammals are found on all but the most barren and unvegetated rocks. As a general rule, the number of species present is proportional to the size of the island, and inversely related to the distance from the nearest major land mass (Medway, 1966a).

A few larger islands support mammals that are apparently survivors from a period of more extensive exposure of dry land. But in general the faunas of neighboring islands are similar in composition and clearly related to those of the nearest mainland. Both features suggest that colonisation has taken place mainly by invasion across the sea, followed by local intra-archipelago movements.

Certain species of Sundaic mammal very frequently occur on islands. Predominant among these are members of the riparian forest and mangrove community, which as noted above, also aggressively colonize clearings and cultivated land.

In the heyday of zoological exploration of this region (1880-1930), mammals of the small islands of the continental shelf were energetically collected and were often described as new forms. Some of these island 'subspecies' fail to stand up to reassessment by modern statistical methods. Others are taxonomically valid, distinguished by color or metrical characters. On several islands, however, the island fauna contains only one member of a genus-group which is represented by two or more species on the nearby mainland. Character displacement may then occur, and it is not uncommon for the island population to be intermediate between its mainland relatives; indeed sometimes it is difficult to assign such an animal confidently to either species. Examples are provided by the mouse-deer (Chasen, 1940) and the field rat (Medway and Lim, 1966).

4. ALTITUDINAL ZONATION

The altitudinal zonation of the vegetation noted above (2.2.) is paralleled among the animals. On the high mountains of the Sundaic hinterlands a distinctive mammal fauna is generally found at elevations above 3000 ft (910 m). In Malaya a significant proportion of this fauna comprises species with Himalayan affinities, whose range extends down the central mountain chain. On isolated, peripheral mountains the specialized fauna may be absent, and normally lowland species occur up to high elevations (Yong, 1969a). In the three large islands some highland mammals also show Himalayan affinities, but the majority are local endemics. On both Mt Kinabalu (Borneo) and Mt Loeser (Sumatra) at still higher elevations, in the upper montane zone above circa 7000 ft (2134 m), there is an endemic species of rat.

Although absent from the intervening lowlands, several montane species have extensive discontinuous ranges in each of the larger islands of the Sundra shelf. Colonisation in these cases is thought to have occurred during periods of lowered temperature, probably associated with the Pleistocene Ice Ages, which would have lowered the altitudinal limits of montane conditions. In several instances localized subspecies have subsequently differentiated on isolated mountain peaks.

Interpretation of distribution patterns, however, can be made difficult by the tendency for morphologically similar forms to evolve in the montane habitat on different land masses. 'Lumpers' among taxonomists have united these into joint species. Reassessment of the one such group, the medium-sized, mountain spiny rats of Sumatra, Malaya and Borneo, has shown that the similarities are due to convergence (Medway, 1964c). The same is likely to be true in other cases, including *Rattus baluensis* as defined by Chasen (1940).

5. ECOLOGICAL BARRIERS IN THE LOWLANDS

Several lowland mammals are represented by more than one subspecies on a single land mass. In Malaya, clinal changes normally distinguish northern populations from those in the south. In Java the general direction of such clines is east-west. In Sumatra similar clines are found from north to south; the main mountain chain also provides a boundary between population of the east and west coasts. In Borneo, the largest of the islands, both the central mountains and also the larger rivers may be barriers to the distribution of species or subspecies.

Several mammals are represented by phenetically indistinguishable populations in northeast Sumatra and western Malaya. In these cases similarities of the habitat may have favored the selection of convergent phenotypes. It is more likely, however, that the Malacca Straits must be recognized as a relatively ineffective (and presumably geologically recent) barrier to gene exchange.

The very variable Prevost's squirrel (*Callosciurus prevostii*) on the other hand exhibits unmistakable convergence. In southern and central parts of eastern Sumatra a black, white and red population occurs, which closely resembles the populations in Malaya and in west Borneo at the same latitudes. Northwards in Sumatra, progressively darker populations are found culminating in a wholly black and red subspecies. In Malaya, northern populations are not markedly different from those to the south. But in Borneo, a great diversity of localised color patterns is found (Banks, 1931a); in the north again there is a progressive elimination of white, ultimately leading to a black and red subspecies phenetically like, but genetically unrelated to the north Sumatran race.

6. ECOLOGICAL EXCLUSION AND SIBLING SPECIES

In the rich and diversified fauna of Sundaland, at the present state of ecological knowledge, it is often difficult to distinguish the niches of closely related species occurring side by side. This problem is general in equatorial areas, and it has been suggested that 'niche-overlap' is an important factor in accounting for tropical faunal diversity (Klopfer, 1961). Among the forest rats of the genus *Rattus*, however, in most cases careful analysis of trapping data has demonstrated different ecological preferences (Harrison, 1957).

Recent work on two species of bats of genus *Tylonycteris*, which roost in the internodal spaces of bamboo, has demonstrated that minute differences in the size of the entrance hole to the roost serve to maintain ecological separation (Medway and Marshall, in press).

The bats and rodents also provide instances of sibling species, some of which have been investigated. The two lowland rats *Rattus surifer* (Miller) and *R. rajah* (Thomas), for example, are phenetically very alike. They may occur in the same locality, although minor differences in habitat preference have been demonstrated (Harrison, 1957). Genetically however, they are distinguished by very different karyotypes (Yong, 1969b), which would presumably serve to maintain reproductive isolation. Among sympatric species of treeshrews, *Tupaia* spp., behavioral differences, including mating posture have been demonstrated which would effectively prevent interbreeding (Conoway and Sorensen, 1966).

4. THE MONKEYS OF SUNDALAND

1. GENERAL

According to the classification established by Chasen (1940) and followed by Napier and Napier (1967), twelve cercopithecid species have been recognized within the boundaries of Sundaland, as follows:

Cercopithecinae: *Macaca fascicularis* Raffles, *Macaca nemestrina* Linnaeus, *Macaca arctoides* I. Geoffroy.

Colobinae: *Presbytis rubicunda* Müller, *Presbytis frontata* Müller, *Presbytis aygula* Linnaeus, *Presbytis melalophos* Raffles, *Presbytis cristata* Raffles, *Presbytis obscura* Reid, *Presbytis potenziani* Bonaparte, *Simias concolor* Miller, *Nasalis larvatus* Wurmb.

These taxonomic groupings can be examined in the light of the general patterns of distribution and ecological diversification among Sundaic mammals discussed above.

2. CERCOPITHECINAE

No macaque is wholly endemic to Sundaland, although two (*M. fascicularis* and *M. nemestrina*) are widespread and evidently successful in the region. Like other members of the genus they eat mainly fruit, supplemented no doubt by animal protein, including invertebrates, when available. All three species descend to the ground to forage, but in the wild all retire at night to sleep in trees. All macaques are gregarious, associating in large troops at all stages of their life.

Macaca fascicularis: Although frequently seen on the sea-shore and in the coastal forest of nipa palm or mangrove, the crab-eating macaque is by no means confined to this habitat. It is more accurately categorised as an 'edge' species. Like other members of this ecological group, it readily colonizes disturbed or partially cleared land, including plantations, parks and gardens. Where tolerated, self-maintaining populations have become established in suburban districts where adequate tree cover remains, e.g. Federal Hill, Kuala Lumpur, and the Botanic Gardens, Singapore. In rural areas *M. fascicularis* can be a serious pest of crops, including grain. The crab-eating macaque also ranges into undisturbed habitats, not only following the banks of rivers but also far into the virgin forest where it feeds largely in the canopy. It has been recorded in primary forest from sea level up to elevations at or above the upper limit of the dipterocarps, in Sumatra at 3400 ft (Robinson and Kloss, 1918), in Malaya at 5000 ft (personal observation), Java at 6600 ft (Hooijer, 1962) and Borneo at 4500 ft (Sarawak Museum collections).

Troop size in disturbed habitats may be very large; observations range from 38-57 individuals (Furaya, 1962b). Although comparable counts are not available, I have the impression that in forest the numbers per troop are lower.

This macaque is evidently able to exploit a great variety of habitats. It is also widely distributed on large and small islands throughout Sundaland. Localized subspecies have been described from many small islands (cf. Chasen, 1940; Sody, 1949), but none is greatly divergent and all are clearly conspecific. Examination of larger samples would probably eliminate several subspecies.

Eastwards, the crab-eating macaque has crossed the borders of the Sunda Shelf following the chain of the Lesser Sundra Islands up to Sumba, Flores and Timor, where it may have been introduced by man. 'Monkeys' (undetermined) have occurred among animal remains, including domestic goat, from cave deposits in Timor dated by C^{14} at circa 1960 B.C. They have not been found in earlier strata in which human artifacts but not domestic animals were recovered (I. Glover, pers. comm.). It may be significant that west of Sumatra the crab-eating macaque has extended beyond the 100 fathom line to reach Simalur and Nias, but is not found on the more isolated Mentawi Islands (Chasen & Kloss, 1927). In Java, Middle Pleistocene fossils have been identified, differing very little from the extant population (Hooijer, 1962).

Macaca nemestrina: The pig-tail has a more restricted distribution than the crab-eating macaque. It is found in continental Asia only from Assam through southern Burma and western Thailand to the Malay Peninsula, extending to Sumatra and Borneo. It is absent from intervening islands except the two relatively large and close inshore islands of Batam in the Riau Archipelago and Bangka. It has reached the Pagi islands, but is not recorded from others in the Mentawi group or any of the western islands closer to Sumatra. Eastward it is limited by the Macassar Straits.

Ecologically this macaque is confined to inland forest, ranging widely from coastal swamps through the foothills to over 3000 ft in both Borneo (Medway, 1965) and Malaya (personal observation). In rural areas it may emerge from the forest to feed on crop plants, including grain, but it is not normally resident in deforested land or in plantations (cf. Bernstein, 1967).

In primary forest, troops of pig-tailed macaques have extensive ranges. In the Gombak Valley, Malaya, small troops are attracted to fruiting trees (e.g. fig and wild mangoes, *Mangifera* spp.), and will stay in the vicinity until all fruit is finished. They do not appear to remain permanently in a limited area but presumably travel long distances in search of food (cf. McClure, 1964). They come to the ground freely and when disturbed in the forest descend from the trees to escape (Bernstein, 1967; D.J. Chivers, pers. comm.). The size of troops in inland forest is not large; McClure (1964) reports 6 individuals and I have observed troops of 5, 3 as well as apparently solitary individuals. D.J. Chivers (pers. comm.) has seen troops composed of at least 15 animals.

In many parts of their range, wild-caught *M. nemestrina* are domesticated and trained to harvest coconuts. Thus they tend to be artificially transported across natural zoogeographical barriers. Chasen 1940), for instance, believed that those recorded on the islands of Penang and Singapore had escaped from domestication. The population on the Pagi Islands may also have been introduced in this way. The original specimens were described by Miller as a distinct species, but no recent material is available and no subsequent comparison has been made with specimens from neighboring parts of Sumatra.

All other populations in Sundaland were considered by Chasen (1940) to belong to single subspecies *M. n. nemestrina*, geographically replaced by *M. n. andamanensis* Bartlett in the northern part of the Malay Peninsula. Fossils have not been found in Pleistocene or post-Pleistocene deposits outside its present range and there is no indication that the distribution of this species is relict.

Macaca arctoides: The stump-tailed macaque has invaded Sundaland only as far as the northern border of Malaya, where it has been reported from Perlis, northern Perak and Kelantan (Medway, 1969). Its range lies mainly in higher latitudes in Burma, Thailand, Indochina and parts of southern China, indicating that it is adapted to a climate with a more marked seasonality than that prevailing in Sundaland. In the southern and southwestern parts of its range it is sympatric with the pig-tailed macaque.

M. arctoides is the most terrestrial of the three Sundaic

macaques. It is reported to associate in large troops, and to forage in a variety of habitats from cultivated land to forest (Medway, 1969).

3. COLOBINAE

The colobines have an important ecological role as the principal arboreal shoot- and leaf-eating mammals of the Sundaic rainforests, although probably all species also eat fruit when it is available.

The most widespread Asian genus is Presbytis Eschscholtz, the langurs or leaf monkeys. Representatives in Sundaland have been separated into two taxa, Presbytis sensu stricto and Trachypithecus Reichenbach, variously classed as species-groups, subgenera or full genera (Pocock, 1935, 1939; Washburn, 1944; Hooijer, 1962). The endemic genus Nasalis Wurmb is restricted to Borneo, and Simias Miller is found only on the Mentawi Islands. Eastwards, no colobine has crossed the borders of the Sunda Shelf.

The troop size of P. rubicunda appears to be relatively small. Stott and Selsor (1961) recorded 5 - 8 individuals, and Davis (1962) from two or three individuals to 'eight or more', often associated with P. hosei (see under P. aygula-melalophos group, below). The association however was only temporary however for "when a mixed group is alarmed, rubicundus (sic) makes off at once in a series of crashing leaps, while hosei may not flee until shot at".

The distinctive vocalization is a deep 'chuk-chuk-chuk', uttered as the monkey runs off (cf. Stott and Selsor, 1961). Banks (1931b) also described the call as "a loud series of resonant chuckles". The alarm call, 'noted carefully at Bukit Kretam, is almost bird-like, a single note repeated several times' (Davis, 1962). Other vocalizations include a guttural roar or growl uttered by a dominant male and a scream emitted by young (Stott and Selsor, 1961).

Presbytis frontata: The white-fronted leaf monkey is confined to the mainland of Borneo, where it has been taken in the southeast and east, and in the Rejang and adjacent drainages to the north of this river in central Sarawak. In the south there are no records west of the R. Barito, but the localization of specimens is not sufficiently precise to confirm that the river froms a barrier to distribution (Figure 3).

The species is characterized by a patch of white naked skin on the forehead. Two subspecies have been described, distinguished principally by the presence (frontata) or absence (nudifrons Elliot) of a median line of hairs dividing this bare patch. On this character, topotypes are distinct (Chasen, 1940), but there are no skeletal features of supporting the separation (Pocock, 1935).

P. frontata appears to be confined to lowland forest. The greatest elevation on record is 1000 ft (skin in Sarawak Museum). There are no published field observations.

Presbytis aygula - melalophos group: The banded leaf monkeys of this group occur widely in Sundaland and are phenetically very variable in external appearance. Specimens from many different localities (and several that are imperfectly localized) have been described as distinct taxa, forming the basis for a plethora of subspecies. In distinguishing these, some stress has been placed on the lie of the hair on the head (Thomas, 1893; Pocock, 1935), a character which appears to be inconstant and is probably not of systematic importance. No decisive features of the skull or skeleton have been established, and the main distinctions are based on differences in the coat color of adults.

The coloration of the new-born young is less variable. All known neonates (of the races aygula, siamensis Müller and Schlegel, natunae Thomas and Hartert, hosei Thomas, chrysomelas Müller and cruciger Thomas are pale fawn or white; they are either born either with a dark brown or black spinal stripe extending from at least the nape to the base of the tail (Pocock, 1935) or acquire pigmentation in this area early in the process of assuming adult coloration (specimen of P. cruciger, from Selantik, Sarawak; Sarawak Museum collections).

The adult pelage in populations in the center of Sundaland, i.e. Singapore, southern Malaya and eastern Sumatra, all of which are on the same latitude, is dark brown dorsally, with underparts gray or brown with limited areas of white. Three subspecies are recognised in this area by Chasen (1940):- femoralis Martin, australis Miller and percura Lyon (Figure 4). These dark subspecies are surrounded by geographically separated populations in the islands of Riau archipelago, in eastern Sumatra, in southern and central Malaya, and on Bunguran Island, North Natunas. All subspecies show a convergent graying of the thighs, whitening of the underparts, and various patterns of gray, black and white on the head (rhionis Miller, cana Miller, catemana Lyon, paenulata Chasen, nubigena Elliot, siamensis Müller and Schlegel, and natunae).

In northern Sumatra there is a gap from the northernmost specimens of paenulata at circa 2°N to about 3°N, where the monkeys of this group are more or less uniformly dark smoky gray on the back and outersides of the limbs with white underparts and black hands and feet; two subspecies have been described, thomasi Collett and margae Hooijer.

In northern parts of Malaya and in Peninsular Thailand, the subspecies robinsoni Thomas is apparently dimorphic. The commoner phase is black or dark brown on the upperparts, with a restricted area of white in the venter and insides of the thighs. The alternate phase is largely white, with variable areas of gray or brown wash on back and limbs. Mixed troops have been recorded by Robinson and Kloss (1915a).

In west Sumatra, the subspecies sumatrana Müller and Schlegel and batuana Miller of the Batu Islands superficially resemble the dark phase of robinsoni. In the Padang Highlands, sumatrana intergrades with forms in which the entire pelage is infused with red, so that the upperparts are dark brown or black and light red ('light cinnamon-rufous to ochraceous-tawny' of Robinson and Kloss, 1918) rather than black and gray, and the underparts buffy-white to buff tinged rufous. Hooijer (1962) divided the reddened populations into melalophos Raffles (syn. nobilis Gray), which he recorded from Pariaman, 45 km north of Pedang, Mt Korinchi and Benkulen, and *ferruginea* Schlegel, from the Padang Highlands only. Chasen (1940) described a paler race, *fluviatilis*, from the lowlands inland of Palembang, evidently from a zone of intergradation between *melalophos* and *fuscomurina* Elliot in the extreme southeast. In the last the coat lacks red, being largely white with gray or black in the crown, spinal stripe and outersides of the arms (Chasen, 1940).

The population across the Sunda Straits in western Java (aygula) has an iron-gray colored back, the head and cheeks black and the underparts white. In its present distribution, the easternmost extension of the species in Java is the still darker population on Mt. Slamat, in which underparts are mostly black; this has been described as *fredericae* Sody. Mt Slamat marks the present eastern limit of the range in Java, but Hooijer (1962) has reported Middle Pleistocene fossils of *Presbytis* sensu stricto, morphologically indistinguishable from *fredericae*, from a site further east in Central Java. In the same area, *Presbytis* fossils were not found in early Holocene cave deposits, suggesting that local extinction had already occurred at the close of the Pleistocene (Hooijer, 1962).

In Borneo again erythristic examples recur among the subspecies *cruciger*, which is also dimorphic. Of the two phases, the dark is not unlike *femoralis*, with black underparts, white on the abdomen and insides of the thighs, and gray brown and white on the chest. Dark phase individuals show a cline from northeast (*cruciger*) to southwest where they have been recognised as subspecifically distinct (*chrysomelas*). In the red phase, black is confined to a dark spinal band, dark outersides of the arms and dorsal surface of the base of the tail, with the flanks, thighs and parts of the chest, red. Specimens exhibiting this color pattern have only been recorded in Sarawak, supporting the separation of the two subspecies. Adults, including parous females of the two phases, are found together in integrated troops and there is no doubt that they are conspecific. Specimens of *chrysomelas* have not been taken south of R. Kapuas which appears to form a barrier to the species' distribution. Eastwards, specimens have been taken at Dingai in the headwaters of the R. Mahakam, Kalimantan.

In northeastern Sarawak, *cruciger* meets *hosei*, a distinctive gray and white taxon. There are no skull or tooth characters distinguishing these forms (Hooijer, 1962). There is evidence that their ranges overlap only at Niah, the purported type locality of *hosei*, where *cruciger* has also been taken (Sarawak Museum specimen). Mixed troops have not been reported, and intermediates are unknown.

The head pattern of *hosei* is sexually dimorphic, with more extensive dark pigmentation in old females than in males. In northeastern Sabah, *hosei* is replaced by *sabana* Thomas, in which the head pattern shows less divergent sexual dimorphism. South of the range of *sabana*, *hosei* has been collected from the upper reaches of the Kayan (= Bulungan) River, Kalimantan. Further south and east, the species is represented by a third gray and white subspecies, *canicrus* Miller, which exhibits no sexual dimorphism.

To summarise this long account of distribution and subspeciation, it is seen that representatives of the *aygula* - *melalophos* group occur throughout Sumatra, Malaya and the small islands between them, on Batu Island but not on the deep-water islands lying west of Sumatra. They also occur on Java where the Middle Pleistocene range of the population was more extensive than at present, on Bunguran, Natuna Is. and in Borneo north of the line Kapuas-Mahakam. Many localized taxa are recognized, of which all except *cruciger* and *hosei* appear to replace each other geographically.

Throughout this extensive range, all populations appear to have similar ecology. They are confined to inland forest, ranging from the lowlands and foothills up to elevations above the limits of dipterocarp forest: *melalophos* in Sumatra to 7300 ft (Robinson and Kloss, 1918), *siamensis* in Malaya to 6700 ft (personal observation), *fredericae* in Java to 3300 ft (Hooijer, 1962), and *hosei* in Borneo to 4000 ft (Thomas, 1889).

Observations suggest moderate to small troops. Davis (1962:

62) saw sabana 'often in family parties of three, although troops of a dozen or more are common'. In Malaya, siamensis associates in troops of 6 - 20, although apparently solitary individuals may be encountered (personal observation). In Thailand, Robinson and Kloss saw parties of 5 - 6, and in Sarawak, Banks (1931a) reported 3 to 6 or more.

Few descriptions of vocalization have been published. Allowing for variations in authors' attempts to syllablise animal sounds, they appear to be similar in several races. The alarm call of sabana is said to be 'a series of three or four guttural grunts, almost pig-like, the sequence lasting about half a second' (Davis, 1962). That of siamensis is a harsh dull trill, or purring croak; it is often followed by a loud and distinctive 'chak-chak-chak-chak' (personal observation). The last call is also heard from cruciger/chrysomelas in Sarawak, described as 'a noisy staccato chuckle like that of a large squirrel' (Banks, 1931b)

Presbytis cristata: The range of the silvered leaf monkey in Sundaland includes the three Greater Sunda Islands, Bali, Sirhasan I. in the south Natuna group, Bangka, Belitung, most islands of the Riau-Lingga group and narrow strip of the west coast of Malaya. The typical pelage is dark gray, with paler tips to the hairs giving a silvered effect. In Sumatra, Malaya, Borneo and intervening islands, individual variation in coat shade is great and no clear subspecific differences have been established; all are referred to the single subspecies P. cristata by Chasen (1940). In the population of eastern Sabah there is a balanced dimorphism with two adult phases, the normal silvered gray and the alternate red-haired phase. (Allen and Coolidge, 1940).

In western Java an almost entirely black form is found, which has been named sondaica. This is replaced in eastern Java by pyrrha which again occurs as a silvered gray phase and a red phase. The silvered leaf monkeys of Bali have also been separated, as kohlbruggei Sody, on the basis of skull measurements (Pocock, 1935). Fossil teeth and skulls of Middle Pleistocene age found in Java are slightly larger than recent specimens from the same area, but otherwise identical in specific characters (Hooijer, 1962).

The young of all subspecies are born with orange fur all over, and pale skin of face and extremities. The gray adult coat spreads inwards from the hands, feet, head and tail, starting at under three months of age.

The habitat of the silvered leaf monkey on Java is inland forest, ranging from the lowlands to high elevations -- *sondaica* to 5000 ft and *pyrrha* to 5700 ft (Hill, 1960). In Sumatra the greatest elevation recorded is 2450 ft, on the shores of L. Korinchi (Robinson and Kloss, 1918); there is also a specimen from Kotatjane, Mt. Loeser, height unspecified (Bogor Museum). In Borneo, localization of the single highland specimen, said to have been taken at 3000 ft on Mt. Dulit, has been doubted by Chasen (1940). All other records are from lowland swamp forest and mangrove (cf. Banks, 1931b). In Sabah, the habitat includes secondary growth, even in the immediate environs of a village (Stott and Selsor, 1961). In Malaya this monkey is restricted to the junction of mangrove and terrestrial forest along a limited stretch of the west coast. On the outskirts of Kuala Selangor town, towards the south of its range, several troops are permanently resident (Bernstein, 1968). This population appears to subsist off the flowers, fruit and young shoots of plants including ornamental trees and native figs.

Observed troop size in Malaya ranges from 13 (personal observation) to 48 (Furuya, 1962a); apparently solitary individuals may also be met with.

The voice has been analysed by Gibson-Hill (1949) and Furuya (1962a). The species is normally quieter than both the *aygula*-group (above) and *obscura* (below).

Presbytis obscura: The dusky leaf monkey occurs only on the Malay Peninsula, throughout its length from Johore to Tenasserim. The species has reached most of the larger islands off the northern part of the peninsula, on the east coast (exterminated on the latter according to Robinson and Kloss, 1915b), and on the west coast from Penang, Langkawi, Terutau, the Phuket group (Chasen, 1940) to several islands of the Mergui archipelago (Pocock, 1935).

The principal distinguishing feature of the species is the pied color of the face, which is black with a wide interrupted ring of white around each eye and a white patch over the center of the mouth involving both lips. In the adult pelage, the mid-dorsal region from occiput to tail is gray, the forehead and sides of the face are blacker, the limbs are gray and the extremities black. Although individual variation is great, the mainland population has been divided by color differences into a northern subspecies P. O. flavicauda Elliot and a southern P. O. obscura (Chasen, 1940; Ellerman and Morrison-Scott, 1956). Many island populations have also been separated subspecifically from the mainland forms.

The color of the new born young is uniform pale orange brown.

The habitat of obscura in Malaya is terrestrial forest, ranging from the lowlands to 6000 ft (personal observation). It will enter rubber plantations adjacent to forest, especially poorly kept peasant holdings, and in the coastal lowlands of Perak has been found associated with P. cristata in this habitat (D. Labang, personal comm.). Troop size in Malaya varies from 6 - 20, but apparently solitary animals may also be encountered. The distinctive call of this species is a loud snort-like hoot composed of four syllables.

Presbytis potenziani: The Mentawi Islands leaf monkey is restricted to the islands of that name lying in deep water off the Sunda Shelf to the west of Sumatra. The upperparts are black; the forehead, cheeks and chin, white; and the underparts red. The fur of the newborn is orange (Washburn, 1944). Two subspecies are recognised, P. p. potenziani on Sipora and North and South Pagi, and P. p. siberu on Siberut. Nothing is known of the ecology of these leaf monkeys.

Nasalis larvatus: The proboscis monkey exhibits considerable sexual dimorphism, the male being much larger than the female. Both sexes have a prominent fleshy nose, which is greatly enlarged and elongated in the male.

The species is restricted to Borneo, where it has been found in the mangrove forest of coast and estuaries, and the riparian forest of the lower and middle reaches of major rivers in most localities visited by collectors. There is also a record from Quoin Hill, in eastern Sabah, at 700 ft elevation, 10 miles (16 km) inland and near no major river (Medway, 1965). The range of N. larvatus extends to at least two inshore islands, Berhala in Sandakan Bay and Sebatik I. on the eastern Sabah-Kalimantan border. Although it can swim well (Davis, 1962) the proboscis monkey has not succeeded in crossing any major reach of sea to colonize more distant lands.

This monkey is characteristically arboreal, but frequently descends to the ground (Banks, 1931b). Its diet consists of young leaves and shoots of the trees of the mangrove and riparian forest associations in which it occurs. Troop size is 12 - 27 (Napier and Napier, 1967).

The distinctive call is 'deep snorking honk', more resonant in the male than the female (Davis, 1962). Although the nose may move during vocalization, it is not inflatable and apparently has no associated function

Simias concolor: The pig-tailed langur is recorded only from South Pagi and adjacent small islands, Sipora and Siberut, in the Mentawi Islands. It is unlikely that S. concolor is absent from North Pagi.

This langur is convertent with certain macaques (including M. nemestrina) in the reduction of the tail, and in having limbs of more or less equal length. It is also distinguished by a prominent snub nose, which is partly supported by elongated nasal bones. The species is dimorphic, occuring throughout its range in a commoner dark gray phase and a rarer creamy buff phase (Chasen and Kloss, 1927). The habitat is apparently lowland forest, but there is no information is available on its biology or behavior.

5. DISCUSSION AND CONCLUSIONS

1. GENERAL

The monkeys, with the gibbons and orang-utan, are the dominant, and only large leaf-, shoot- and fruit-eating vertebrates of the Sundaic rain forests. The ecology of all species ties them closely to forest, and only one (Macaca fascicularis) is able to thrive in habitats that are artificial but not totally deforested. Although the Dipterocarpaceae dominate in the terrestrial lowland forest, there is no indication that any monkey is specialized to exploit this plant family. Limited available ecological observations indicate rather that all Sundaic monkeys utilize a wide variety of plant species within an equally wide range of forest types.

To some extent, interspecific competition must be reduced by dietic specialization and by vertical separation of activity space. The macaques are primarily fruit-eaters or omnivores, occupy- the lower fruit-eaters, frequenting the middle and upper canopy. Although distribution maps based on museum specimens are probably less reliable for monkeys (which are difficult to collect and laborious to prepare) than for any other group of mammals, it is clear that cercopithecines and colobines coexist throughout Sundaland.

Of the general factors affecting the evolution and distribution of Sundaic mammals, isolation has clearly been important among the monkeys. The distinctiveness of the Bornean fauna is reflected in the presence of three well defined endemic species, and the isolation of the Mentawi Is by two. Two species, Macaca arctoides and Presbytis obscura, which extend to Sundaland only in southern Thailand and Malaya, emphasise the continental connections of the Peninsula.

The family is also widely represented on the smaller islands of the Sunda Shelf. Macaca fascicularis is outstanding and apparently well adapted to crossing the sea barrier to reach small islands. Presbytis cristata and members of the P. aygula-melalophos group are widely distributed; probably in these species colonization followed land connections, which are now submerged, rather than a sea crossing. The presence of *P. obscura* on only the northern islands of the Malaya Peninsula suggests that it invaded from continental Asia relatively recently, after the more southerly east coast islands had already separated.

Altitudinal factors appear to be unimportant. No Sundaic monkey is exclusively montane, and most of them range from the lowlands to elevations above the limits of dipterocarp forests. Of the many races of the *P. melalophos-aygula* group, only *ferruginea* has a purely upland range. The strictly lowland species, *Nasalis larvatus*, is confined to edaphically determined forest types, i.e. mangrove and river bank vegetation, which clearly set a limit to its altitudinal distribution. It is suggested below that interspecific competition, rather than ecological factors related to altitude, may restrict *Presbytis cristata* to the subcoastal lowlands of Malaya.

Within each major land-mass of Sundaland, the directions of clines and other patterns of subspeciation conform to those of mammals in general and it is presumed that similar barriers operate. Except in the north of the island, populations of *melalophos-aygula* on the west coast of Sumatra are racially distinct from those in the eastern lowlands. The R. Kapuas of western Borneo appears to have prevented southward movement of *P. chrysomelas*, and the Barito may have been effective in limiting westward spread of *P. frontata*. On the other hand, on the two shores of the Malacca Straits at its narrowest part, the populations of *Macaca fascicularis*, *M. nemestrina*, *Presbytis aygula-melalophos* group and *P. cristata* in each case are phenetically alike, despite the intervening sea.

The precise ecological niche of each species is not always clear, particularly among *Presbytis*. In Malaya, troops of the *P. melalophos-aygula* group occur in tall forest alongside *P. obscura* but I am unable to detect any difference in the sector of the habitat exploited by the two species. They may mix temporarily, but permanently integrated troops have never been observed. The two species are clearly distinguished by vocalization, facial coloration and juvenile pelage. If inter-breeding is prevented by social, behavioral or other mechanisms reflecting a relatively distant phylogenetic relationship, then slight differences in ecology, so far unidentified, may suffice to prevent competitive exclusion.

In Java, populations of P. aygula-fredericae and of P. sondaica-pyrrha similarly coexist in inland forests apparently without competing. By contrast, P. cristata in Malaya has a very restricted range at the junction of terrestrial forest and mangrove on the west coast. Mixed troops of P. cristata and P. obscura have been reported from secondary forest and poorly kept plantation where their ranges overlap. It is possible that cristata has only recently invaded the Malay Peninsula, and is now excluded from inland forests by the presence of other leaf monkeys. In both Borneo and Sumatra, cristata apparently has a more extensive range, which perhaps reflects of the absence of a closely related competitor.

In Borneo, in addition to P. cristata, twelve other recognized taxa of the genus Presbytis occur. The distribution maps show that the distinctive endemic P. rubicunda, represented by five subspecies, is most widespread and occurs alongside all others. The remainder fall into three groups (chrysomelas-cruciger, frontata-nudifrons and hosei-sabana-canicrus) which are largely exclusive in their ranges. The main habitat, occupied by all, is the dipterocarp forest of lowlands and foothills; but two forms, at least, occur at elevations high enough to show that no mountain range in Borneo would be an impassable barrier to them. The wide altitudinal tolerance of these monkeys implies that they are not adapted to any particular plant community. It is possible that here again interspecific competition is responsible for the interdigitating pattern of ranges. More intensive field work may show that features of the ecology of P. rubicunda, so far unrecognized, serve to separate it from congenerics.

2. SYSTEMATIC CONCLUSIONS: CERCOPITHECINAE

Of the three macaques in Sundaland, M. arctoides clearly evolved outside the area and is adapted to different ecological conditions. The absence of _M. nemestrina_ from Java, both recent and in the Middle Pleistocene, suggests that this species is also a late immigrant from continental Asia, and presumably reached Borneo via Sumatra and a Bangka-Belitung land bridge. The relatively small troops encountered in inland forest, widely scattered and apparently ranging over large areas, suggest that the pig-tailed macaque is only marginally successful in the predominantly forested habitats of Sundaland.

Still more clearly _M. fascicularis_ is adapted to the 'edge' habitat, rather than the deep forest. Its ubiquity and its large and relatively static troops attest to the successful exploitation of this widespread habitat. Fossils demonstrate that the species was already in Sundaland by the Middle Pleistocene. A distinct subspecies has evolved on Java; but on the other mainland masses allopatric populations are not morphologically distinguishable. In this mobile and ecologically opportunistic species, migration and genetic exchange may occur readily, even across sea barriers. If _M. fascicularis_ proves to be conspecific with _M. mulatta_ (Fooden, 1964) and to have evolved on continental Asia, its ecological preferences would clearly preadapt it to rapid invasion and colonization of Sundaland at some period of lowered sea-level early in the Pleistocene.

3. SYSTEMATIC CONCLUSIONS: COLOBINAE

Morphological features support the division of Sundaic langurs into the two genus-group taxa, _Presbytis_ and _Trachypithecus_ (Washburn, 1944; Hooijer, 1962). What is known of their vocalizations also accords with this division, although a description of the call of _P. frontata_ is lacking. The orange to brownish color of the neonates is a further character linking species assigned to _Trachypithecus_ (Washburn, 1944). No common neonatal color pattern, however, is apparent among all members of _Presbytis_ (_sensu stricto_). The cruciform pattern is restricted to the _aygula-melalophos_ group which, as shown below, can reasonably be recognized as a single species or superspecies. Even among these it is not universal; a skin of _cruciger_ in the Sarawak Museum lacks the typical pattern. The neonates of _P. rubicunda_ are pale red all over; they show incipient cruciform pattern only in the assumption of the adult shade by progressive darkening from the spinal region. The color of the neonate of *P. frontata* is unknown. As yet, in sum, there is no clear indication that all taxa conventionally associated in *Presbytis* sensu stricto are phylogenetically closely related, and the assemblage may ultimately prove to be artificial.

P. rubicunda is considerably more widespread in Borneo than *P. frontata*. The two species occur together over a large proportion of the range of the latter and there are no obvious natural barriers (except possibly ther R. Barito) limiting the range of either species. Unless this is a stage in replacement of one by the other, ecological differences presumably prevent competitive exclusion.

P. rubicunda can apparently also live alongside representatives of the *aygula-melalophos* group. The ranges of the latter and of *P. frontata*, however, are largely complementary, suggesting that they may be ecologically exclusive and cannot have evolved in sympatry. The *aygula-melalophos* group has representatives outside Borneo and presumably invaded the island after the evolution of *P. frontata*, which has perhaps been partially supplanted.

The taxonomy of what I have so far non-committally called the *melalophos-aygula* group is not easy to interpret. The pronounced geographical variation and the evidence of polymorphism independently evolved at least twice suggest that this group is genetically plastic, and perhaps actively evolving at present. There may also be ecological or behavioral mechanisms (e.g., troop endogamy) that reduce the rate of genetic exchange between populations.

Pocock (1935) classified known forms into seven allopatric species under the following names:- *P. femoralis* in Malay Peninsula, Singapore, the Riau archipelago, central Sumatra and the Natunas; *P. thomasi* north of this group in Sumatra, and *P. melalophos* to its south; *P. aygula* in Java; and *P. chrysomelas*, *P. hosei* and *P. sabana* in Borneo. Subsequently Chasen (1940) 'lumped' all taxa into two species, in an arrangement explitic1y rejected by Pocock (*op. cit.*) which united populations with grey upperparts in northwest Sumatra, Java and north and east Borneo into one species (*P. aygula*) and all others into a second species, for which *P. melalophos* is the prior name.

These two conflicting classifications imply totally different interpretations of the evolutionary history of the group. If Chasen was correct, the gray-backed species is relict in its present disjunctive distribution and represents the initial colonizer of Sundaland. Pocock's more cautious classification assumed that the similarities of *thomasi*, *aygula*, *sabana* and *hosei* represent convergence and not genetic affinity (Pocock, 1935: 918).

Both men had access to the same or similar material, and no conclusive new evidence has since become available. If modern techniques (e.g. comparative karyology or serology) cannot be applied, neither interpretation can be wholly refuted. In the main Pocock's approach is more acceptable. Within the group unambiguous examples of convergence are provided by the obviously unrelated alternative white polymorph of *robinsoni* vs. *fuscomurina*, and the red polymorph of *cruciger* vs. *melalophos* and *ferruginea*. Among other mammals, similar situations have been attested (3.4, 3.5 above) including the convincing case of *Callosciurus prevostii*.

The number of species recognised by Pocock however, is centainly too large. In Sumatra there are intermediates linking *femoralis* and *melalophos*; *natunae* and *chrysomelas*-*cruciger* are also clearly conspecific. The taxa *thomasi* and *margae* can either be considered conspecific with others in Sumatra, or (in the absence of intermediate specimens) as forming a local polytypic species. Only where *hosei* meets *cruciger* in the Baram and Niah drainages of Sarawak is there evidence of reproductive isolation. This can be recognized by separating *hosei* with *sabana* and *canicrus* as subspecies. The Javan representatives can again be treated either as conspecific with the Malayan and Sumatran populations, or as allopatric species.

At its most radical, this interpretation classifies the group into two species: *P. hosei* including *sabana* and *canicrus*, and *P. aygula* (the prior name) comprising all other recognized subspecies. The more cautious alternative separates P. hosei as above, P. aygula on Java, and P. melalophos to include the remainder, with or without thomasi (including margae) as a fourth species.

The species P. cristata has been held to include two continental Asian taxa, namely atrior Pocock and germani Milne-Edwards, which have a combined range from southeastern Burma through southern (non-peninsular) Thailand to Indochina (Ellerman and Morrison-Scott, 1956). Between these and the Sundaic populations a wide gap intervenes, consisting of virtually all the Malay Peninsula. Here, as indicated above, P. obscura replaces cristata; the latter has a foothold only on the west coast of Malaya, where it is probably a geologically recent invader from Sumatra across the Malacca Straits, and is perhaps excluded from the inland habitat by competition. It is possible that Indochinese and Bornean populations formerly had contact along the now drowned northern coast of Sundaland. Meanwhile the affinities of the two groups may need reconsideration.

P. obscura, as at present defined, occurs only in the Malay Peninsula. Here it has a wide ecological range from mountain tops to coastal forest, and has also colonized many offshore islands. To the north P. o. nubigena is apparently replaced without intergradation by crepuscula Elliot. The latter is at present considered to be the representative of a related but distinct species which includes barbei Blyth, shanicus Wroughton and phayrei Blyth. This classification was first proposed by Pocock (1935), who clearly had a limited concept of species (see above), and later followed by Ellerman and Morrison-Scott (1965). The resemblances between certain taxa in these two groups are said to be close (Pocock, 1935), and in captivity a female P. obscura subsp. has sucessfully hybridized with a male that may be P. phayrei (Badham, 1967). Once again there seems to be a case for re-examination of available evidence and, if possible, the application of modern techniques.

The two subspecies of P. potenziani are clearly related, and presumably descend from a single initial colonization from Sumatra, followed by intra-archipelago movement. The deep water barrier between these islands and the Sumatran mainland has prevented reinvasion and further genetic exchange. In isolation, a number of distinctive characters have evolved. The anatomy of the skull and the orange-furred young, however, link *P. potenziani* with the *Trachypithecus* group (Washburn, 1944) rather than with *Presbytis aygula* (Chasen, 1940). The red color of the underparts of the adult is not unique in the subgenus, since *P. cristata* in Borneo and Java exhibits erythristic dimorphism.

Chasen (1940) included the proboscis monkey in his list of Bornean mammals with Indochinese affinities, implicitly suggesting relationship with the snub-nosed langurs *Rhinopithecus* (*Presbytiscus*) of northern Vietnam. Ecological evidence provides no support for this hypothesis, but suggests instead that *Nasalis* evolved in Borneo as specialized offshoot of the ancestral colobine stock. Davis (1963) conjectured that the long nose of the male had no selective value, but could have originated by allometry as an associated effect of the exaggerated sexual dimorphism characteristic of the species.

The affinities of *Simias* are equally uncertain, but again there seems to be no ecological grounds to associate it with the other langurs with prominent noses and nasal bones such as *Nasalis* (but see Groves, this volume).

ACKNOWLEDGEMENTS

I am grateful to Mr. Benedict Sandlin, Curator, Sarawak Museum, for permission to examine the collections in his care, and for other facilities kindly afforded; to Dr. S. Somadikarta for supplying details of the connections in the Museum Zoologicum at Bogor. Mr. E.R. Alfred, Director, kindly made available material in the Singapore National Museum, and the distributional details were collated by Mr. Richard Lim. Above all, I wish to express my thanks to the Trustees of the Wenner Gren Foundation and Dr. J.R. Napier, whose invitation to join the Symposium provided the initial stimulus to draft this paper.

REFERENCES

Allen, G.M. & H.J. Coolidge, Jr. (1940). Asiatic primate expedition collections: Mammals. Bull. Mus. comp. zool. **87**: 131-166.

Badham, M. (1967). Breeding the spectacled langur. Int. Zoo Yb. **7**: 89.

Banks, E. (1931a). The races of Sciurus prevostii in Borneo. Proc. zool. Soc. Lond. 1931: 1335-1348.

Banks, E. (1931b). A popular account of the mammals of Borneo. J. Malay. Br. roy. Asiatic Soc. **9**(2): 1-139.

Banks, E. (1935). Seasonal rainfall in Borneo. Sarawak Mus. J. **2**: 364-370.

Bernstein, I.S. (1967). A field study of the pig-tailed monkey (Macaca nemestrina). Primates **8**: 217-228.

Bernstein, I.S. (1968). The Lutong of Kuala Selangor. Behaviour **32**: 1-16.

Chasen, F.N. & C.B. Kloss. (1927). Spolia Mentawiensa: mammals. Proc. zool. Soc. Lond. 1927 (53): 797-839.

Chasen, F.N. (1940). A handlist of Malaysain mammals. Bull. Raffles Mus. **15**: 1-209.

Conaway, C.H. & M.W. Sorenson. (1966). Reproduction in tree shrews. Symp. zool. Soc. Lond. **15**: 471-492.

Dale, W. L. (1963). Surface temperatures in Malaya. J. trop. Geogr. 17: 57-71.

Davis, D.D. (1958). Mammals of the Kelabit Plateau, Northern Sarawak. Fieldiana: Zool. 39(15): 119-147.

Davis, D.D. (1962). Mammals of the lowland rain forest of North Borneo. Bull. natn. Mus. Singapore 31: 1-129.

Ellerman, J.R. & T.C.S. Morrison-Scott. (1956). Checklist of Palaearctic and Indian Mammals. London: British Museum (Natural History).

Fooden, J. (1964). Rhesus and crab-eating macaques; intergradation in Thailand. Science 143: 363-365.

Furuya, Y. (1962b). Ecological survey of the wild crab-eating monkeys in Malaya. Primates 3: 75-76.

Gibson-Hill, C.A. (1949). Calls of the silvered leaf-monkeys. Malay. Nat. J. 4: 40.

Harrison, J.L. (1957). Habitat of some Malayan rats. Proc. zool. Soc. Lond. 128: 1-21.

Harrison J.L. (1962a). Mammals of Innisfail. Aust. J. Zool. 10:

Harrison, J.L. (1962b). The distribution of feeding habits among animals in a tropical rainforest. J. anim. Ecol. 31: 53-64.

Harrison, J.L. (1969). The abundance and population density of mammals in Malayan lowland forests. Malay. Nat. J. 22: 174-178.

Hill, J.E. (1960). The Robinson collection of Malaysian mammals. Bull. Raffles Mus. 29: 1-112.

Ho, R. (1964). The environment. In Malaysia: a survey, (Ed. Wang Gungwu). London and New York: F.A. Praeger.

Hooijer, D.A. (1948). Prehistoric teeth of man and of the orang-utan. Zool. Meded. Mus. Leiden. 29: 175-301.

Hooijer, D.A. (1960). Quaternary gibbons from the Malay Archipelago. Zool. Verh. Mus. Leiden. 46: 1-42.

Hooijer, D.A. (1962). Quaternary langurs and macaques from the Malay Archipelago. Zool. Verh. Mus. Leiden. **55**: 1-64.

Klopfer, P.H. (1961). On the cases of tropical species diversity: niche overlap. Amer. Nat. **95**: 223-226.

Lönnberg, E. & E. Mjöberg. (1925). Mammals from Mt. Murud and the Kalabet country. Ann. Mag. nat. Hist. **16**: 508-516.

McClure, H.E. (1964). Observations of primates in climax diptocarp forest near Kuala Lumpur, Malaya. Primates **5**: 39-58.

McClure, H.E. (1966). Flowering, fruiting and animals in the canopy of a tropical rainforest. Malay. Forester **29**: 182-203.

Medway, Lord. (1960). The Malay tapir in late Quaternary Borneo. Sarawak Mus. J. **9**: 356-360.

Medway, Lord. (1964a). The late Upper Pleistocene fauna of Borneo. Stud. Speleol. **1**: 33-37.

Medway, Lord. (1964b). The fauna. In Malaysia: a survey. (Ed. Wang Gungwu) London and New York: F.A. Praeger.

Medway, Lord. (1964c). Comments on the status of Rattus inas (Bonhote), with observations on the distribution of this and related rats in the Sunda subregion. Fedn. Mus. J. **9**(N.S.):

Medway, Lord. (1964d). The marmoset rat, Hapalomys longicaudatus Blyth Malay. Nat. J. **18**: 104-110.

Medway, Lord. (1965). Mammals of Borneo: field keys and an annotated checklist. J. Malay. Br. roy. Asiatic Soc. **36**(3): 1-193.

Medway, Lord. (1966a). The fauna of Pulau Tioman: Mammals. Bull. natn. Mus. Singapore. **34**: 9.32.

Medway, Lord. (1966b). The Ulu Gombak field studies centre. Malay. Scientist **2**: 1-15.

Medway, Lord. (1969). The wild mammals of Malaya. Kuala Lumpur: Oxford University Press.

Medway, Lord. (in prep.). Phenology of Upper dipterocarp rain forest.

Medway, Lord & B.L. Bim. (1966). The specific relations of Rattus tiomanicus (Miller). Bull. natn. Mus. Singapore. 34: 33-38.

Medway, Lord & A.G. Marshall. (in press). Roost-site selection among the flat-headed bat, Tylonycetria spp. J. Zool. Lond.

Napier, J.R. & P.H. Napier. (1967). A handbook of living primates. London and New York: Academic Press.

Nisbet, I.C.T. (1968). Utilization of mangroves by Malayan Birds. Ibis 110: 348-352.

Pocock, R.I. (1935). On monkeys of the genera Pithecus (or Presbytis) and Pygathrix found to the east of the Bay of Bengal. Proc. zool. Soc. Lond. 895-961.

Pocock, R.I. (1939). The fauna of British India, including Ceylon and Burma: Mammalia I. London: Taylor and Francis.

Poore, M.E.D. (1964). Vegetation and Flora. ch. 2 in Malaysia: a survey. (Ed. Wang Gungwu) London and New York: F.A. Praeger.

Robinson, H.C. & C.B. Kloss. (1915a). On a collection of mammals from the Siamese province of Bandon. J.fed. Malay States Mus. 5: 111-127.

Robinson, H.C. & C.B. Kloss. (1915b). Zoology of Koh Samui and Koh Pennan: Mammals. J. fed. Malay States Mus. 5: 130-139.

Robinson, H.C. & C.B. Kloss. (1918). Mammals (of Korinchi). J. fed. Malay States Mus. 8: 1-18.

Sody, H.J.V. (1949). Notes on some primates, carnivora and babirusa from the Indo-Malayan and Indo-Australian regions. Treubia 20: 121-190.

Stott, K. & G.J. Selsor. (1961). Observations of the maroon leaf monkey in North Borneo. Mammalia 25: 184-189.

Thomas, O. (1889). The mammals of Mt. Kina Balu, North Borneo. Proc. zool. Soc. Lond. 228-236.

Thomas, O. (1893). Descriptions of two new North Bornean Mammals. *Ann. Mag. nat. Hist.* (6)**12**: 230-232.

Washburn, S.L. (1944). The genera of Malaysian langurs. *J. Mammal.* **25**: 289-294.

Yong, H.S. (1969a). Rats from Medah Peak. *Malay Nat. J.* **22**: 53-56.

Yong, H.S. (1969b). The systematic status of Malayan *Rattus rajah* and *Rattus surifer*. *Bull. Brit. Mus. (Nat. Hist.) Zool.* (in press).

Figure 1: A map of Sundaland to show the limits of the continental shelf and principal areas and islands mentioned in the text.

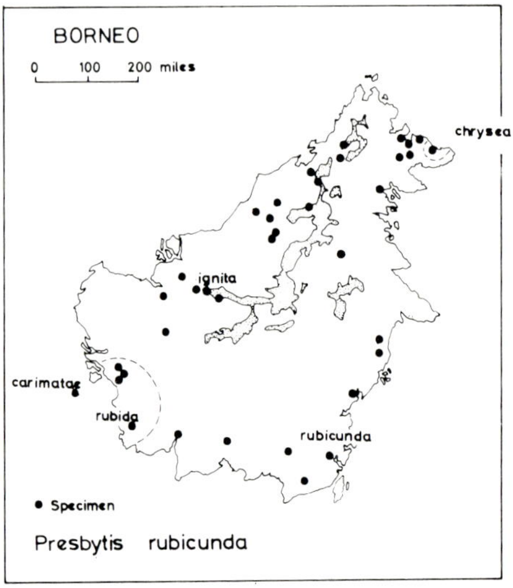

Figure 2: Distribution of specimens referred to the species *Presbytis rubicunda*.

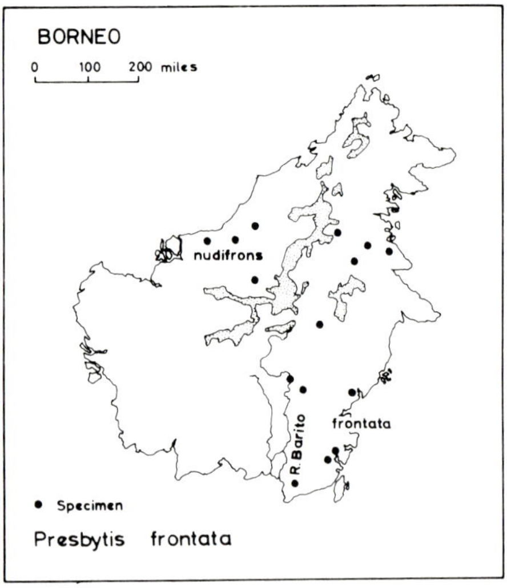

Figure 3: Distribution of specimens referred to the species *Presbytis frontata*.

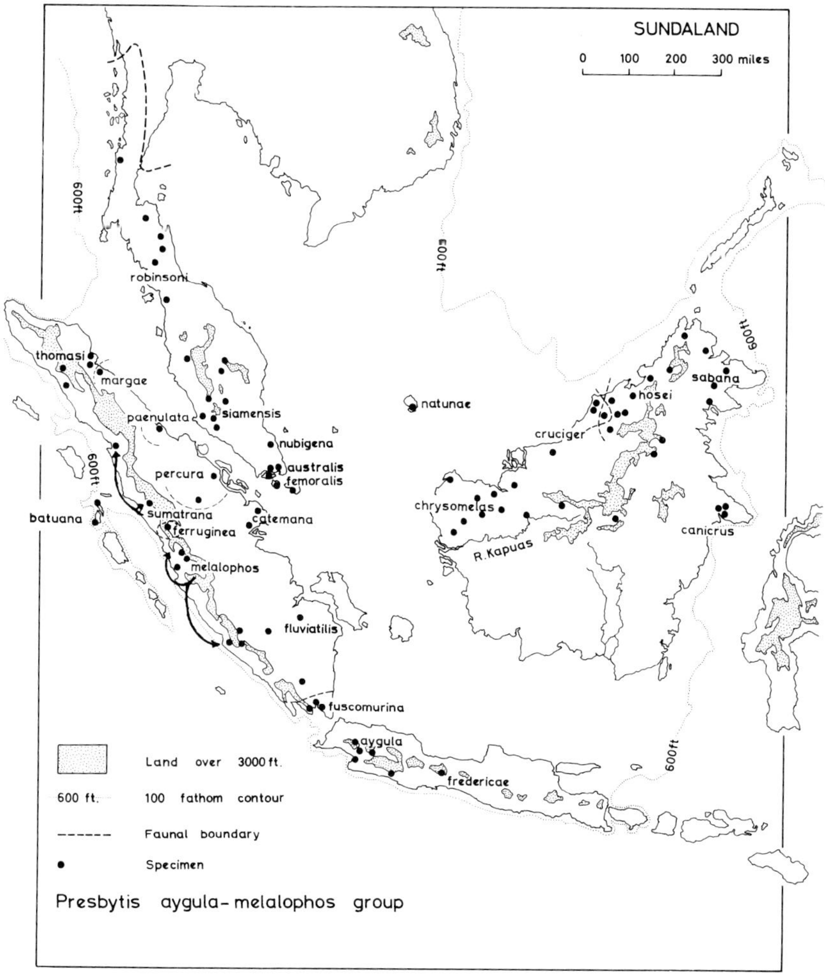

Figure 4: Distribution of specimens, and approximate boundaries of recognized subspecies, of the Presbytis aygula-melalophus group.

THE FORGOTTEN LEAF-EATERS, AND THE PHYLOGENY OF THE COLOBINAE

C. P. Groves

INTRODUCTION

This paper will attempt to clarify the neglected taxonomy of the subfamily Colobinae. The species of restricted regions have been dealt with by Hill (1936), Pocock (1934) and Hooijer (1962) among others; but there seems never to have been an attempt to consider all forms together. The result of this regionalism can be appreciated with a few examples. The problem of the generic status of the common langurs of India and Southeast Asia - whether they should be put into one genus or into four - has remained unsolved largely because no author who has discussed the question has been able to treat all four putative genera at once. Hooijer (1962) divides the Indonesian langurs into two full genera, *Presbytis* and *Trachypithecus*, and these two groups are unquestionably natural ones with clear-cut differences between them. However, the Indian langurs, which Hill (1936) refers to the genera *Semnopithecus* and *Kasi*, to some extent bridge the gap between the two Indonesian genera.

The same may be said of the African leaf-eaters. Verheyen (1962) in his study of the skills of the different species unites them into a single genus, *Colobus*, with three subgenera, stating that the subgenus *Procolobus* (including only the Olive colobus) may have to be raised to generic rank. On the other hand Kuhn (1967), while recognizing *Procolobus* as a full genus, includes Red Colobus as well as the Olive; this opinion was based on a consideration of the digestive apparatus, and other factors (see below).

While much discussion has thus taken place over the two widespread groups, the *Colobus* and *Presbytis* complexes, the remaining species, all Asiatic, have been largely ignored. Chasen (1940) and Ellerman and Morrison-Scott (1951) between them recognize five species, each occupying its own genus or subgenus: *Nasalis larvatus*, *Simias concolor*, *Rhinopithecus* (*Rhinopithecus*) *roxellanae*, *Rhinopithecus* (*Presbytiscus*) *avunculus* and *Pygathrix nemaeus*. Considering that these two works recognize between them only one further genus, *Presbytis*, with a dozen species, this makes for a very top-heavy classification. It is intended to examine in this paper the interrelationships of the "forgotten leaf-eaters" and compare them as far as possible with the other members of the subfamily.

HISTORY OF CLASSIFICATION

The first of the species to be made known to science was the Douc, described by Linnaeus in 1771 as *Simia nemaea*. The description is recognizably of the northern (red-legged) race of the Douc; the type locality Cochin-China, is not necessarily in error as the territory known as Cochin-China in the 18th century covered essentially the whole of the present-day South Vietnam. The southern (black-legged) Douc was made known in 1871.

The Proboscis monkey was described as *Cercopithecus larvatus* by van Wurmb in 1781, from specimens collected at Pontianak, Borneo, and in 1812 was separated into a distinct genus *Nasalis* by Etienne Geoffroy St. Hilaire. Vigors and Horsfield in 1828 proposed a second species, *Nasalis recurvus*, which however was based on a young male of the same species and, according to Chasen (1940), might also have been from Pontianak. Chasen himself in the same place described a subspecies *Nasalis larvatus orientalis*, based mainly on the lighter coloration. As the present author intends to show (in a future paper) is the case with the gibbons of Borneo, there are no clear-cut subspecies of the Proboscis monkey in Borneo, merely differing frequencies of color-types; it happens that the pale type is predominant, though not universal, in Eastern Borneo.

The first snub-nosed monkey was described in 1870 by Milne-Edwards, as *Semnopithecus roxellanae*, named after the beautiful snub-nosed Roxellana, concubine - and later the wife - of Suleiman the Magnificent. The species was discovered by Père David, also the discoverer of the Giant Panda and of Père David's Deer. In a subsequent paper (1872) Milne-Edwards raised the species to generic rank under the name *Rhinopithecus*. Further specimens were collected by Berezovski (reported by Büchner, 1891), Styan (reported by de Winton, 1899), the Kelley-Roosevelts expedition (Osgood, 1932) and Williston and the Brooke-Dolan expedition (Allen,

1938). These, and the animals now living in various Chinese zoos (Jarvis, 1966), seem to be the only recorded specimens. A second species, Rhinopithecus bieti, was described by Milne-Edwards in 1897 on specimens collected by Biet, Bonvalot and Prince Henri d'Orléans; this animal is in effect a starter, less brightly colored version of R. roxellanae, and ranks as one of its subspecies. The same status is assigned by Ellerman and Morrison-Scott (1952) to the third described form, R. brelichi Thomas, 1903, from the Van Gin Shan mountains of Kweichow; but in this case the differences seem of rather a different order and its status as a full species is implied.

A fourth species of Rhinopithecus, described by Dollman (1912) as R. avunculus, was not a montane form like the three previous ones but came from the tropical forests of Tonkin (now North Vietnam). Pocock, on the basis of its longer fingers and toes, created a genus Presbytiscus for it, a taxon which seems altogether unnecessary in view of the close relationship to the other species, especially R. brelichi. This last-named species is also known from only a very few specimens.

The fourth of the snub-nosed monkeys, the Pig-tailed langur from the Mentawai Islands, was described as a new genus and species Simias concolor by Miller in 1903. A subspecies, S. c. siberu, was erected by Chasen and Kloss in 1927; since then, nothing more has been recorded. However, as noted by Chasen (1940), the fairly frequent cream-and-tawny mutants of the species resemble the skins of Nasalis larvatus.

SYSTEMATICS

All these monkeys have a marked nasal pecularity. In Rhinopithecus and Simias the nose is prominent and upturned; in the latter the prime direction is forward, but in the former it turns up towards the forehead, the nostrils each having a flap of skin partially covering them from the distal-external margin. In Pygathrix the nose is flat but the little flaps of skin are present. The nose of the juvenile Nasalis is like that of Simias, but with age becomes more prominent; in the adult male it is markedly bul-bour and pendant, overhanging the mouth.

The limb proportions of all four genera are almost identical. In particular the intermembral index is higher than in any other Colobine, being approached only by the Red Colobus (Table 6).

It is from the skulls however (Tables 1 and 2) that the most convincing evidence for the interrelationships of the four genera can be derived. The skulls of _Nasalis_ and _Simias_ are narrow, and long-faced; those of _Rhinopithecus_ on the other hand are markedly broad, with wide square orbits and short-faces; in _Pygathrix_ these features are only slightly less marked. The interorbital space of the first two is very narrow, quite unlike the second two. The braincase is longer in the second group, shorter and more rounded in the first. The second group have robust jaws with out-turned angles and high ascending rami. The nasal bones of _Nasalis_ and _Simias_ are long and narrow while those of _Rhinopithecus_ and _Pygathrix_ are short, imperfectly formed and sometimes partially or even, in the former, completely suppressed on one or both sides.

Externally, the _Nasalis_-_Simias_ group can be distinguished from the _Rhinopithecus_-_Pygathrix_ group by the characters of the nose, as described above, and by the backswept, closely adpressed hairs of the cheeks, contrasting with the shorter, more untidy, hair of this region of the latter. The _Nasalis_-_Simias_ group also have a differential hair-pattern, the hair being more fully banded on the proximal half of the back then on the distal.

At the same time, it should be pointed out that the relationship between _Pygathrix_ and _Rhinopithecus_ is less close than that between _Nasalis_ and _Simias_. The color-pattern is quite different, although it is true that within _Rhinopithecus_ there are also prominent differences in color-pattern. The nose is not upturned in _Pygathrix_, although it resembles a flattened version of that of _Rhinopithecus_. There are also a few differences in the skull, mainly concerned with the more powerful jaws and teeth of the latter.

The question of speciation within the genera recognised is a vexed one. In case of allopatric species, such as those of _Rhino-_

pithecus, there is no clear biological criterion which can be employed to decide whether the geographically discrete forms which are recognisable should be classed as species or subspecies; thus one must fall back on purely morphological criteria. The phenotypic differences between the forms under consideration have to be assessed to try to determine what magnitude of difference they represent in the genotype. Currently, differences of color within the limits of a single color-pattern are assessed as of lesser value than differences in the color-pattern itself, (see, for example, Booth, 1955, on speciation within the Cercopithecus mona group); by the same token, mere size differences will be of lesser significance than differences of shape and proportion, unless of course these latter are purely allometric characters.

In Rhinopithecus, the differences between roxellanae and bieti concern color (within a single pattern type), size, elongation of hairs on the back of the thigh, and the nature of the hair-patterns on the proximal part of the tail. Of these, only the last is possibly of complex genetic origin. The two forms are obviously very closely related, although equally obviously distinct as far as the limited material goes, and rank as subspecies of a single species.

However, the differences between the roxellanae/bieti complex and brelichi are not so simple. They concern tail length (shorter than head and body in the first comples, much longer in the second), absence of noticeable differentiation in body hair type, the nature of hair on the tail, the color-pattern (white tail-tip, absence of white in ano-genital region, white patch on withers), and possibly a different sutural pattern in the skull. The conclusion is that brelichi is a distinct species. It is noteworthy, in addition, that brelichi is intermediate between roxellanae and avunculus (ranking as good species even in the "lumpers" classification of Ellerman and Morrison-Scott).

In Pygathrix, the differences between nemaeus and nigripes, striking though they are, involve only color-substitutions and different forms of the same pattern. There is no difficulty in seeing them as races of a single species.

It is felt that these interrelationships are best recognised by the following arrangements:

Genus Nasalis. Synonym: Simias.

Species: larvatus, concolor.

Genus Pygathrix.

Subgenus Pygathrix.

Species: nemaeus.

Subgenus Rhinopithecus. Synonym: Presbytiscus.

Species: roxellanae, brelichi, avunculus.

A detailed consideration of these taxa follows:

GENUS Nasalis E. Geoffroy, 1812. Synonym: simias Miller, 1903.

Colobine monkeys, with arms nearly as long as legs; nose prominent, tip somewhat upturned at least in females and young; face red or black and naked. Facial portion of skull, long, narrow, and with elongated nasal bones; interorbital space very narrow; braincase short, rounded. Jaw angle somewhat expanded; hypoconulid present on M_3 with a tuberculum sextum in 7-10% of cases; no concavity behind supra-orbital torus. Palate comparatively short. Hair of cheeks brushed back towards midline of neck; hair banded with light and dark on forepart of back, more uniformly colored on hindpart. In males of both species, the ischial callosities are united across the midline. Large, stoutly built monkeys.

Distribution: East Indian Archipelago.

1. Nasalis larvatus Wurmb, 1781. Proboscis monkey.

This species is confined to Borneo. The back is brick-red in color: individual hairs are yellow, becoming red at the tips. Black bands on the hairs become increasingly prominent anteriorly. On the arms this band is strong, and each color is diluted, the yellow base becoming straw-colored and the red tip, yellowish. The overall effect on the arms is gray; the same color occurs on the legs. On the shoulders there is no zone of solid color, but on the hips and upper thighs there is a transitional zone of pure orange between the back and legs. The back-swept cheek hairs are pale reddish in the male, but more yellowed in the female; a creamier tint of the same color extends over the whole under-surface. On the top of the head is a cap of solid dark-red, extending back between the cheek-pads to the withers. Finally, the tail is a striking creamy-white, as is a triangular area on the croup immediately in front of the tail. The head and body length is 66-76 cm in males and 53-61 cm in females, with a tail 95-110 percent of head and body length. Weights: Males, 35-50 lbs; females, 15-24 lbs.

The skin of the face is entirely brick-red. The nose is somewhat elongated even in females and young; in males is enormous and pendulous.

Variations on this pattern concern mainly the intensity of coloration, which may be much lighter and yellower than described above. There may be much less differentiation between the line of the cheek-pads and that of the cap and its posterior extension. Such individuals do indeed predominate in eastern Borneo, as surmized by Chasen (1940), who referred them to his subspecies _N. l. orientalis_; but also they occur elsewhere, together with animals showing an intermediate state of coloring.

Table I shows that three morphological demes are recognizable, but again there is nothing to indicate subspecific differences. The localities are as follows:

Western: R. Sampang, Pontianak, Mankol, R. Kapuas.

Eastern: R. Baai, Labuan Bini, R. Birang, R. Domaring, Karan Tigan, R. Berau, R. Segah, Buntok.

South-eastern: Klumpang Bay, Balikpapan Bay, R. Karang Mumus, Pamukang Bay, R. Djambajan.

Note that the "South-eastern" deme is coastal only. The specimens from Buntok on the R.Barito at 1°40'S, 114°50'E belong geographically to the southeast but occur further inland; they are identical to the "Eastern" deme (topotypes of orientalis) in skull morphology, but none of the nine skins is of the pale orientalis type.

2. Nasalis concolor. Miller, 1903. Pig-tailed langur. Confined to the Mentawai Islands: Siberut, Sipura, Pagai Utara and Pagai Selatan.

The back is dull brownish-black in color; solid at the hind-parts but on the foreparts the hairs are banded with cream and brown-black. This banding does not extend to the arms, unlike N. larvatus. The hair on the cheeks is brushed back forming a thick mat over the ears, as in N. larvatus, but is not lighter than the general body-color. The tail is very short and quite naked except for a few hairs at its tip.

Miller (1903) and Chasen and Kloss (1929) both record creamy-buff mutants, said to resemble skins of N. larvatus.

The head and body length is 450-525 mm, and the tail is 23-35 percent of the head and body length. Weight of a single female specimen is given as 15 lbs 2 oz.

The Pig-tailed langur is said to be a robustly built, thickset animal. The skin of the face is black; the nose is not as elongated as in N. larvatus.

The ascending ramus of the mandible is higher than in N. larvatus.

Although too few specimens are available for accurate judgement Chasen and Kloss's (1929) Simias concolor siberu may be a valid race. The type skull is significantly smaller and broader than specimens from Pagai Selatan and Sipura (Table I); this is due largely to its lesser degree of prognathism. Females are not smaller but, again, are broader.

The two species of this genus have diverged considerably in isolation, but their relationship is clear to see (but scc Mcdway, this volume). Essentially, N. larvatus has become larger, with a much greater degree of sexual dimorphism, and its coloration has become dominantly erythristic. On the other hand, N. concolor has become dominantly melanistic with a remarkable specialization of the tail. It may be noted that the prime habitat of both seems to be swampy coastal forest. The resonant, honking alarm call of N. larvatus, made by the male through his exaggerated nose (Kern, 1965), would seem to be unnecessary in an environment like the Mentawai Islands where there are no large predators.

GENUS Pygathrix E. Geoffroy, 1812.

Colobine monkeys with arms nearly as long as legs; nose (often upturned towards forehead) with a small flap on upper rim of each nostril; face naked and dark colored (yellowish to blue) in circumocular region, more furry and white around mouth. Facial portion of skull short and broad; with nasal bones reduced, sometimes obsolete on one side or both; interorbital space very broad; orbits prominent with strong bony torus above and outside them. Palate very long, often extending back far beyond third molar. Jaw angle slightly or not expanded. Hypoconulid always present on M3, with a tuberculum sextum in 20-100 percent of cases according to species. Sutures fuse rather late in life, being still easily visible when M3 is well worn. Color-patterns vary from species to species. Ischial callosities are always relatively small. Large, robust species: China and Indo-China. The four species replace each other geographically.

SUBGENUS Pygathrix E. Geoffroy, 1812.

Nose flat; black blue-gray, with 7-10 alternate black and white bands on individual hairs. Hands and feet black; underside, throat and cheeks white; red collar, bordered behind with black, on throat; a red forehead band, bordered behind with black. Tail and perineum white, extending in a triangle in front of tail-root. In skull, face not quite so broad; palate not as long as distance between staphylion and basion.

1. Pygathrix (Pygathrix) nemaeus Linnaeus, 1771. Douc.

The only species of this subgenus. There are two well-marked subspecies:

a. Pygathrix (Pygathrix) nemaeus nemaeus Linnaeus 1771.

Central Vietnam and southern Laos. The legs are maroon from the knees down, but the thighs are black. The arms are white from the elbows to the wrists, the upper arms being colored like the body. The white hairs on the cheeks are longer, and on the bi-color frontal band, red predominates over black. The face around the eyes is reddish-yellow. The skull is larger than in the second race; 2 percent (9 skulls) show a tuberculum sextum on M_3; 21.4 percent (14 skulls) show a persistent metopic suture.

Localities: Nape (Laos), Mt. Sontra and Col des Nuages (S. Vietnam). There is no evidence for the existence of the Douc on Hainan.

b. Pygathrix (Pygathrix) nemaeus nigripes Milne-Edwards, 1871.

The hindlimbs are entirely black; the forelimbs (except for the hands) are colored entirely like the body. The cheek-hairs are shorter; on the forehead band, black predominates over red. The circumocular region is black. The skull is smaller than the nominate rate; 60 percent (of 10 skulls) have a tuberculum sextum; there is no metopism. A note on one of the labels in the British Museum (Nat. Hist.) states that the penis and scrotum are blue.

Kloss (1926) described a supposedly new race of Douc

as Pygathrix nemaeus moi. The characters on which this form was founded fall within the range of variation of the subspecies as a whole; indeed they are weakly marked even in their type skin. It is possible that skulls from the northern part of the subspecies range average larger than those from the southern part, tending towards the nominate rate.

Localities: Saigon, Bienhoa, Trangbom, Djiring, Dalat, Langbian, Nhatrang (South Vietnam).

SUBGENUS Rhinopithecus Milne-Edwards, 1872.

Nose raised, "snubbed"; color variable, but always with underside and innerside of limbs light-colored. In skull, face very broad; palate longer than distance from staphylion to basion.

1. Pygathrix (Rhinopithecus) roxellanae Milne-Edwards, 1870, Golden Snub-nosed monkey.

Size medium to large; tail shorter than head and body length (86-95 percent); body hair long (50-80 mm), dark grayish, with long (120-160 mm) yellow guard-hairs intermingled; tail gray brown, darkening towards tip, its hairs generally straight but sometimes curly towards base; a white area on backs of thighs and in genital region; top of head blackish in male; underside whitish or golden; no white on whithers. Digits short, stubby. Skull large, with bregma indented posteriorly. The nostril flaps are well shown in a photograph on p. 30 of Van Gulik (1967).

The distribution encompasses part of the mountainous region on the south-eastern slopes of the Tibetan plateau, in Szechwan, Kansu and Yunnan. There are two well-marked subspecies.

a. Pygathrix (Rhinopithecus) roxellanae roxellanae Milne-Edwards, 1870.

The eastern flanks of the Szechwan mountains. The

smaller of the two, head and body length 680-760 mm in male; circumocular region pale bluish; tail hairs not curly; hands and feet golden; underparts reddish gold; brightest in old males. Hairs not elongated in ischial region. 21.4 percent (14 skulls) show a tuberculum sextum.

Localities: Moupin, Kwanhsien, Wen Chuan Hsien, Yang liu pa (Szechwan); Ssigou (Kansu). The supposed occurrence of this form in Assam (Gee, 1952), more likely refers to the subsequently discovered *Presbytis geei*.

b. *Pygathrix* (*Rhinopithecus*) *roxellanae bieti* Milne-Edwards, 1897.

Between the Yangtse and Mekong rivers in Yunnan. Larger (perhaps the largest of the Cercopithecoidea), head and body length 830 mm in male; circumocular region greenish; tail hairs curly towards the base in the male; hands and feet black; underside whitish-grey. Hairs very elongated in ischial region and down backs of upper part of thighs in male, forming a fringe 200-210 mm long. A little crest, drooping forward with hairs 97-120 mm long in male, on crown. All three adult skulls show a tuberculum sextum.

Localities: Atentse, Tsekou (Yunnan). Restricted to the eastern side of the Mekong Gorge. The only known specimens are those of the type series in Paris, consisting of two adult males, two adult females, two juveniles and an infant.

It is noteworthy that in the three large museum series of the nominate race, the sex ratio for adults is very disproportionate, as follows:

B.M. (N.H.)	One male, four females. One male not in the same series.
U.S.N.M.	One male, nine females.
A.M.N.H.	One male, four females and one specimen, sex unknown.

This makes altogether 4 males and 17 females in collections, a most unusual proportion, especially in view of the more brightly colored and conspicuous pelage of the male. The dominantly female element in collections suggests that the species may perhaps live in one-male groups.

2. Pygathrix (Rhonopithecus) brelichi Thomas, 1903.

Size medium, head and body length 730 mm., tail longer than head and body (133 percent); body hair very long (70-80 mm), brownish-gray, without long guard hairs; tail blackish with white tip, with hairs curly, parted above for most of tail length, in male; no white area on back of thighs or genital region; top of head light gray with a dark crown mark; underside dirty gray white; a white patch on withers; skull unknown in adult, but in the juvenile in the British Museum the bregma is not indented. To judge from this juvenile the skull would on average be smaller than P.(R) r. roxellanae (Table 4). The digits are short as in the previous species.

Localities: Van Gin Shan range, south of Middle Yangtse. No other records exist for this species, of which only the type and one other specimen are known.

Rhinopithecus tingianus Matthew and Granger, 1923, from the Middle Pleistocene of Yenchingkou, China, is considered by Colbert and Hooijer (1953) as a race of P. roxellanae. The locality concerned is only about 110 miles NNE of the Van Gin Shan; consequently it would be more probably a race of the present species, if the subgenus was divided into species by that time. Certainly it would be ancestral to P.(R) brelichi. The skull, as kindly measured and estimated by Mr. Eric Delson (Table 2), would have been a little smaller than a female P.(R) roxellanae, and considerably smaller than a male - perhaps comparable with that suggested above for brelichi - but the teeth are much larger than those of the only available specimen of the latter: a not unknown trait of Pleistocene ancestral forms. It is provisionally proposed to class tingianus as a subspecies of brelichi; to judge from the figure in Colbert and Hooijer, the bregma - as in brelichi - is not indented

3. Pygathrix (Rhinopithecus) avunculus Dollman, 1912. Tonkin snub-nosed monkey. Size small, head and body length 560-620 mm; tail much longer than head and body (143-148 percent); body hair short (45-50 mm), black, without long guard hairs; tail black with straw-colored hairs intermixed, and white tip, hairs curly and parted above in the male; a white area on backs of thighs and in genital region; top of head ashy-brown with a black crown mark; underside white; no white on withers. Bregma of skull not indented. Digits long, slender, like those of P. (P) nemaeus, Nasalis and Presbytis. All of the seven skulls show a tuberculum sextum on M_3.

The face of this species is stated on a label to be as follows: "face pale bluish white, a stripe across nose to hairly part of upper lip, violet, 13 mm deep". In another specimen, the label states, "lips terracotta". Another has, "chin-flesh-colored". Localities: Bac-kan and Yen-bay, N. Vietnam. The species is known only from these localities, and has only been collected by two expeditions.

Of remaining Colobines studies, only the following species show a tuberculum sextum: Colobus guereza, 33.3 percent (24), Colobus verus, 90.9 percent (22), Presbytis phavrei, 13.0 percent (23), Presbytis entellus, 8.7 percent (23).

CLASSIFICATION AND PHYLOGENY OF THE COLOBINAE

Table 5 compares the average figures for some important indices in the Colobinae. It is noteworthy that in the facial and face-braincase indices, Nasalis has the highest values in the subfamily, and Pygathrix has some of the lowest. Thus in four out of five indices, selected for their ability to discriminate, the two genera stand at opposite ends of a spectrum.

The only things which Pygathrix and Nasalis have in common

are the upturned nose (in some species) and the high intermembral index (Table 6). Since in skull features and external characters they are so far apart, the possibility must be seriously considered that these two features are characters of common inheritance (symplesiomorph) among Colobines. That this is quite likely the case with the limb proportions, is borne out by the similarity of the Pontian colobine *Mesopithecus* to *Nasalis* and *Pygathrix* (Table 6). The question of the nose is not so simple. It is a highly plastic organ within the Colobinae (Dandelot, 1968), and in any case the structure is not entirely identical in the two genera. *Nasalis* lacks the prominent "nose-flaps" of *Pygathrix*. However, the similarity in general outline is unmistakable and recalls also *Theropithecus* among the Cercopithecinae.

The classification of the other Colobines presents problems which only a detailed study of all groups together can solve. The following remarks may help to identify some of the problems.

Within the genus *Presbytis*, the relationships seem more complex than is expressed by the usual four subgenera or genera. The four "species" *melalophos*, *aygula*, *frontata* and *rubicunda* form a natural group with similar skull shape, limb indices and hair patterns. When their distributions are plotted on a map, they appear approximately to replace one another, with the exception of *rubicunda*, which is sympatric with each of the other three in Borneo. Detailed study and examination of locality records may reveal that only two species are involved. For the present, as suggested by Medway (this volume) a number of species may be recognized; certainly *aygula* is a heterogeneous taxon as it stands and provisionally should be broken into its constituent parts, allowing *aygula* of Java, *thomasi* of North Sumatra, and *hosei* of North-East Borneo each to rank as a separate species.

Presbytis johni and *P. senex*, probably conspecific, are intermediate between *entellus* and *cristatus*. In skull shape they are rather nearer the latter; in limb indices, the former. During a lecture in 1966, Frank Poirer described the infant of *johni* as golden, like *cristatus*.

Presbytis phayrei and P. obscurus, again perhaps conspecific, are as different from cristatus (with which they are usually combined in the genus or subgenus Trachypithecus) as from johni. The golden color of the neonate does, however, provide a link between them.

P. potenziani has a golden infant, but is otherwise quite intermediate in both skull and postcranial skeleton between P. cristatus and the P. melalophos-group.

It would probably be best to recognize no subgenera in the genus Presbytis until it can be ascertained what they would contain. The assemblage does form a natural group, even though there are few obvious characters to define it.

In all the nose is flat, the face comparatively broad, the zygomata bowed, the braincase long compared to the face, the arm 74-85 percent of the leg length. The hair patterns are complex, often throwing up crests on the crown of the head.

In the African colobines, the Olive Colobus (P. verus) tends to be the "odd man out" in its skull indices, as indeed Verheyen (1963) notes, but in limb indices it is the mainland Red Colobus (badius/rufomitratus) which stand out. Dandelot (1968) shows that there are several highly distinctive color types in this Red Colobus group; by the criteria for allopatric species discussed above, some of these would certainly stand as full species, but detailed study has not been undertaken as yet. The genus Colobus is probably a natural group distinguished by the suppression of the thumb. In both skull and limbs it is broadly intermediate between Presbytis and Nasalis. In line with the primitive status of the Olive Colobus as pointed out by Verheyen (1936) and Hill (1952), Procolobus is here recognized as a valid genus. Kuhn (1967) includes the Red as well as the Olive Colobus in the genus Procolobus, citing the lack of sexual dimorphism, the absence of gross age-changes in color, the difference in jaw musculature, the form of the stomach and the presence of sexual swelling in females and adolescent males: these being characters in which the Red and Olive Colobus show similarity and differ from the Black-and-White group. A good case could, however, be made out for regarding most or all of these characters as "characters of primitive retention", or African colobine symplesiomorphs - perhaps even symplesiomorphs of the Colobinae as a whole. At any rate, the Olive Colobus shows such clear-cut primitive characters in such regions as skull, larynx, pelage etc., regions in which the Red and Black-and-White groups show similar specializations, that its generic status should surely be upheld. It could even be maintained that the lack of the thumb has occurred independently (Kuhn, 1967) and that *Procolobus verus* is a relict form whose origins pre-date the separation of the African and Asiatic lines of the Colobinae.

In conclusion then, the subfamily Colobinae would consist of five genera: *Nasalis*, *Pygathrix*, *Presbytis*, *Procolobus* and *Colobus*. The first two stand at opposite poles of the subfamily, although members of both possess similar nose shape (probably superficial) and similar limb indices (probably an ancestral character). *Presbytis* has some groups of closely related species but no clearly separated subgenera, at least as far as can be seen at the moment. *Colobus* and *Pygathrix* on the other hand each have two well-marked subgenera. The genera *Simias*, *Rhinopithecus* and *Presbytiscus* are sunk

APPENDIX: A NOTE ON *PARACOLOBUS CHEMERONI* R. LEAKEY 1969

A new genus and species of Colobine (*Paracolobus chemeroni*) has recently been described from a skull and skeleton from the Chemeron Beds, Kenya, a Lower Pleistocene deposit (R. Leakey, 1969). The completeness of the remains makes this form the best-known fossil Colobine after *Mesopithecus*. Since, in the original description, it was compared only with *Colobus polykomos*, it seems worthwhile to try to assess its position within the subfamily as a whole.

In the first place, this appears to be a monkey of relatively large size. The braincase is incomplete, but the facial skeleton is entire. Since in living Colobines the facial length is never more than 62 percent of the braincase length (and is as low as 39 percent in *Presbytis rubicunda*) it seems reasonable to assume that in the Chemeron form with a facial length of 70 mm the braincase would measure not less than 113 mm (if the face was as long as a Proboscis monkey) and possibly as much as 180 mm (if the face was as short as a Maroon Langur). Either way, it was larger than any living member of the Colobinae.

As estimated from the photographs (Plate I in the description) as well as the measurements given by R. Leakey in his Table I, the facial index would be approximately 67 percent - closest to *Colobus* among the living genera. Palate index and skull breadth index cannot be estimated. The interorbital index, however, is high 36.5, resembling *Pygathrix* rather than any other genus.

As regards non-metrical features, the skull seems to be most similar to *Colobus*. The subraorbital torus is very well-marked, with a distinct concavity behind it, as in Colobus and *Pygathrix* alone. There is a distinct but weak supraorbital notch, as in *Colobus*. If one assumes, as is most likely on geographical grounds, that it belongs to the African section of the subfamily (unfortunately the manus is missing, so this cannot be checked), then one notices further features: the face is strongly prognathous with a sloping premaxillary region, the jaw angle is somewhat expanded, the bregma appears to to be indented, the upper incisors are rather high-crowned, and there is no tuberculum sextum on M_3. These features align it with *Colobus* rather than *Procolobus*, and would seem to indicate that the two African genera had already become separate, unless the Olive colobus has undergone a great deal of "secondary simplification" in its cranial anatomy. Since no photograph or detailed description of the palate and basicranium is given in the original description, no idea of the status of *Paracolobus* can be formed from a consideration of this highly diagnostic region.

It is also of interest to try to assess the intermembral index of the Chemeron form. In living Colobines (Table 6) the brachial index varies from 90 to 116; assuming that the brachial index of

Paracolobus fell within this range, one can obtain estimates of the radius which with the ulna is the only major long bone that is incomplete in the type skeleton. Estimates would range from 230 to 297 mm and an intermembral index, using these figures, would range from 93.6 to 106.5. This puts it in the putative "primitive Colobine" range of Nasalis and Pygathrix (unless it is actually above 100, a proportion unknown in living Cercopithecoidea), and not too far from the living Colobus (Piliocolobus) badius-group. This would seem to be a very plausible result.

In conclusion, then, the available data implicate Paracolobus chemeroni as already recognizably closer to the modern Colobus than to any other genus (including Procolobus), but with characters that set it off as fairly primitive genus, recalling in particular Pygathrix among the Asiatic group.

ACKNOWLEDGEMENTS

Many thanks are due to Dr. Charles O. Handley, Chairman of the Division of Mammals, Smithsonian Institution, Washington, D.C. and Dr. J.R. Napier until recently the Director of the Primate Biology Program at the Smithsonian; to Dr. G.B. Corbet and Mr. J.E. Hill, of the Mammal Section, British Museum (Natural History), London, and to Monsieur F. Petter and Monsieur M. Roch of the Department des Mammifères et Oiseaux, Musée Nationale d'Histoire Naturelle, Paris, for permission to study the collection under their charge, and help in doing so. I also wish to state my indebtedness to Dr. J.R. and Mrs. P.H. Napier, and Monsieur P. Dandelot, Mr. T.F. MacIntyre, Mr. P. Ryan and Dr. H.-J. Kuhn, for stimulating discussions, although they are in no way responsible for the conclusion here presented. I am also very grateful to Mr. Eric Delson and Dr. Cliff Jolly for supplying me with important information and measurements on material in the American Museum of Natural History (New York); and to Mr. John Osborne, of the Faculty of Archaeology and Anthropology, Cambridge University, for drawing the figures.

REFERENCES

Allen, G.M. (1938) Mammals of China and Mongolia, Part I. New York: American Museum of Natural History.

Booth, A.H. (1955). Speciation in the Mona monkeys. J. Mammal. 36: 434-449.

Buchner, E. (1891). Die Säugethiere der Ganssu-Expedition (1884-87). Mel. Biol. Bull. Acad. Imp. Sci. St. Petersb. 13: 146.

Chasen, F.N. (1940). A Handlist of Malaysian Mammals. Bull. Raffles Mus. 15: 1-209.

Chasen, F.N. and Kloss, C.B. (1927). Spolia Mentaweiensia. Proc. zool. Soc. Lond. 797-840.

Colbert, E.H. and Hooijer, D.A. (1953). Pleistocene Mammals from the Limestone Fissures of Szechwan, China. Bull. Amer. Mus. Nat. Hist. 102: Art. 1.

Dandelot, P. (1968). Primates: Anthropoidea. Preliminary Identification Manual for African Mammals 24. Washington: Smithsonian Institution Press.

De Winton, W.E. (1899). On Chinese Mammals, principally from Western Szechwan. Proc. zool. Soc. Lond. 572-578.

Ellerman, J.R. and Morrison-Scott, T.C.S. (1951). Checklist of Palaearctic and Indian Mammals, 1758 to 1946. London: British Museum (Nat. Hist.).

Gee, E.P. (1952). Possible occurrence of the Snub-nosed monkey (Rhinopithecus roxellanae) in Assam. J. Bombay Nat. Hist. Soc. 51: 264-5.

Hill, W.C.O. (1936). Supplementary observations on Purple-faced leaf monkeys. Spolia Zeylano 20: 115-133.

Hill, W.C.O. (1952). The external and visceral anatomy of the Olive Colobus monkey (Procolobus verus). Proc. zool. soc. Lond. 127-186

Hooijer, D.A. (1962). Quaternary langurs and macaques from the Malay Archipelago. Zool. Verh. Leiden 55: 1-64.

Jarvis, C. (1966). International Zoo Yearbook Vol. 6 (Ed. Jarvis, C.). London: Zoological Society.

Kern, J.A. (1965). Observations on the habits of the Proboscis monkey, Nasalis larvatus (Wurmb), made in the Brunei Bay area, Borneo. Zoologica 49: 183-192.

Kuhn, H.-J. (1967). Zur Systematik der Cercopithecidae. In Neue Ergenbnisse der Primatologie (Progress in Primatology), (eds. Starck, D., Schneider, R. and Kuhn, H.J.) Stuttgart: Gustav Fischer.

Leakey, R.E.F. (1969). New Cereopithecidae from the Chemeron Beds, Kenya. In Leakey, L.S.B., ed., 1969, Fossil Vertebrates of Africa 1: 53-70.

Miller, G.S. (1903). On ninety-two new mammals from Malaysia. Smithsonian misc. Coll. 45: 1-73.

Osgood, W.H. (1932). On Mammals from Indo-China from the Kelley-Roosevelts and Delacour Expeditions. Field Mus. Nat. Hist. Zool. 18: 191-339.

Pocock, R.I. (1934). The monkeys of the genera Pithecus (or Presbytis) and Pygathrix found to the East of the Bay of Bengal. Proc. Zool. Soc. Lond. 895-961.

Van Gulik, R.H. (1967). The Gibbon in China. Leiden: E.J. Brill.

Verheyen, W.N. (1962). Contribution a la craniologie comparee des primates. Musee Roy. Afr. centrale Tervuren Ser. 8. Sci. Zool. 105: 1-256.

KEY TO TABLES 1-3

Facial index	-	Nasion prosthion x 100/Biorbital breadth
Palate index	-	Staphylion-basion x 100/Palate length
Skull breadth index	-	Biorbital breadth x 100/Bizygomatic breadth
Interorbital index	-	Interorbital width x 100/Postorbital breadth
Face braincase index	-	Nasion-prosthion x 100/Nadion inion

TABLE 1

SKULL MEASUREMENTS IN THE GENUS *NASALIS*

MEANS AND STANDARD DEVIATIONS

Male	Prosthion-inion	Prosthion-basion	Bizygomatic breadth	Nasion-prosthion	Toothrow	Interorbital width	Ascending ramus ht.
N. larvatus							
Western (8)	129.2 ± 1.76	92.3 ± 1.38	92.1 ± 1.69	58.6 ± 1.95	32.6 ± 1.12	11.9 ± 0.62	49.6 ± 3.68
Eastern (5)	127.3 ± 1.73	91.3 ± 2.47	85.9 ± 2.12	56.4 ± 1.56	32.6 ± 1.12	11.2 ± 0.90	47.8 ± 2.74
Southeastern (13)	131.1 ± 2.32	94.8 ± 3.00	91.0 ± 1.80	57.6 ± 3.54	32.7 ± 0.87	12.3 ± 1.32	48.0 ± 3.39
N. concolor							
S. Pagai, Sipora (4)	104.8 ± 1.73	74.8 ± 1.00	72.8 ± 2.65	44.5 ± 0.87	28.7 ± 1.06	8.0 ± 1.00	44.5 ± 4.50
Siberut (1)	96.5	66.0	72.5	40.0	28.0	7.0	44.0
Female							
N. larvatus							
Western (2)	104.5	72.5	74.7	40.5	29.0	8.8	39.5
Eastern (7)	108.3 ± 2.92	75.8 ± 2.32	75.6 ± 1.41	43.6 ± 2.49	30.1 ± 1.41	8.5 ± 1.14	40.3 ± 2.65
Southeastern (5)	106.2 ± 3.16	73.3 ± 3.32	73.7 ± 2.18	42.8 ± 1.32	29.9 ± 2.76	8.1 ± 0.90	41.4 ± 3.00
N. concolor							
S. Pagai, Sipora (5)	95.0 ± 2.24	66.2 ± 1.41	66.6 ± 1.64	38.6 ± 2.92	27.1 ± 1.23	6.7 ± 0.76	39.4 ± 1.71
Siberut (2)	95.9	65.5	71.7				

TABLE 2

SKULL MEASUREMENTS IN THE GENUS PYGATHRIX

MEANS AND STANDARD DEVIATIONS

	Prosthion-inion	Bizygomatic breadth	Palate length	Staphylion-basion	Toothrow	Interorbital width	Symphysis ht.
Male							
P. (R.) r. bieti (1)	135.0	103.0	54.0	-	34.0	18.0	39.0
" " r. roxellanae (3)	127.3 ± 3.24	98.7 ± 0.71	48.7 ± 1.23	41.0 (1)	33.8 ± 1.00	18.5 ± 1.50	39.0 ± 3.00
" " avunculus (4)	117.0 ± 2.82	94.0 ± 2.45	43.5 ± 1.00	39.0 ± 2.00	30.0 ± 1.00	16.0 ± 1.41	34.3 ± 1.30
" (P.) n. nemaeus (6)	116.8 ± 1.48	82.5 ± 3.26	40.7 ± 1.67	44.3 ±1.79	29.9 ± 1.10	15.2 ± 0.78	24.5 ± 1.55
" " " nigripes (8)	111.8 ± 3.63	80.8 ± 1.90	36.7 ± 1.23	43.0 ± 2.65	29.7 ± 1.32	13.0 ± 0.87	23.8 ± 0.79
Female							
P. (R.) r. bieti (1)	120.0	84.0	45.0	36.0	33.0	18.0	29.0
" " r. roxellanae	110.5 ± 2.74	84.9 ± 1.79	40.9 ± 2.07	34.7 ± 1.92	31.9 ± 0.77	15.6 ± 1.00	31.7 ± 2.59
" " avunculus (2)	104.5	78.5	35.5	34.0	29.0	12.5	22.9
" (P.) n. nemaeus (3)	107.7 ± 2.12	74.7 ± 3.61	35.7 ± 1.00	38.7 ± 2.55	28.0 ± 0.00	12.7 ± 0.71	20.3 ± 0.71
" " n. nigripes (2 to 3)	103.5	75.2 ± 4.04	33.3 ± 2.51	38.0	27.3 ± 0.58	10.8 ± 2.17	20.3 ± 1.58
? " (R.) tingianus (1)	(107.5)	(75.5)	36.4	(33.0)	-	11.5	-

TABLE 3

SKULL MEASUREMENTS OF SUBGENUS *RHINOPITHECUS* IN JUVENILE I STAGE (DECIDUOUS DENTITION + M_1^1)

	Prosthion-inion	Bizygomatic breadth	Palace length	Staphylion-basion	Interorbital	Symphysis
P. (*R.*) *r. roxellanae*						
B.M.8.10.9.3	95	68	31	28	10	21
8.10.9.5	98	71	34	26	12	17
P. (*R.*) *brelichi*						
B.M.5.3.24.1	89	65	29	24	10	21

TABLE 4

FIRST MOLAR MEASUREMENTS IN THE SUBGENUS RHINOPITHECUS

	Upper		Lower	
Male	Length	Width	Length	Width
P.(R.) r.bieti (2)	8.0 - 8.6	7.6 - 8.0	7.5 - 8.4	6.3 - 6.5
" " " roxellanae (5)	7.5 - 8.8	7.4 - 8.0	7.5 - 8.0	6.2 - 6.5
" " avunculus (2)	6.3 - 6.4	6.0 - 6.4	6.2 - 6.5	5.0 - 5.2
" " tingianus (5)	7.3 - 8.4	7.4 - 7.7	8.0	7.3
Female				
P.(R.) r.bieti (1)	7.7	7.5	7.3	6.0
" " " roxellanae (14)	7.0 - 8.5	6.4 - 7.7	6.9 - 8.1	4.7 - 6.7
" " brelichi (1)	6.3	5.8	6.2	4.6
" " avunculus (3)	6.0 - 6.4	6.0 - 6.2	5.8 - 6.5	4.9 - 5.0

(including figures from Colbert and Hooijer, 1953)

TABLE 5

AVERAGE SKULL-INDICES OF SELECTED SPECIES OF COLOBINAE

	Braincase length		Facial Index		Palate Index		Skull Breadth Index		Interorbital Index		Face Braincase Index	
	♂	♀	♂	♀	♂	♀	♂	♀	♂	♀	♂	♀
NASALIS												
larvatus (13/7)	91.8	83.1	86.6	74.3	108.4	111.4	75.9	77.7	24.9	19.3	61.4	52.4
concolor (3/5)	78.3	74.6	76.5	67.7	107.1	104.6	81.3	85.6	20.4	17.1	56.8	51.7
PYGATHRIX												
r. bieti (1/1)	105.0	98.0	52.2	51.2	-	80.0	89.3	94.4	34.6	35.3	45.7	43.9
r. roxellanae (3/10)	99.7	90.1	54.7	52.0	85.4	84.8	87.4	88.3	36.1	31.3	47.4	43.3
avunculus (4/2)	91.5	86.5	56.5	54.4	89.6	95.8	86.2	86.6	32.0	26.4	50.1	42.8
n. nemaeus (6/3)	88.3	82.7	59.0	57.4	108.8	108.4	93.7	97.3	30.2	24.7	50.6	50.4
n. nigripes (8/3)	86.2	84.0	58.9	56.3	117.2	114.1	90.3	90.4	26.1	21.6	49.6	45.6
PRESBYTIS												
entellus (7/11)	89.4	84.4	59.2	59.0	88.9	85.5	76.9	80.0	25.7	22.6	46.3	46.0
johni (8/5)	80.3	77.2	58.9	58.7	97.9	95.0	76.0	77.0	21.3	20.0	45.5	44.7
cristatus (9/8)	76.3	72.4	62.6	61.0	102.9	104.7	75.3	79.5	24.6	24.4	48.1	47.2
phayrei (11/16)	80.5	77.2	59.1	59.1	110.5	113.1	81.1	82.9	21.7	21.1	46.2	44.9
potenziani (7/4)	79.0	77.0	52.9	47.5	102.1	109.9	81.5	85.0	20.0	19.3	44.3	37.7
rubicundus (13/15)	77.5	77.9	50.5	48.4	123.0	117.4	85.1	94.2	17.0	17.7	39.2	37.9
COLOBUS												
polykomos (28/19)	86.4	82.9	67.2	64.4	104.1	103.9	74.4	75.5	25.2	23.1	50.0	46.3
badius (64/48)	85.5	78.4	67.8	67.7	98.7	97.1	76.8	79.9	26.9	24.0	51.8	50.3
verus (6/8)	73.4	71.6	61.3	59.1	110.6	119.7	77.4	80.5	20.8	21.1	44.7	43.9

(including figures from Verheyen, 1962)

TABLE 6

LIMB INDICES IN SUBFAMILY COLOBINAE

	Brachial index	Intermembral index	
Mesopithecus pentelici	98	88	
Nasalis spp.	97 - 107	90 - 98	(1)
Pygathrix spp.	97 - 107	92 - 96	(11)
Presbytis aygula/melalophos	103 - 111	74 - 80	(16)
" chrysomelas/frontata/ rubicunda	108 - 116	74 - 77	(14)
" senex	98 - 106	74 - 78	(10)
" johni	100 - 106	76 - 81	(4)
" entellus	98 - 106	77 - 84	(8)
" potenziani	106	81	(1)
" pyrrhus/cristatus	92 - 99	79 - 82	(12)
" obscurus/phayrei	95 - 101	78 - 85	(11)
" geei	96 - 101	77 - 80	(9)
Colobus guereza	90 - 97	75 - 80	(7)
" polykomos	96 - 100	76 - 80	(9)
" badius/rufomitratus	94 - 103	83 - 89	(13)
" kirkii	92 - 97	78 - 79	(2)
" verus	103	80	(1)

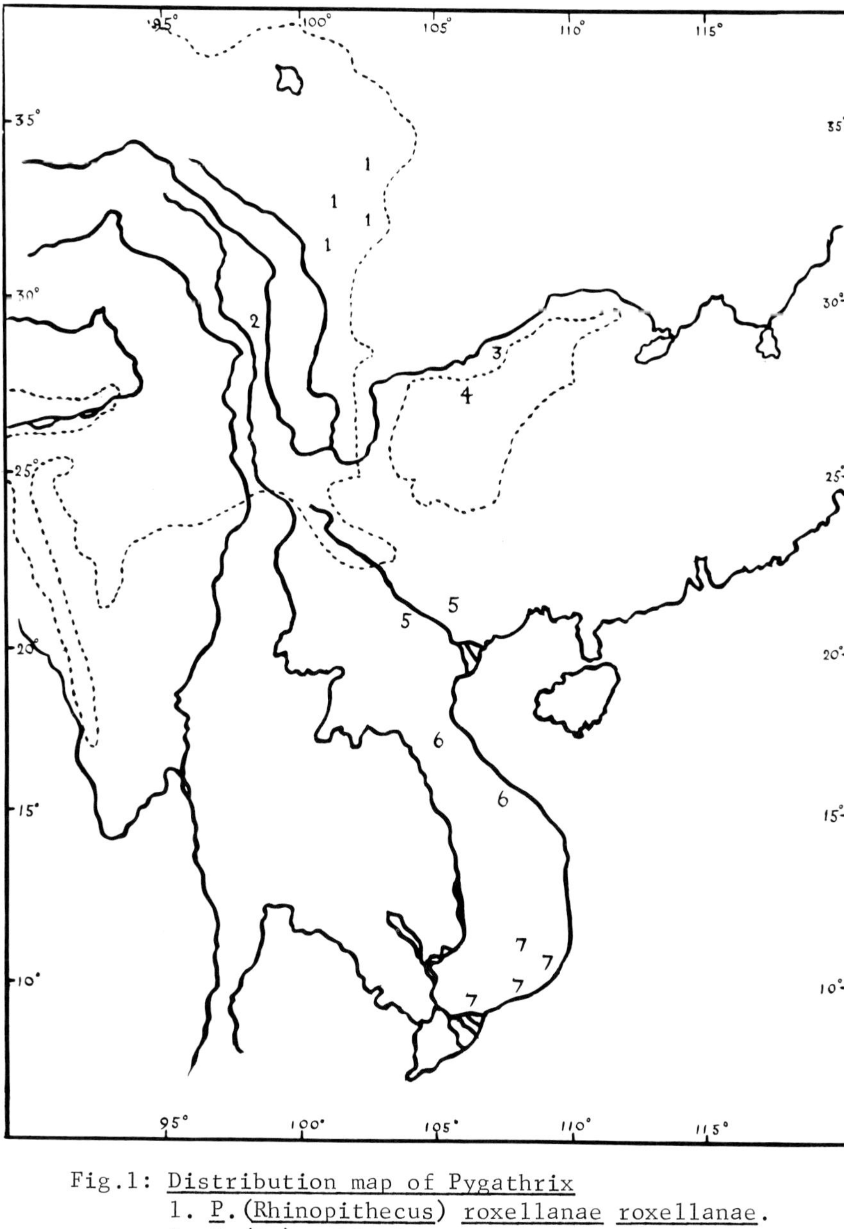

Fig.1: Distribution map of Pygathrix

1. *P.* (*Rhinopithecus*) *roxellanae roxellanae*.
2. *P.* (*R.*) *r. bieti*.
3. *P.* (*R.*) *brelichi tingianus*.
4. *P.* (*R.*) *b. brelichi*.
5. *P.* (*R.*) *avunculus*.
6. *P.* (*Pygathrix*) *nemaeus nemaeus*.
7. *P.* (*P.*) *n. nigripes*.

Fig. 2: Color pattern of P. (R.) avunculus (not previously figured).

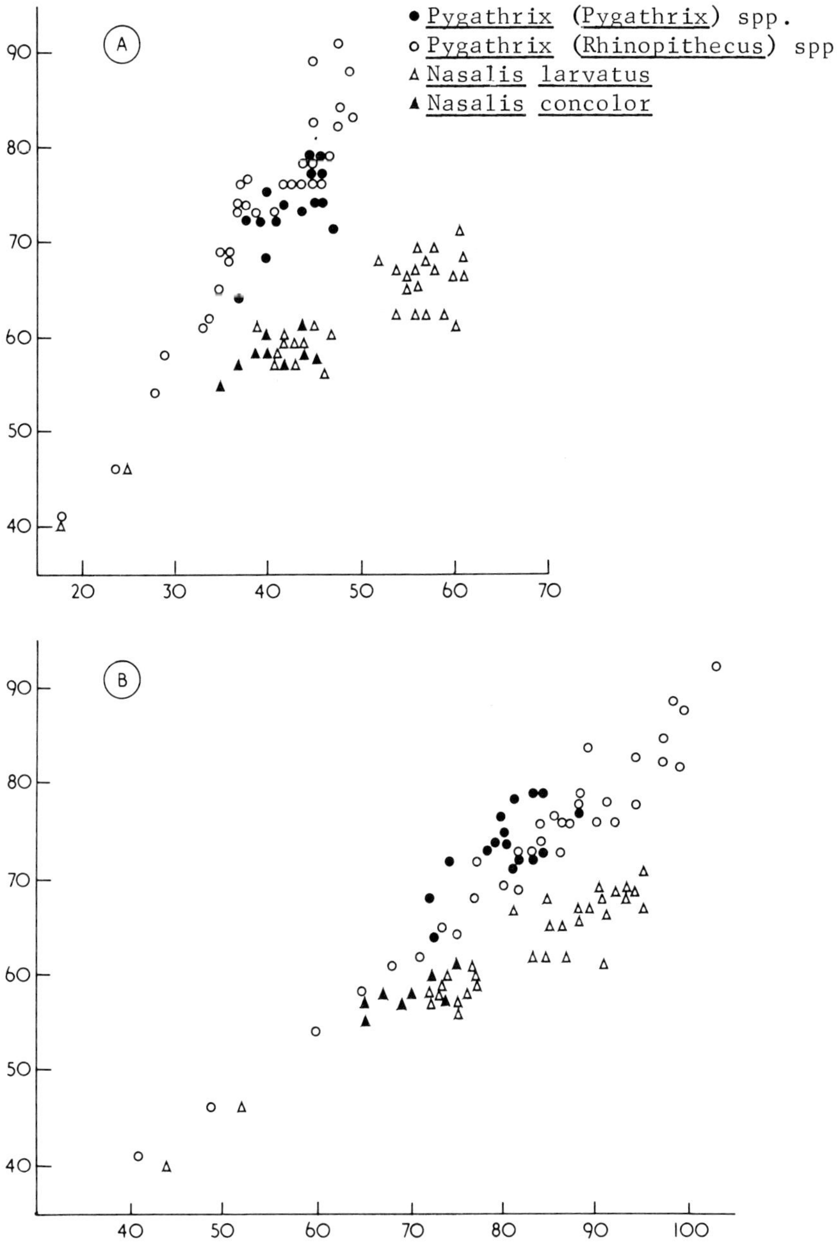

Fig.3: Graphs of skull measurements for *Pygathrix* and *Nasalis*.

(a) Ordinate - Biorbital breadth; Abscissa - Nasion-prosthion length (facial length).

(b) Ordinate - Biorbital breadth; Abscissa - Bizygomatic breath.

ECOLOGY, ZOOGEOGRAPHY, AND SYSTEMATICS OF SOME AFRICAN FOREST MONKEYS

Urs H. Rahm

INTRODUCTION

This paper is an attempt to establish how far ecological and geographical barriers can contribute to the solution of some outstanding problems of primate systematics. The present survey is limited to certain forest monkeys of Africa, with special reference to Central and East Africa, the habits of which are well known from personal experience. A brief summary of the present and past distribution of the forest would appear to be necessary.

The African rain forest, as it exists today, begins in Sierra Leone as a relatively narrow band running along the coast of West Africa until interrupted by the so-called Dahomey Gap, where the savanna reaches the coast. It continues through Equatorial Africa and covers most of the Congo Basin and the northern and eastern Congo extending as far as the mountain chain of the Central African Rift Valley. In many regions, especially in the southern Congo, the forest makes fingerlike penetrations of the savanna to form gallery forests, an important primate habitat. Forest islands or "outliers" may be found in the savannas close to the main forest block. In Uganda and western Kenya there exist several forest islands which are botanically and ecologically similar to the equatorial rain forest. In East Africa, equatorial conditions are limited to a coastal belt where residual forest is mixed with woodland extending far south (e.g. Knysna forest in the Cape region). On the mountain ridge through the Central African Rift Valley and on isolated mountains in Ethiopia, Kenya, Tanzania and South Africa, many relics of montane forest are to be found, most of them geographically separated from each other (Fig. 1).

We have evidence that the equatorial forest belt was subjected to periodic contraction and expansion but maintained a permanent core in some areas. The tectonic changes and the climatic oscillations which began at the end of the Miocene reached a climax in the Pleistocene. There are reasons to believe that the four major glaciations in the northern hemisphere had their counterpart in the southern hemisphere. During the glaciation at higher altitudes the tropical region had a cooler and wetter climate and montane conditions were to be found as low as 500 meters.

In the Pleistocene the forest covered a much greater area and the now isolated montane forests in East Africa were more closely linked at that time. Lowland forest also had a much wider distribution. This situation created ideal living conditions for most of the forest animals in East Africa and elsewhere.

In considering the present distribution of the forest animals, particularly those inhabiting montane forest, we have strong reason to believe that the mountains of the Central African Rift Valley were linked by highland forest to those of the Cameroon region and that the mountains in Angola were connected by montane forest, through Katanga, with forest bordering L. Tanganyika. A similar, more or less continuous connection, probably existed from the Ethiopian highlands down to the Cape passing through East Africa. On the other hand, the dryer and hotter conditions during the interglacial periods (especially between the Mindel and Riss glaciations) were responsible for the drying out of lakes and the contraction of the forest which survived only on some of the higher mountains. The most important montane forest blocks persisted, probably in the region of the Ruwenzori, the Virunga Volcanoes and Mt. Cameroon. There is reason to believe that even during the worst interpluvial periods, Uganda and the region around L. Victoria, was covered with lowland forest. At present, this forest only survives in form of forest islands (Budongo, Bugoma, Kabira, Bukoba, Kakamega and others). The relatively recent regression and disappearance of the forest in this area is mainly due to human activity. The forest has been cut for firewood and to create pasture for cattle. An identical situation is to be found on the slopes of mountains in East Africa. These forest islands still harbour many species of mammals belonging to the fauna of the Congo forest, most of them belonging to the same subspecies (Table 1). _C. ascanius_, for instance, occurs as far as the Mau Escarpment. Further eastwards, fewer typical rain forest mammals are found, having been replaced geographically by other subspecies or even species. Above species level nearly every family of the forest mammals has one or two representatives that can be found in the forest islands of east and southeast Africa. These animals show a more or less wide distribution and form a number of species and subspecies. Among the most important species and genera are Cephalophus caerulus, Heliosciurus gambianum, Cricetomys gambianus, Anomalurus, Dendrohyrax. For example, Cephalophus caerulus, has a very similar distribution to that of Cercopithecus mitis.

It is not the purpose of this paper to discuss mammals other than monkeys in any detail, but the problem of speciation and subspeciation is essentially the same. When a forest area is cut into two, either by climatic changes or human activity, the fauna of the two blocks may evolve in different ways and ultimately give rise to new subspecies. If for instance two forest areas previously continuous are still loosely linked by woodland, two populations may occasionally meet in this area and interbreed. When the two forests are definitely isolated for a long period by an ecological barrier new species may eventually arise. Such separations and isolations of forest areas, due to human activity, are commonly seen in many regions in East Africa and if these forest remnants were not under constant protection they would rapidly disappear. There is little chance that these isolated forests will join up again unless a drastic change in climate occurs. It is possible therefore to theorize that subspecies will continue to evolve in these forest islands as they have done in the past. Furthermore it is evident that the differences between two subspecies depend on the length of time that the populations have been isolated from each other. The following examples may illustrate this in respect of forest monkeys.

SYSTEMATICS OF FOREST LIVING MONKEYS

BLUE MONKEYS

In spite of the wide distribution of this monkey and the differences of the color pattern, C. mitis forms a homogenous group; we can therefore follow Schwarz (1928) and G. M. Allen (1939) in recognizing one species with several subspecies. The species C. mitis includes the Pocock's leucampyx-group and Sykes' albogularis-group. Hill (1966) employs the taxon of superspecies for groups in the genus Cercopithecus. Within the superspecies C. mitis he admits three species: C. mitis, C. albogularis and C.

nictitans. The justification for uniting C. nictitans and C. mitis is unclear at the present time. If the localities are correct, C. m. maesi and C. nictitans nictitans are sympatric in the region of Budjala in the Congo-Oubangui District. It is also possible that C. m. stuhlmanni meets C. n. nictitans near the headwaters of the R. Rubi (Uele District).

C. mitis ranges far to the east and south-east and is the only forest Guenon which has survived there. At one time the distribution area included probably large regions of East Africa when it was largely covered with forest. Subsequently the different subspecies of C. mitis became adapted to different kinds of forests (swamp forest, montane forest, coastal forests, savanna woodland and forest islands of different types). Generally speaking, the C. (mitis) mitis-group of Hill (1966) includes the western subspecies, and the C. (mitis) albogularis-group, the eastern one. Booth (1968) emphasises that the East African Rift Valley acts as a geographical barrier between the two groups but this is questionable. C. mitis stuhlmanni inhabits the whole forest region in the eastern and northeastern Congo. Setzer (1956) attributes the C. mitis specimens from the southern Sudan to this subspecies. C. m. neumanni is considered as a synonym of C. m. stuhlmanni and this subspecies therefore occurs in Uganda and western Kenya in forest islands as far as the Mau Escarpment. The C. mitis specimens from L. Manyara and the Ngorongoro region are listed by Swynnerton and Hayman (1951) as C. m. neumanni; these authors observe that it is "doubtfully separable from C. m. stuhlmanni." Booth, on the other hand, mentions her specimens from the same region as a hybrid swarm intermediate between C. m. stuhlmanni and C. m. albogularis. Concerning the localities and the distribution of C. m. schoutedeni, C. m. kandti and C. m. doggetti, a confusing situation is to be found in the L. Kivu area according to the map published by Schouteden (1947). This is probably due to the fact that skins, prepared by natives, were bought by collectors and that some localities are not precisely defined. Comparing my own material from this area with the Tervuren collection it was possible to draw some conclusions concerning the distribution and the evolution of C. mitis in the L. Kivu area (Fig. 2).

Firstly we have to consider the enormous changes that have occurred in the pattern of vegetation during the last hundred years, due to deforestation by the indigenous population. Furthermore, the formation of L. Kivu is, geologically speaking, a recent event that took place some 15,000 - 20,000 years ago. *C. m. stuhlmanni* occupies the lowland forest of the eastern Congo and is also found in the montane forest of the mountain chain, west of L. Kivu (Kahuzi). In Uganda it has been seen in the Budongo, Bugoma, Kibale, Kalinzu and other forests. *C. m. kandti* is a typical species of the montane (cloud and bamboo) forests of the Virunga Volcanoes which define the border between the Congo, Rwanda and Uganda. In Rwanda *C. m. kandti* is to be found in the montane forest east of L. Kivu, a continuation of the Virunga forest to the south (Rugege forest). In Uganda it occurs, for example, in the Chuya bamboo forest. *C. m. doggetti* lives in Rwanda and Burundi where it inhabits gallery forests, swamp forest and papyrus swamps around the many lakes. In Uganda *C. m. doggetti* is found in forests near Sango Bay, along the R. Kagera and in the Kayonza forest. It is likely that *C. m. doggetti*, following the gallery forests, comes into contact with *C. m. kandti* on the eastern edge of the Rugege forest. One *C. mitis* skin is known from this region which appears to be an intermediate between the two subspecies. *C. m. schoutedeni* was described from Jdjwi Island in L. Kivu where it lives in the remaining montane forest (approx. 35 sq km) in its pure form. My 17 specimens from this island are quite uniform in color. Schouteden (1947) mentions two localities (Bobandana, Rubengera) on the shore of L. Kivu and one, north of L. Kivu, at Burunga. It is quite possible that *C. m. schoutedeni* occupied the whole valley before L. Kivu was formed and that, separated from the Jdjwi-population by the lake, certain small groups survived in the surrounding areas. It is very probable that such groups, living on the west shore of L. Kivu, came into contact with *C. m. stuhlmanni*. I have three skins from this region which are clearly intermediate between *C. m. stuhlmanni* and *C. m. schoutedeni*.

Ansell (1960) reports a *C. mitis* from the eastern parts of Lusaka (Rural) District but, until more material becomes available, Ansell provisionally regards the single specimen as an intermediate between *C. m. moloneyi* and *C. m. erythrarchus*.

Schwarz (1928) regarded C. m. francescae as a synonym of C. m. moloneyi. Ansell (1960), basing his opinion on his own material, writes "there can no longer be any doubt that this is a valid subspecies and indeed a very distinct one". On the other hand the same author estimates C. m. stevensoni "as only weakly separable perhaps not really worth distinguishing from C. m. erythrarchus". It is quite possible that this C. mitis population on Mt. Silinda was cut off from C. m. erythrarchus in a relatively recent period. Roberts (1954) gives an accurate description of C. m. schwarzi and it seems reasonable to accept it as a valid subspecies. Dandelot (1928) does not mention C. m. schwarzi, neither as a subspecies nor as a synonym. G. M. Allen (1954) considered C. albogularis zammaranoi as a synonym of C. a. albotorquatus and Dandelot (1968) reintroduced it as a good subspecies of C. (m.) albogularis. The status of C. m. phylax as a valid subspecies is doubtful; Schwarz (1928) regarded it as a race of C. m. kolbi and Dandelot (1968) considers it as a synonym of C. (m.) a. albotorquatus. As far as we know, no cranial comparisons have been made for the C. mitis group, nor are any biochemical studies available. Due to their complex and variable coloration some of the specimens are difficult to classify. In some areas intermediate forms or hybrids are to be found, but for most of the subspecies we can conclude that they are ecologically or geographically isolated at the present time. The distribution map (Fig. 3), compared with that of the present pattern of forests (Fig. 1), shows a clear correlation and it is justifiable to postulate that the present degree of isolation of the different C. mitis subspecies will remain substantially as it is, or will even become more pronounced due to further deforestation.

a. C. mitis mitis Wolf 1822. Synonyms: Simia leucampyx Fischer 1829; C. diadematus Geoffroy 1834; C. dilophos Ogilby 1838; C. pluto Gray 1848; (C. stuhlmanni) nigrigenis Pocock 1907. Distribution: Angola, (Dondo, Columbo, Loanda etc.). Habitat: gallery forest along the R. Cuanza.

b. C. mitis maesi Lönnberg 1919. Synonym: C. leucampyx maesi Lönnberg 1919. Distribution: Congo Basin (no specimens are known from this center), southern limit: R. Kasai and R. Sankuru; also present in the Congo-Oubangui District.

Habitat: rain forest and swamp forest.

c. *C. mitis stuhlmanni* Matschie 1839. Synonyms: *C. octoleus* Sclater 1902; (*C. stuhlmanni*) *carruthersi* Pocock 1907- *C. princeps* Elliott 1909; *C. leucampyx schubotzi* Matschie 1913; *C. leucampyx elgonis* Lönnberg 1919; *C. neumanni* Matschie 1905; *Lasiopyga leucampyx mauae* Heller 1913. Distribution: northeastern and eastern Congo, mountains west of L. Kivu, Budongo forest, Mt. Elgon, Kakamega forest, forest islands.

d. *C. mitis schoutedeni* Schwarz 1928. Synonym: *C. leucampyx schoutedeni* Schwarz 1928. Distribution: Jdjwi Island in L. Kivu. Habitat: montane forest.

e. *C. mitis kandti* Matschie 1905. Synonym: *C. insignis* Elliot 1909. Distribution: Virunga Volcanoes and adjacent mountains. Habitat: montane and bamboo forest.

f. *C. mitis doggetti* Pocock 1907. Synonym: *Lasiopyga leucampyx sibatoi* Lorenz 1913. Distribution: parts of Burundi and Rwanda, southern Uganda, western Tanzania. Habitat: montane forest, forest islands, gallery forest and papyrus swamps.

g. *C. mitis opisthostictus* Sclater 1894. Distribution: Southern Congo (Katanga) and Zambia. Habitat: forest islands and gallery forest (the white dots in Fig. 3 indicate, schematically, that in spite of the large distribution area, this subspecies only occurs in specific areas).

h. *C. mitis boutourlinii* Giglioli 1887. Synonym: *C. omensis* Thomas 1901. Distribution: Charada forest (Kaffa), Gimma, R. Omo. Habitat: forest islands and gallery forest.

i. *C. mitis zammaranoi* de Beaux 1924. Synonym: *C. (Insignicebus) albogularis zammaranoi* de Beaux 1924. Considered as a synonym of *C. m. albotorquatus* by some authors. Distribution: Bidi Scionde (lower R. Juba), Uebbi Schebeli.

Habitat: gallery forest.

j. C. mitis albotorquatus Pousargues 1896. Synonym: C. rufotinctus Pocock 1907. Distribution: Ola Uagere (Somalia), R. Tana (Kenya). Habitat: forest islands and gallery forest.

k. C. mitis phylax Schwarz 1927. Considered as a synonym of C. m. albotorquatus by some authors. Distribution: Patta Islands.

l. C. mitis kolbi Neumann 1902. Synonyms: C. kolbi hindei Pocock 1907; C. kolbi nubilus Dollman 1910. Distribution: Mt. Kenya, Aberdare Mts., Kedong Escarp., Fort Hall, Nairobi forest. Habitat: montane forest, forest islands and gallery forest.

m. C. mitis kibonotensis Lönnberg 1908. Synonyms: Lasiopyga albogularis maritima Heller 1913; Lasiopyga albogularis kima Heller 1913. Distribution: Mt. Kilimanjaro, Mt. Meru, Pare Mts., Usambara Mts., Magrotto Mts., Mt. Mbololo, Chyulus Hills, Taita Hills, Mazeras. Habitat: montane forest, gallery forest, coastal forest.

n. C. mitis albogularis (Sykes 1831). Synonyms: Semnopithecus ? albogularis Sykes 1831; Semnopithecus albigularis Lesson 1840. Distribution: Zanzibar and Mafia Islands.

o. C. mitis monoides Geoffroy 1841. Synonym: [C. albogularis] rufilatus Pocock 1907. Distribution: R. Rufiji, Kilosa forest, Morogoro, Mikindani etc. Habitat: gallery forest, montane and coastal forest.

p. C. mitis moloneyi Sclater 1893. Distribution: Karonga, Uzungwe Mts., Rungwe Mts., Ukinga Mts., Poroto Mts., Rukwa etc. Habitat: montane forest.

q. C. mitis nyasae Schwarz 1928. Distribution: Fort Lister, Vipyo Plateau, Lychenya Plateau, Fort Johnson, Chirinda Forest, Namuli Mts. Habitat: montane and relict forests.

r. *C. mitis francescae* Thomas 1902. Recently re-introduced as a valid subspecies by Ansell (1960). Listed by G. M. Allen (1954) as a synonym of *C. m. moloneyi*. Distribution: Mt. Waller, Nkata Bay, Nchisi Mts., Chinteche. Habitat: montane forest.

s. *C. mitis erythrarchus* Peters 1852. Synonyms: *C. stairsi* Sclater 1892; *C. albogularis beirensis* Pocock 1907; [*C. stairsi*] *mossambicus* Pocock 1907. Distribution: Inhambane, Umtali, Beira, Gorongoza, littoral northwards to lower Zambesi. Habitat: coastal and gallery forests.

t. *C. mitis stevensoni* Roberts 1948. Distribution: Mt. Selinda, Ansell (1960) writes: "I regard *C. m. stevensoni* as only weakly separable, perhaps not really worth distinguishing from *C. m. erythrarchus*". Habitat: montane forest.

u. *C. mitis schwarzi* Roberts 1931. Distribution: Pilgrims Rest District and Woodbush. Habitat: larger forests, local in distribution.

v. *C. mitis labiatus* Geoffroy 1842. Synonym: *C. samango* Sundevall 1844. Distribution: Pondoland, King Williams Town, Stutterheim, from Kaffraria eastwards to Natal and Zululand. Habitat: montane and coastal forest.

BLACK AND WHITE COLOBUS

In grouping the different black and white colobus monkeys, the classification of Schwarz (1928) and Dandelot (1968) is here followed. Dandelot accepts four species (*C. polykomos*, *C. satanas*, *C. abyssinicus* and *C. angolensis*). G. M. Allen (1939) established four sections but only one species (*C. polykomos*). To distinguish four species is also defensible from the point of view of zoogeography. The three *C. polykomos* subspecies inhabit the forests of West Africa; Booth (1954) thinks the R. Sassandra forms an effective barrier between C. p. polykomos and C. p. dollmani. On the other hand, C. p. vellerosus is also to be found on the right bank of the R. Bandama and Booth (1954) mentions interbreeding between C. p. vellerosus and C. p. dollmani in the Bouaflé Forest Reserve. I possess a skin from Mankono which belongs to C. p. dollmani. All three subspecies have white tails and the white pattern of the face recalls somewhat that of C. abyssinicus. The skins examined in the British Museum (Nat. Hist.) show white or grayish shoulder patches (best developed in C. p. polykomos; only the presence of some gray hairs in C. p. vellerosus) recalls the epaulettes of C. angolensis. C. satanas is entirely black. Thanks to the black and white color pattern, the other two species, C. abyssinicus and C. angolensis are better distinguishable. Both species must have had a much wider distribution in East Africa during pluvial maxima, when most of eastern Africa was covered with forest. The distribution map shows that C. abyssinicus constitutes the northern group whereas C. angolensis represents the southern one. C. angolensis probably reached the east coast, following the montane forests of the Central African Rift Valley along L. Tanganyika and north of L. Nyasa, reaching finally the Usambara Mts. Today subspecies of C. angolensis survive in montane forest islands and coastal forests (see map). The same is true for C. abyssinicus which came probably to East Africa moving eastwards north of L. Victoria. Another group of this species went as far as the Ethiopian Highlands. This species too survives nowadays in the forest islands, having developed different subspecies. For these two Colobus species, C. abyssinicus and C. angolensis, the same problem arises as for C. mitis, namely to what extent we can, and should, distinguish subspecies. Due to the black and white color pattern this task is somewhat easier than in C. mitis. In establishing subspecies we can also take into consideration the present isolation of the different populations living in forest islands. In the angolensis-group, C. a. angolensis is separated from the other subspecies by the R. Lualaba-Congo. In the eastern Congo we find five subspecies, two of them inhabiting lowland forest (C. a. cottoni, C. a. cordieri) three living in montane forest (C. a. prigoginei, C. a. adolfi friederici, C. a. ruwenzorii). C. a. cottoni, C. a. cordieri and C. a. prigoginei are closely related forms in which the white shoulder hairs are poorly developed. As far as we know at present,

C.a.cordieri inhabits the forest between the R. Lowa and the R. Elila. *C.a.prigoginei* was probably cut off from *C.a.cordieri* in relatively recent time by montane grassland formations which are mostly of secondary origin. I do not agree with Dandelot (1968) who considers *C.a.adolfi friederici* as a synonym of *C.a.ruwenzorii*. The latter subspecies has much better developed shoulder hairs and only some short grayish-white hairs at the end of the tail. Similar subspeciation between the Ruwenzori and the Kivu mountains is to be found in other groups of mammals (Muridae, Sciuridae). We face the same problem in Tanzania with *C.a.sharpei* and *C.a.palliatus*, the former is considered by some authors as a synonym of the latter. Taking the *C.abyssinicus*-group we encounter an identical problem concerning subspeciation. *C.a.occidentalis* is separated from *C.a.uellensis* by the R. Oubangui and the swamps in the north-western Congo. The two subspecies *C.a.dodingae* and *C.a.matschiei* are very close to *C.a.uellensis* but are here regarded as valid subspecies. The three subspecies *C.a.percivali*, *C.a.kikuyensis* and *C.a.caudatus*, living in isolated montane forests are geographically well separated. The shape of the bushy tail and its color pattern develops gradually from the northern to the southern subspecies. They are related to *C.a.abyssynicus*. We keep therefore *C.a.percivali* as a valid subspecies (the differences between *C.a.abyssinicus* and *C.a.percivali* are not less than that between *C.a.kikuyuensis* and *C.a.caudatus*).

a. *C.polykomos polykomos* (Zimmermann 1780). Synonym: *Cebus polykomos* Zimmermann 1780; *Simia regalis* Kerr 1792; *Simia tetradactyla* Link 1795; *Simia polycomos* Schreber 1800; *Simia comosa* Shaw 1800; *Col(obus) ursinus* Ogilby 1835. Distribution: Gambia, Guinea, Sierra Leone, Liberia, western Ivory Coast. Habitat: montane forest and rain forest.

b. *C.polykomos dollmani* Schwarz 1927. Perhaps a hybrid form of *C.p.polykomos* and *C.p.vellerosus*. Distribution: between the R. Sassandra and the R. Bandama. Habitat: rain forest.

c. *C. polykomos vellerosus* (Geoffroy 1834). Synonyms: *Semnopithecus vellerosus* Geoffroy 1834; *Semnopithecus bicolor* Wesmael 1835; *C. leucomeros* Ogilby 1838. Distribution: Ivory Coast east of R. Bandama, Dahomey, Togo, Ghana, western Nigeria. Habitat: rain forest.

d. *C. satanas* Waterhouse 1838. Synonyms: *Semnopithecus anthracinus* Leconte 1857; *Stachycolobus municus* Matschie 1917; *Stachycolobus limbarenicus* Matschie 1917; *Stachycolobus zenkeri* Matschie 1917. Distribution: Southern Cameroons, Rio Muni, Gabon, Fernando Po, probably in some areas in the Congo Republic. Habitat: rain forest.

e. *C. angolensis angolensis* Sclater 1860. Synonyms: *C. a. sandbergi* Lönnberg 1908; *C. palliatus weynsi* Matschie 1913; *C. maniemae* Matschie 1914; *C. benamakimae* Matschie 1914. Distribution: Northeastern Angola, Congo Basin. Habitat: rain forest, swamp forest, forest galleries (southern Congo).

f. *C. angolensis cottoni* Lydekker 1905. Synonyms: *C. palliatus cottoni* Lydekker 1905; *C. palliatus mawambicus* Matschie 1913; *C. mawambicus nahani* Matschie 1914. Distribution: Northeastern Congo (right bank of R. Congo). Habitat: rain forest.

g. *C. angolensis cordieri* Rahm 1959. Synonym: *C. polykomos cordieri* Rahm 1959. Distribution: eastern Congo (see map). Habitat: rain forest.

h. *C. angolensis prigoginei* Verheyen 1959. Distribution: Mt. Kabolo (west of L. Tanganyika). Habitat: montane forest.

i. *C. angolensis ruwenzorii* Thomas 1901. Distribution: Mt. Ruwenzori. Habitat: montane forest.

k. *C. angolensis adolfi friederici* Matschie 1914. Distribution: Virunga Volcanoes, mountains east and west of L. Kivu, Minziro forest (Tanzania).

l. C. angolensis sharpei Thomas 1902. Synonym: C. langheldi Matschie 1914. For the type locality see Ansell (1959). Distribution: North of L. Nyasa (Rungwe Mts., Poroto Mts., Livingstone Mts., Uzungwe Mts.). Habitat: montane forest.

m. C. angolensis palliatus Peters 1868. Distribution: R. Pangani, Amboni, Amani, Usambara Mts., Uluguru Mts., near Dar-es-Salaam etc. Habitat: montane forest, forest islands, gallery forest, coastal forest.

n. C. abyssinicus occidentalis (Rochebrune 1886). Synonyms: Guereza occidentalis Rochebrune 1886; C. escherichi Matschie 1914. Distribution: eastern Nigeria, southern Cameroons, Gabon, Congo Republic, southern Sudan. Habitat: rain forest, forest islands and gallery forest in the Sudan.

o. C. abyssinicus uellensis Matschie 1913. Synonyms: C. matschiei uellensis Matschie 1913; C. matschiei ituricus Matschie 1913; C. matschiei dianae Matschie 1913; C. matschiei brachychaites Matschie 1913; C. a. terrestris Heller 1913; C. occidentalis ituricus Lorenz 1914; C. occidentalis rutschuricus Lorenz 1914. Distribution: northern and north-eastern Congo, Budongo forest. Habitat: rain forest, forest galleries, forest islands.

p. C. abyssinicus dodingae Matschie 1913. Synonmy: C. (Guereza) matschiei dodingae Matschie 1913. Distribution: Dodinga Mts., Lotti forest, Mt. Lotuke, Torit. Habitat: montane forest.

q. C. abyssinicus matschiei Neumann 1899. Synonyms: C. a. roosevelti Heller 1913; C. a. elgonis Granvik 1925. Distribution: Kavirondo, R. Lukosa, Kisumu, Mau forest, Mt. Elgon, Nandi forest, Mt. Debasien, Speke Golf etc. Habitat: montane forest, forest islands, gallery forest.

r. C. abyssinicus abyssinicus (Oken 1816). Synonyms: Lemur abyssinicus Oken 1816; C. guereza Rüppell 1835; Guereza rüpellii Gray 1870; C. a. poliurus Thomas 1901;

C. poliurus managaschae Matschie 1913. Distribution: south L. Tana, Sidano, near Dongila, L. Abaya, west of R. Awash etc. Habitat: montane forest.

s. C. abyssinicus gallarum Neumann 1902. Distribution: Djaffa Mts. Habitat: montane forest.

t. C. abyssinicus percivali Heller 1913. Distribution: Mt. Gargues (Matthews Range). Habitat: montane forest.

u. C. abyssinicus kikuyuensis Lönnberg 1912. Synonyms: C. caudatus thikae Matschie 1913; C. caudatus laticeps Matschie 1913. Distribution: Mt. Kenya, Thomson's Falls, Kijabe, Nanyuki, Nyeri. Habitat: montane forest, gallery forest.

v. C. abyssinicus caudatus Thomas 1885. Synonym: C. albocaudatus Lydekker 1906. Distribution: Mt. Kilimanjaro, Mt. Meru. Habitat: montane forest.

THE RED COLOBUS

The present situation concerning the Colobus badius is confusing. We agree with Dandelot (1968) that no definite answer can be given from the point of view of species and subspecies. G.M. Allen (1939) divided the C. badius-group into sections, including C. verus, and Dandelot (1968) established a subgenus Piliocolobus with 6 species and 9 to 11 subspecies. Verheyen (1957) proposed, mainly for the Congo forms, an arrangement based on his craniometrical studies. The range of the C. badius-group goes from Senegal to the Zanzibar Islands but is not a continuous one. We can distinguish four main areas: C. b. temminckii, C. b. badius and C. b. waldroni occupy the forests of West Africa from Senegal to Ghana. Dandelot (1968) considers C. b. waldroni as probably a valid species. After Booth (1954), C. b. waldroni has not been found west of the R. Bandama or the R. Nzi; but on the other hand C. b. badius coexists on the left bank of the R. Nzi with C. b. waldroni. Booth (1954) did not find any proof of hybridization and considers, as does Hayman, C. b. waldroni to be a subspecies of C. b.

badius. *C.b.pennantii* and *C.b.preussi* belong, after G.M. Allen (1939), to the *C.badius*-group of West Africa. A confusing situation reigns in Equatorial Africa and as already noted by Malbrant and Maclatchy (1949), very little material is available from this region. *C.b.bouvieri* from the R. Sanga and the R. Likoula is regarded by Dandelot (1968) as a subspecies of *C.b.pennantii*. *C.b.nigrimanus* from Liranga, located in the distribution area of *C.b.oustaleti*, has to be considered as a synonym of the latter. *C.b.tholloni* is a good subspecies which occupies the Congo Basin. We ignore the possibility of this form existing on the right bank of the R.Lomami. Here we follow Dandelot (1968) in admitting *C.b.powelli* and *C.b.brunneus* as synonyms of *C.b.oustaleti*. It is possible that the R. Aruwimi-Ituri form a barrier between *C.b.powelli* (now *C.b.oustaleti*) and *C.b.ellioti*, but it is possible that the two come together at the headwaters of these rivers and in the region west of L.Albert. Concerning *C.b.powelli*, Verheyen concludes through his craniometrical studies that this form is an intermediate between *C.b.brunneus* and *C.b.ellioti* but there is no ecological justification for this belief. To the south, the limit of the area occupied by *C.b.ellioti* is the R. Elila. In the region south of this river lives *C.b.foai* which is quite different from *C.b.ellioti* judging from the skins of both subspecies. The southernmost limit of *C.b.foai* is still uncertain. Regarding *C.b.graueri*, I consider this form as a synonym of *C.b.foai*, at least for the moment, until more material is available. *C.b.graueri* was described after specimens coming from the mountains north-west of L.Tanganyika (Fizi, Baraka). Specimens of the Red Colobus I observed and shot in the Kahuzi region, just west of L. Kivu are very much like the *C.b.foai* specimens. It is interesting that these specimens are quite different from the *C.b.ellioti* material I collected in the equatorial forest some 100 km west of the Kahuzi mountains. The *C.b.graueri* population in the mountain was probably cut off from the *C.b.foai*-group in relatively recent times due to the formation of a man-made secondary montane grassland belt. Two subspecies of *C.badius* occur in Tanzania: *C.b.tephrosceles* inhabits forest islands and gallery forest in north-western Tanzania, *C.b.gordonorum* is restricted in the montane forest on the Uzungwe Mts. *C.b.kirkii* is quite different from the other Red Colobus, due to its isolation on Zanzibar Island; but to underline its relationship I prefer to keep it as a subspecies of *C. badius* as G.M. Allen (1939) does.

a. *C. badius badius* (Kerr 1792). Synonyms: *Simia (Cercopithecus) badius* Kerr 1792; *Simia ferruginea* Shaw 1800; *C. ferruginosus* Geoffroy 1812; *C. rufoniger* Ogilby 1838. Distribution: Eastern Sierra Leone, forest areas of Guinea, Liberia, Ivory Coast to the right bank of R. Bandama and, according to Booth (1958), on the left bank of the R. Nzi. Habitat: rain forest.

b. *C. badius temminckii* Kuhl 1820. Synonyms: *C. temminkii* (sic) Kuhl 1820; *C. fuliginosus* Ogilby 1835; *C. rufo-fuliginus* Ogilby 1838. Distribution: Senegal, Gambia, Guinea. Habitat: gallery forest, montane forest.

c. *C. badius waldroni* Hayman 1936. Distribution: Ivory Coast (left bank of the R. Bandama and the R. Nzi), Ghana. Habitat: rain forest.

d. *C. badius preussi* (Matschie 1900). Synonym: *Piliocolobus preussi* Matschie 1900. Distribution: Nigeria (left bank of the R. Cross), Cameroons. Habitat: rain forest.

e. *Colobus badius pennantii* Waterhouse 1838. Distribution: Fernando Po Island. Habitat: rain forest.

f. *C. badius bouvieri* (Rochebrune 1887). Synonyms: *Piliocolobus bouvieri* Rochebrune 1887; *C. likualae* Matschie 1914. Distribution: R. Sanga, R. Likoula, N'Gabe. Habitat: rain forest.

g. *C. badius oustaleti* Troussart 1906. Synonyms: *C. (Tropicolobus) umbrinus* Matschie 1914; *C. b. nigrimanus* Troussart 1906; *C. (Piliocolobus) powelli* Matschie 1913; *C. (Tropicolobus) schubotzi* Matschie 1914; *C. (Piliocolobus) brunneus* Lönnberg 1919. Distribution: Northern Congo (left bank of R. Oubangui, from the junction of the R. Oubangui and R. Congo to L. Albert). Southern limit: R. Congo and probably R. Aruwimi-Ituri. Habitat: swamp forest, rain forest, forest galleries.

h. C. badius elliotti Dollman 1909. Synonyms: Piliocolobus anzeliusi Matschie 1914; Piliocolobus ellioti melanochir Matschie 1914; C. variabilis Lorenz 1914; C. multicolor Lorenz 1914; C. b. langi J.A. Allen 1925. Distribution: eastern Congo (right bank of R. Congo). Northern limit probably R. Aruwimi-Ituri, southern limit R. Elila. Habitat: rain forest.

i. C. badius foai Pousargues 1899. Synonyms: C. graueri Dollman 1909; Piliocolobus kabambarei Matschie 1914; Piliocolobus lulindicus Matschie 1914. Distribution: eastern Congo (right bank of R. Congo) between R. Elila and R. Lukuga. Mountains west of L. Tanganyika and L. Kivu. Habitat: rain forest and montane forest.

k. C. badius tholloni Milne-Edwards 1886. Synonym: C. (Piliocolobus) lovizettii Matschie 1913. Distribution: Congo Basin (left bank of R. Congo). Southern limit probably R. Kasai-Sankuru. Schouteden (1947) reports two localities in the north-western Congo, the area of C. b. oustaleti. Habitat: swamp and rain forest.

l. C. badius tephrosceles Elliot 1907. Synonym: Tropicolobus gudoviusi Matschie 1914. Distribution: Burundi (Muhinga), Uganda (R. Ruahara, west and southwest of L. Victoria, near L. Rungwe? Habitat: forest islands and gallery forest.

m. C. badius gordonorum (Matschie 1900). Synonym: Piliocolobus gordonorum Matschie 1900. Distribution: Uzungwe Mts. Habitat: montane forest.

n. C. badius rufomitratus Peters 1879. Distribution: R. Tana. Habitat: gallery forest.

o. C. badius kirkii Gray 1868. Distribution: Zanzibar

Island.

OTHER GUENONS

Cercopithecus hamlyni Pocock 1908. Synonym: *Cercopithecus leucampyx aurora* Thomas and Wroughton 1910. Still very little is known about this species and relatively few specimens are to be found in museum collections. The Tervuren Museum possesses 12 skins from localities such as Epulu, Kisangani, Kiloboze, Shabunda, Ruhengeri etc. The British Museum (Nat.Hist.) possesses a single skin from the Congo and I have 11 skins from the eastern Congo. J.A. Allen (1922) writes: "The American Museum Congo Expedition collected several thousand mammals in the Ituri forest region, from Stanleyville northwards, including some four hundred Primates, without either Lang or Chapin meeting with this species". Eight of my specimens come from the montane forest west of L. Kivu. Either *C. hamlyni* is more common in this habitat than in the lowland forest or it is more easily detected in the less dense montane forest. On the other hand it is quite possible that this species was originally confined to these mountains and that later it spread into the lowland forest between the westward running rivers. Schaller (1963) postulates a similar ecological history for the mountain gorilla. At present, and probably for several hundred years, the mountain population of *C. hamlyni* is separated from the lowland population by an almost uninterrupted belt of montane prairies and grassland. The specimens of the mountain population are somewhat different from those occurring in the lowland forest and it might be that they will have to be considered as a new subspecies (the most striking difference is the length of the tail which, in the mountain specimens, is as long as the headbody length; in the lowland animals the tail is about one third longer). Concerning the biology of *C. hamlyni*, my observations provide no evidence at all that this monkey is nocturnal. Most of the primate literature is wrong in this respect, nor have we yet established the origin of this belief. *C. hamlyni* lives in small groups of ten or less and is often to be found on the ground. Fig. 7 shows the known distribution. The northernmost locality is Epulu, which was a collecting station for animals and we do not know if the specimens labelled "Epulu" were caught north or south of the R. Ituri-Epulu or if therefore this river is a barrier or not. The southern limit is the R. Lukuga and to the west *C. hamlyni* does not cross the R. Congo. The most eastern specimen comes from the bamboo forest near Kabale (Uganda).

Cercopithecus lhoesti lhoesti Sclater 1898. Synonyms: *Cercopithecus thomasi* Matschie 1905; *Cercopithecus thomasi rutschuricus* Lorenz 1915. *C. lhoesti* has approximately the same distribution as *C. hamlyni*, but is much more common than the latter, expecially on the mountains around L. Kivu (Fig. 8). This species was probably also once confined to montane forest and penetrated the lowland forest in the same way as *C. hamlyni*. It has a wider range in Rwanda and Burundi, where it also occurs in some gallery forests. Only slight individual variations in color are to be found and the specimens from the lowland and the montane forest are very much alike. Schwarz (1928) states that *C. hamlyni* is probably a subspecies of *C. lhoesti* and G. M. Allen (1939) mentions both under the *C. lhoesti* group. However as they share the same habitat and have the same distribution, they should be considered as two good species. The present author does not agree with Tappen (1960) who suggests that the most important factor limiting the distribution of *C. lhoesti* is food in the undergrowth. Tappen also suggests that *C. lhoesti* might be unable to compete with other monkeys beyond the range of distribution of certain favored food plants growing close to the ground. G. M. Allen (1939) and Dandelot (1968) consider *C. preussi* as a subspecies of *C. lhoesti* (*C. lhoesti preussi* Matschie 1898). This subspecies inhabits the montane forests in south-eastern Nigeria (Fig. 8) and Fernando Po. We have some evidence that the mountains of the Central African Rift Valley were linked with Mt. Cameroon by montane forest during pluvial maxima. The distribution pattern of other animals, including mammals, supports this theory.

We have already mentioned on previous pages that rivers can constitute geographical barriers and promote subspeciation. Two of such rivers, which are efficient barriers, are the R. Congo and the R. Oubangui. They not only form a dividing line between subspecies, but are also an important limit for the distribution of certain other mammalian species. Rivers such as the R. Kasai and the R. Sankuru and several smaller ones in the eastern Congo, are less important but also constitute barriers for subspecies. Cercopithecus ascanius and Cercopithecus mona subspecies have a similar distribution in the Congo.

Cercopithecus ascanius (Audebert 1799). Synonym: Simia ascanius Audebert 1799. The subspecies C.a.whitesidei Thomas 1909 inhabits the rain forest and swamp forest of the Congo Basin. The limit to the west, north and east is formed by the R.Congo. The R.Kasai and the R.Sankuru, together with the limit of the tropical forest, constitute the southern limit. C.a.schmidti Matschie 1892 lives on the right bank of R.Congo. The northern boundary is formed by the R.Oubangui and the R.Bomu and the distribution in the north-eastern Congo is also limited by the edge of the forest. The subspecies C.a.montanus Lorenz 1914 has to be considered as a synonym of C.a.schmidti, therefore the distribution to the east extends to the western mountain ridge of the Central African Rift Valley (C.a.schmidti does not occur in Rwanda and Burundi). On the other hand C.ascanius schmidti is present in several forest islands in Uganda, which re-affirms that these forest relics belong to the rain forest of the Congo. C.a.schmidti is not to be found in the higher montane forest and the altitudinal limit lies between 1800 and 2000 m (the author has not observed this monkey in the bamboo forest). C.a.katangae Lönnberg 1919 inhabits the gallery forests and forest islands in the southern Congo. The western limit is the lower R.Kasai but C.a.katangae extends across this river in the south-western Katanga; it also occurs there in a small area in Angola. Following the gallery forests, C.a.katangae is to be found as far south as northern Zambia. C.a.ascanius (Audebert 1799) inhabits forests and gallery forest of the south western Congo and in Angola. From the Lunda district in Angola, de Barros Machado (1965) described recently a new subspecies, C.a.atrinasus. In the same region he found C.ascanius specimens which have to be regarded as intermediate forms between C.a.atrinasus and C.a.katangae, and others which are intermediate between C.a.atrinasus and C.a.ascanius.

CONCLUSION

From the facts mentioned above we can draw the following conclusion. It is evident that the edge of the forest is an ecological barrier for forest-dwelling primate species which can only penetrate savanna country following gallery forests; however they can also be found in suitable forest islands located in savanna regions. In a few cases only, the limit between lowland and montane forest seems to constitute an ecological barrier for subspecies; among the primates, there is no example of a species that is restricted to montane forest regions. For several species and subspecies rivers constitute an effective geographical barrier. We can therefore conclude that geographical and ecological barriers can be helpful in defining the limits of distribution of a subspecies; furthermore, such barriers may even be decisive in considering whether or not a population is a subspecies. Never-the less it is as well to be aware of the fact that, under certain conditions, these barriers can be by-passed. It is in such circumstances that intermediate forms or hybrids are found.

ACKNOWLEDGEMENTS

I am indebted to the Director and Staff of the British Museum (Nat. Hist.), London and the Musée Royal D'Afrique Centrale, Tervuren, who placed their collections at my disposal. I am also happy to acknowledge the hospitality of the Wenner Gren Foundation for Anthropological Research and the assistance with the manuscript given me by the editors of this volume.

REFERENCES

Allen, G.M. (1939). A Checklist of African Mammals. Bull. Mus.Comp.Zool.Harv. 83: 1-763.

Allen, G.M. and Lawrence, B. (1936). Scientific results of an expedition to rain forest regions in E.Africa. III Mammals, Bull.Mus.Comp.Zool.Harv. 79: 31-126.

Allen, G.M. and Loveridge, A. (1933). Reports on the scientific results of an expedition to the southwestern highlands of Tanganyika Territory, II Mammals. Bull.Mus.Comp.Zool. Harv. 75: 47-140.

Allen, G.M. and Loveridge, A. (1942). Scientific results of a fourth expedition to forested areas in east and central Africa, I Mammals. Bull.Mus.Comp.Zool.Harv. 89: 147-214.

Ansell, W.F.H. (1959). The type localities of Colobus angolensis sharpei and Colobus angolensis sanbergi. Rev.Zool.Bot.Afr. 60: 168-171.

Ansell, W.F.H. (1960). Contributions to the mammalogy of N. Rhodesia. Occasional Papers Nat.Mus.Southern Rhodesia. No. 24: 351-398.

Barros Machado de, A. (1965). Acerca do achado em Angola de una forma racial inédita do Cercopithecus ascanius. Boletin da Acad. das Ciências Lisboa 37: 182-198.

Barros Machado de, A. (1969). Mamiferos de Angola ainda não citados ou pouco conhecidos. Publ.cult.Co.Diam.Ang. Lisboa. No.46: 93-232.

Booth, A.H. (1954). Note on the Colobus Monkeys of the Gold and Ivory Coast. Ann.Mag.Nat.Hist. 7: 857-860.

Booth, A.H. (1958). The zoogeography of West-African primates: A Review. Bull.Inst.Fr.Afr.Noire 20: 587-622.

Booth, C.P. (1968). Taxonomic studies of Cercopithecus mitis Wolf (East Africa). Nat.Geogr.Soc.Research Reports (1963

Projects): 37-51.

Carcasson, R.H. (1964). A preliminary survey of the zoogeography of African Butterflies. East African Wildlife Journal, 2: 122-157.

Chapin, J.P. (1932). The Birds of the Belgian Congo. Bull.Amer. Mus.Nat.Hist. 65(1): 1-756.

Dandelot, P. (1968). Preliminary identification manual for African Mammals, Primates: Anthropoidea. Washington: Smithsonian Institution Press.

Ellermann, J.R., Morrison-Scott, T.C.S. and Hayman, R.W. (1953). Southern African Mammals. London: British Mussum (Nat.Hist.).

Haddow, A.J. (1956). The blue monkey group in Uganda. Uganda Wild Life and Sport 1: 22-26.

Hill, J.E. and Carter, T.D. (1941). The Mammals of Angola, Africa. Bull.Amer.Mus.Nat.Hist. 78: 1-211.

Hill, W.C.O. (1966). Primates: Vol. VI, Cercopithecoidea. Edinburgh: University Press.

Hollister, N. (1924). East African Mammals in the United States National Museum, part III, Primates etc. Bull.U.S.Nat.Mus. 99: 1-151.

Malbrant, R. et Maclatchy, A. (1949). Faune de l'Equateur Africain Francais Tome II, Mammifères. Encyclopédie Biologique, Paris: Lechevalier.

Moreau, R.E. (1966). The Bird faunas of Africa and its islands. London: Academic Press.

Rahm, U. (1961). Esquisses mammalogiques de basse Côte d'Ivoire. Bull.Inst.Fr.Afr.Noire 22(4): 1229-1265.

Rahm, U. (1966). Les Mammifères de la forêt équatoriale de l'Est du Congo. An.Musée Roy.Afr.Centr.Tervuren,Belgique. Ser. 8 Sci.Zool. 149: 39-121.

Rahm, U. et Christiansen, A. (1960). Notes sur Colobus polykomos cordieri du Congo Belge. Rev.Zool.Bot.Afr. 61: 215-

-220.

Richards, P.W. (1957). The Tropical Rain Forest. Cambridge: University Press.

Roberts, A. (1954). The Mammals of South Africa. Johannesburg: Central News Agency South Africa.

Schaller, G.B. (1963). The Mountain Gorilla. Chicago. Univ. of Chicago Press.

Schouteden, H. (1947). De Zoogdieren van Belgisch-Congo en van Ruanda-Urundi. Ann. Musée Roy. Congo Belge. Tervuren. Sér. II + III: fasc. 1-3.

Schwarz, E. (1928). Notes on the classification of the African monkeys in the genus Cercopithecus Erxleben. Ann. Mag. Nat. Hist. (10) 1: 649-663.

Setzer, H.W. (1956). Mammals of the Anglo-Egyptian Sudan. Proc. U.S. Nat. Mus. 106: 447-587.

Smithers, R.H.N. (1966). The Mammals of Rhodesia, Zambia and Malawi. London: Collins.

Swynnerton, G.H. and Hayman, R.W. (1951). A checklist of the Land Mammals of the Tanganyika Territory and the Zanzibar Protectorate. Jour. East Africa Nat. Hist. Soc. 20: (6 and 7): 274-394.

Tappen, N.C. (1960). Problems of distribution and adaption of the African monkeys. Current Anthropology 1: 91-120.

Ulrich, W. (1961). Zur Biologie und Soziologie der Colobusaffen (Colobus guereza caudatus Thomas 1885). Der Zool. Garten (NF) 25: 305-368.

Verheyen, W. (1957). Bijdrage tot de Craniometrie van Colobus badius. Ann. Musée Roy. Congo Belge, Tervuren. Serie 8°, Sci. Zool. 62: 7-94.

Table 1

	Congo	Mabira	Mibale	Budongo	Kakamega	Mt. Elgon	Mt. Kenya
Cercopithecus ascanius	+	+	+	+	+		
Perodicticus potto	+	+		+	+		
Galago demidovii	+			+			
Protoxerus stangeri	+	+	+	+	+	+	+
Tamiscus emini	+	+	+				
Tamiscus alexandri	+	+		+			
Funisciurus pyrrhopus	+		+				
Anomalurus	+				+	+	+
Atherurus africanus	+	+			+		
Malacomys longipes	+	+	+				
Hybomys univittatus	+	+	+				
Hylomyscus stella	+	+					
Manis tricuspis	+	+					
Hylochoerus	+						

Lowland rain forest mammals found in the Congo forest and in forest islands in Uganda and Kenya.

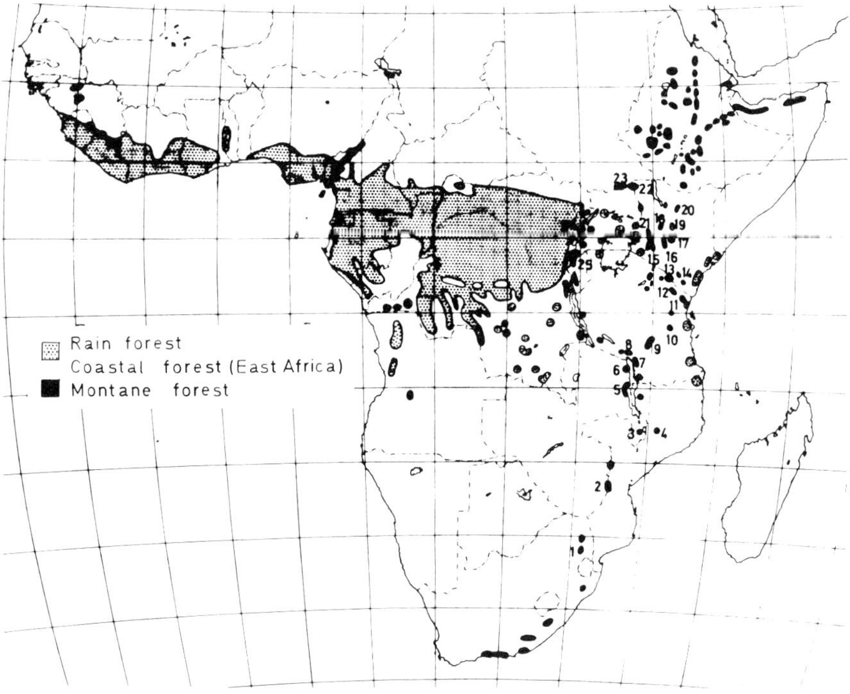

Fig. 1: Distribution of lowland rain forest, coastal forest, (East Africa) and montane forest in Africa. 1. Pilgrims Rest, 2. Mt. Silinda, 3. Mlanje Mts., 4. Namuli, 5. Mt. Waller, 6. Nyika, 7. Rungwe Mts., 8. Poroto Mts., 9. Uzungwe Mts., 10. Uluguru Mts., 11. Usambara Mts., 12. Pare Mts., 13. Mt. Kilimanjaro, 14. Chyulus and Teita Hills, 15. Highlands with Mau Escarpment, 16. Aberdare Mts., 17. Mt. Kenya, 18. Loriyu Plateau, 19. Matthews Range, 20. Marsabit Mts., 21. Mt. Elgon, 22. Dodingas, 23. Imatongs, 24. Mt. Ruwenzori, 25. Virunga and Rugege Mts.

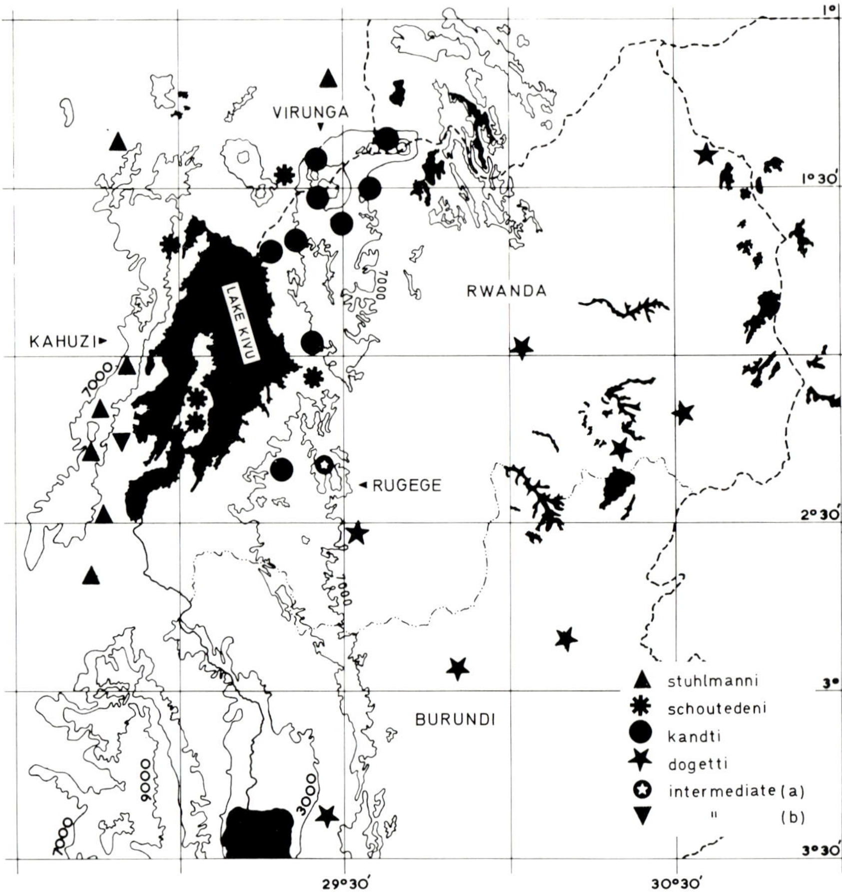

Fig. 2: Distribution of the Cercopithecus mitis subspecies in the L. Kivu area (eastern Congo, Rwanda, Burundi). Intermediate (a) = C. m. kandti/doggetti, intermediate (b) = C. m. stuhlmanni/schoutedeni.

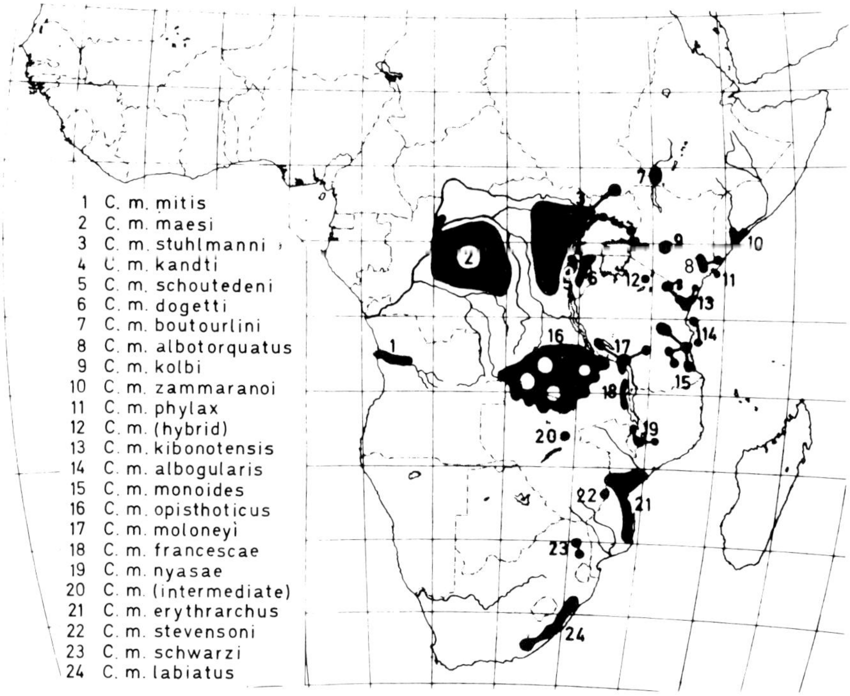

Fig. 3: Distribution of the *Cercopithecus mitis* subspecies.

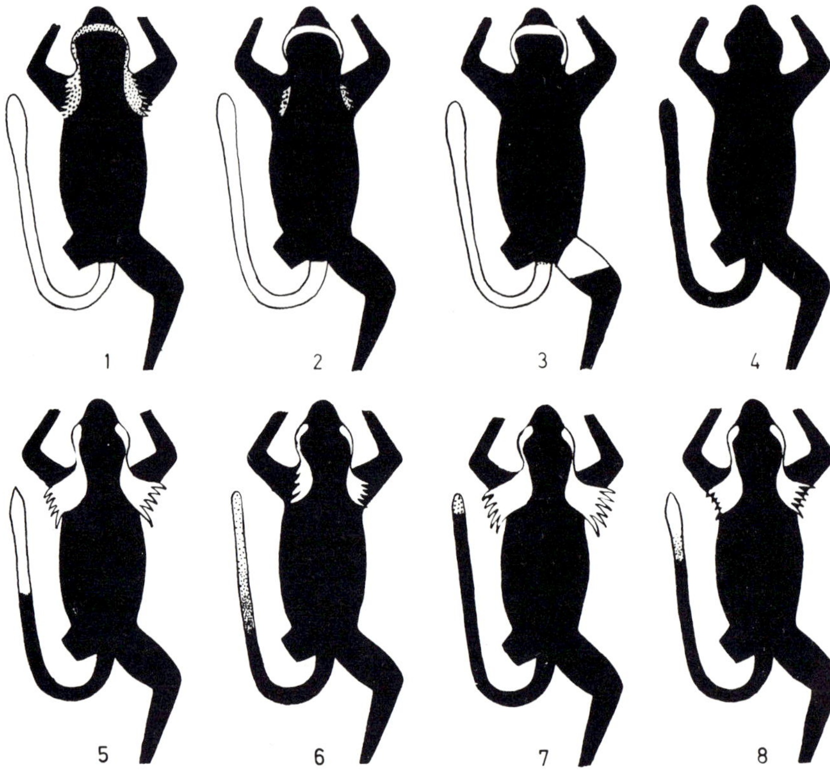

Fig. 4: Black and White Colobus: Variation of the color pattern in the different species and subspecies. (1) C. polykomos polykomos, (2) C. polykomos dollmani, (3) C. polykomos vellerosus, (4) C. satanas, (5) C. angolensis angolensis, (6) C. angolensis cottoni, (7) C. angolensis r wenzorii, (8) C. angolensis adolfi friederici, (9) C. angolensis cordieri (10) C. angolensis prigoginei, (11) C. angolensis sharpei, (12) C. angolensis palliatus, (13) C. abyssinicus caudatus, (15) C. abyssinicus kikuyuensis, (16) C. abyssinicus percivali (17) C. abyssinicus gallarum, (18) C. abyssinicus abyssinicus, (19) C. abyssinicus dodingae, (20) C. abyssinicus uellensis, (21) C. abyssinicus occidentalis.

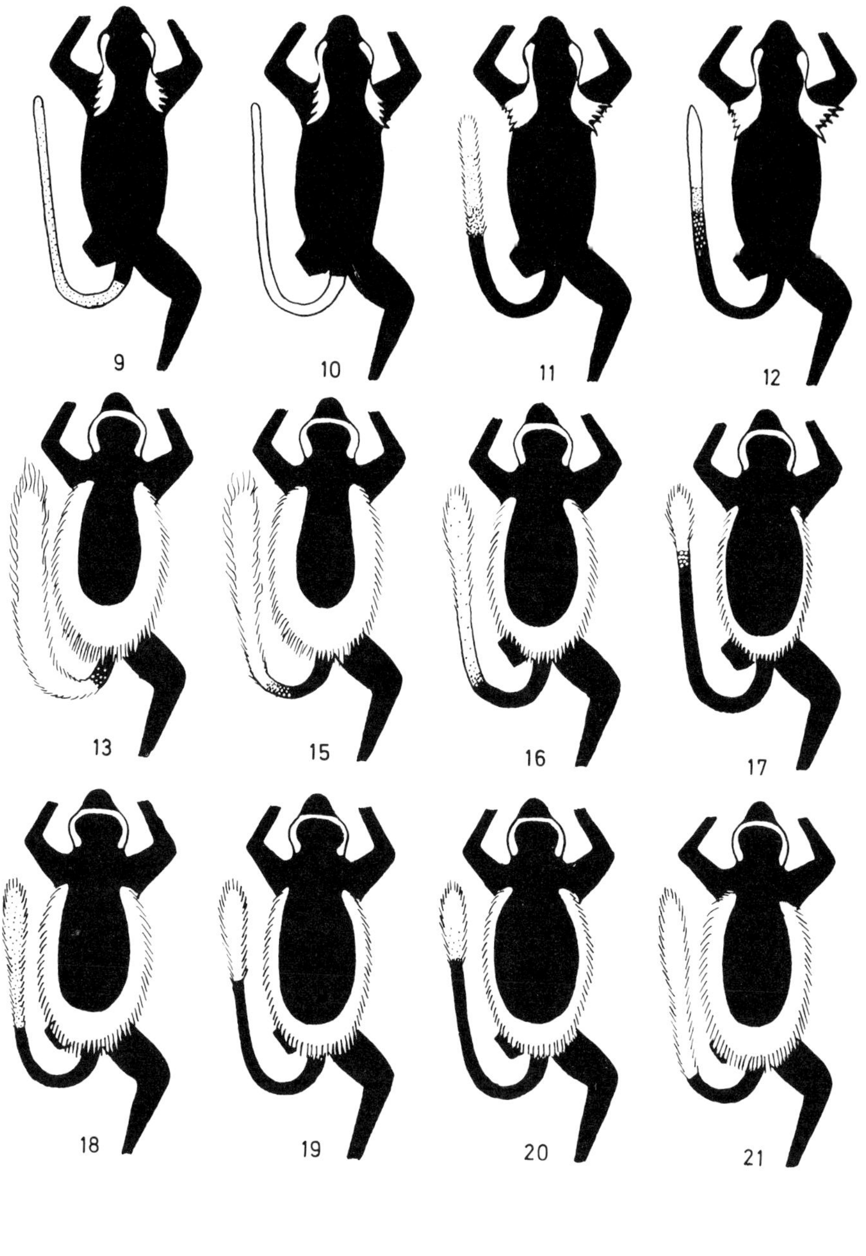
9
10
11
12
13
15
16
17
18
19
20
21

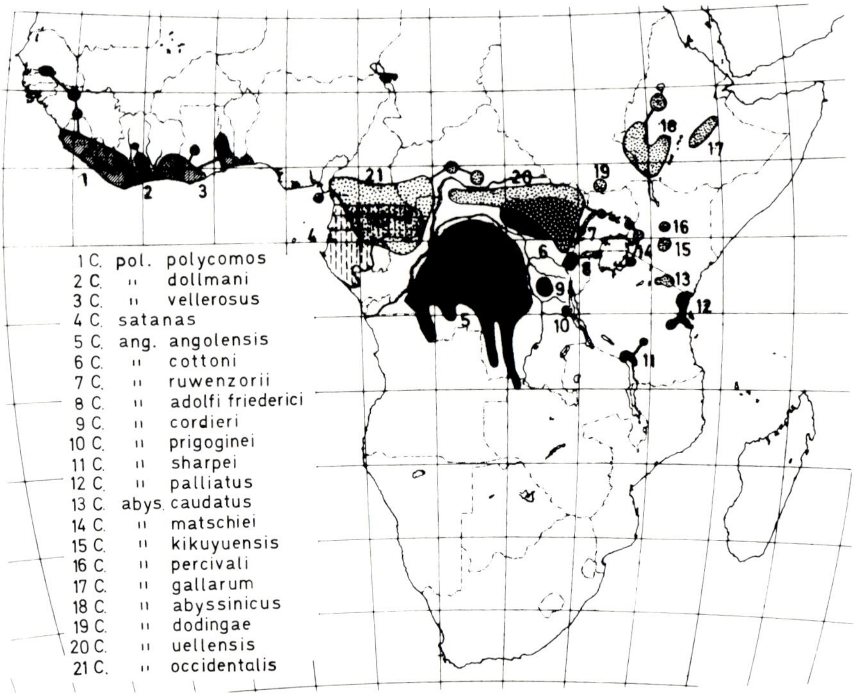

Fig. 5: Distribution of the Black and White Colobus (C. polykomos, C. satanas, C. angolensis, C. abyssinicus).

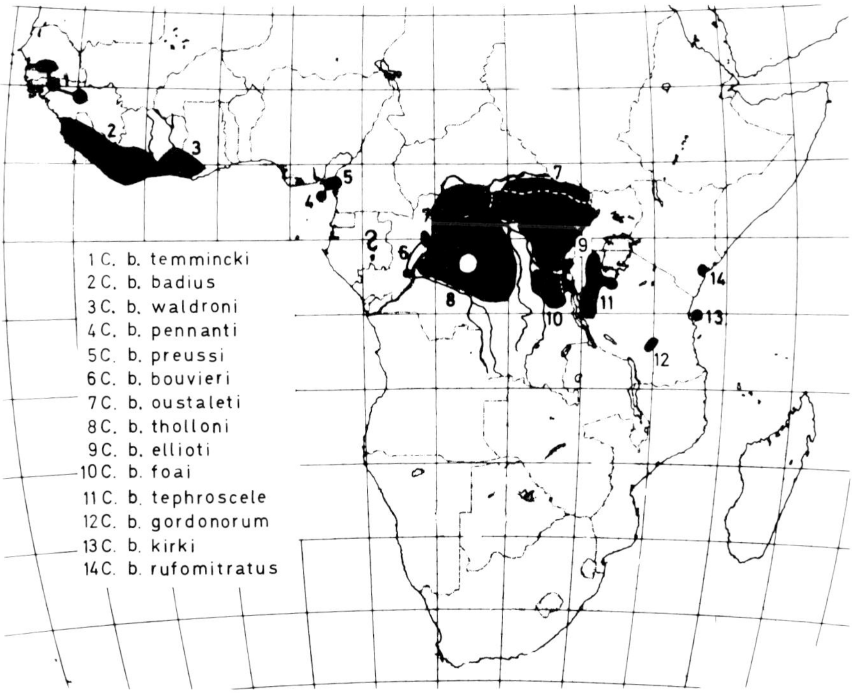

Fig. 6: Distribution of the Red Colobus. The interrupted lines in no 7 (C. badius oustaleti) give the approximate limits of the synonyms C. b. powellii and C. b. brunneus and in no 10 (C. badius foai) the limit of the synonym C. b. graueri.

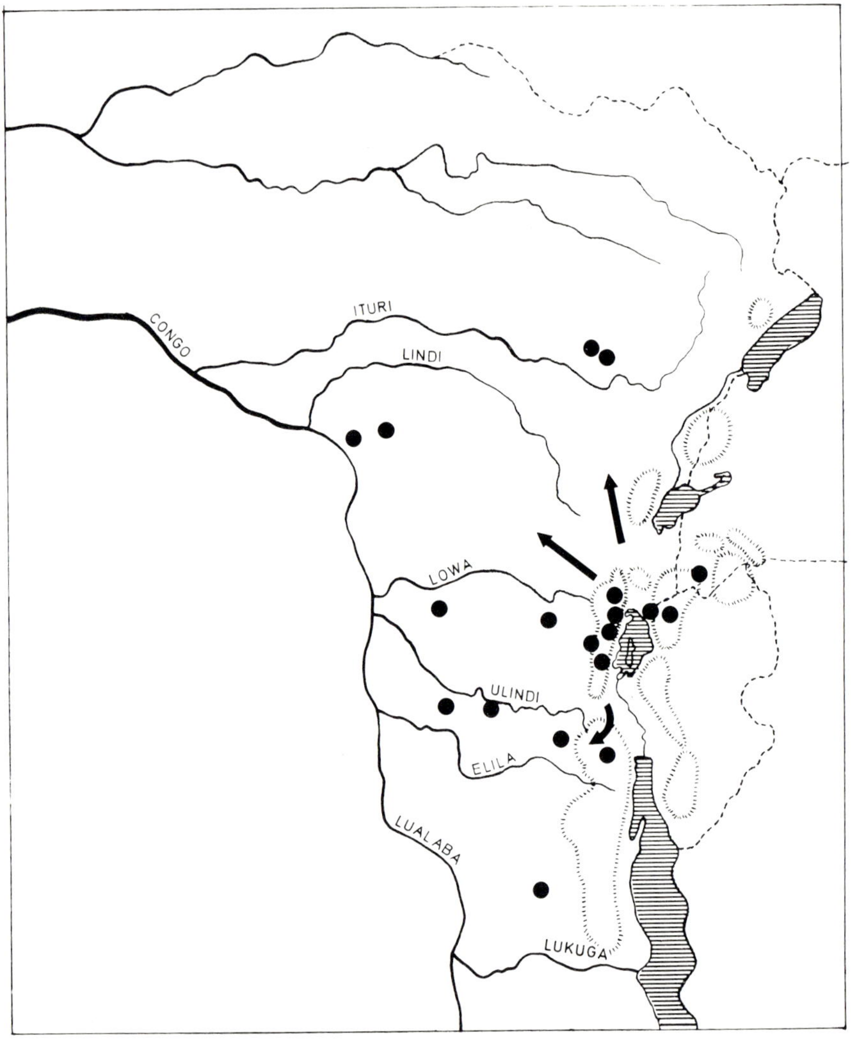

Fig. 7: Distribution of Cercopithecus hamlyni (compare with Fig.8).

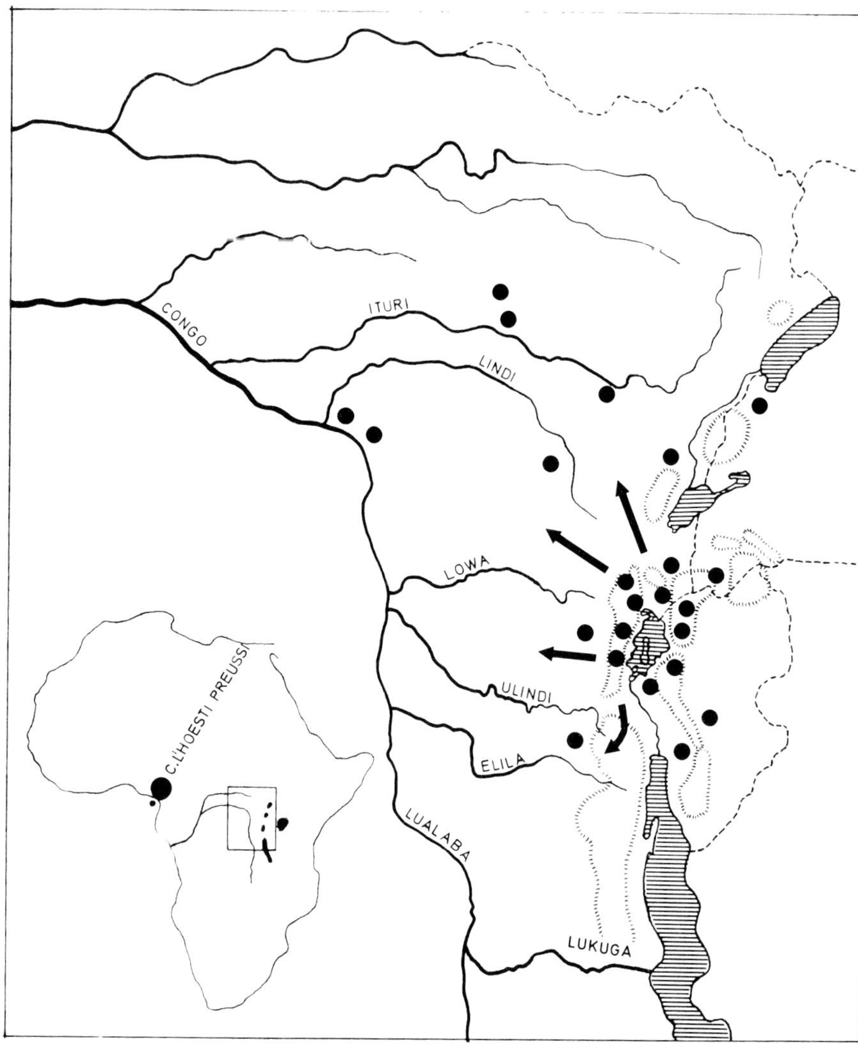

Fig. 8: Distribution of Cercopithecus lhoesti. The square on the map of Africa shows the area of the enlarged map.

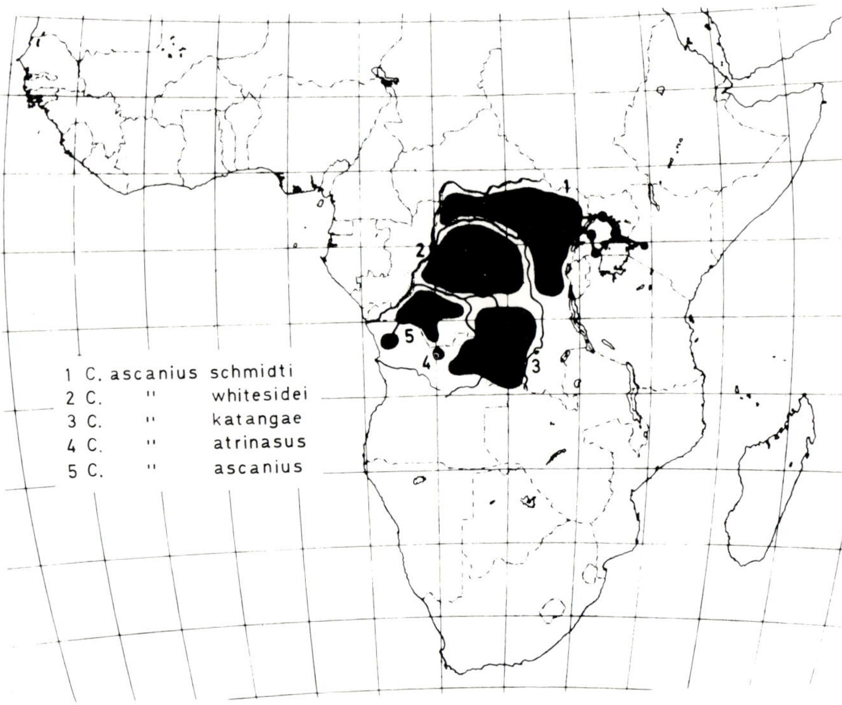

Fig. 9: Distribution of Cercopithecus ascanius.

PART V

SYSTEMATIC CONCLUSIONS

AN ANNOTATED CLASSIFICATION OF THE CERCOPITHECOIDEA

Richard W. Thorington, Jr. and C. P. Groves

Zoological classifications have heuristic value, since they require investigators to focus their attention on the comparison of animals. Such an exercise sharply delineates areas of ignorance from areas of knowledge. We submit and annotate this classification in the hope that it will emphasize the areas of doubt and thus stimulate further investigations into the biology of Old World monkeys.*

We list species names without mentioning subspecies. Except where we have specified otherwise, our usage is the same as that of Kuhn (1967) who itemizes all the included subspecies.

SUBORDER: ANTHROPOIDEA Mivart, 1864

SUBFAMILY: CERCOPITHECOIDEA Gray, 1821

We see no reason to change these higher categories as adopted by Simpson (1945) and many reasons to retain them The inclusion of the Ceboidea, Cercopithecoidea, and Hominoidea in a gingle suborder seems a useful device to convey similarity and is consistent with known phylogeny even with a fairly strict definition of monophyly. The use of Infraorders, Catarrhini and Platyrrhini, is convenient although they make the classification somewhat top-heavy

FAMILY: CERCOPITHECIDAE Gray, 1821

SUBFAMILY: CERCOPITHECINAE Gray, 1821

Schultz' work, as summarized in his paper in this volume, provides ample justification for inclusion of the cercopithecines and

---

* Dr Thorington and Dr Groves were asked to act as taxonomic referees for the 1969 Conference and to write a concluding paper for this volume incorporating the most recent views on the classification of Old World monkeys. (Eds.)

colobines in a single family. Division of the subfamily Cercopithecinae into two tribes, Papionini and Cercopithecini, as done by Kuhn (1967), is probably helpful. However, it is questionable whether the use of tribes in such a small subfamily increases clarity more than it increases nomenclatorial complexity.

GENUS: _Macaca_ Lacepede, 1799

M. sylvanus	Linnaeus, 1758
M. sinica	Linnaeus, 1771
M. radiata	Geoffroy, 1812
M. silenus	Linnaeus, 1758
M. nemestrina	Linnaeus, 1766
M. fascicularis	Raffles, 1821
M. mulatta	Zimmerman, 1780

Fooden (1964) presented evidence of hybridization in the wild between _M. fascicularis_ and _M. mulatta_. There is not sufficient evidence of intergradation between these two to warrant synonymizing them, however.

M. assamensis	McClelland, 1839
M. cyclopis	Swinhoe, 1862
M. arctoides	Geoffroy, 1831
M. thibetana	Milne-Edwards, 1870
M. fuscata	Blyth, 1875

With respect to the last three species, we follow Fooden (1967, a, b).

M. nigra Desmarest, 1822

Inclusion of *Cynopithecus* in the synonymy of *Macaca* follows Fooden (1969). For species however we interpret his evidence rather differently. He showed that all the Celebes forms are allopatric and that they exhibit clinal variation. This suggests to us that these are conspecific forms which have relatively recently interbred and are freely capable of interbreeding at present, in spite of some morphological differences.

GENUS: *Cercocebus* Geoffroy, 1812

C. torquatus Kerr, 1792

C. galeritus Peters, 1879

C. albigena Gray, 1850

C. aterrimus Oudemans, 1890

There was little contributed at the Conference to clarify the systematics of *Cercocebus*. Booth's (1965) suggestion that *C. atys* (Audebert, 1797) is not conspecific with *C. torquatus* need evaluation. We follow Kuhn (1967), since we are not convinced that the pelage difference between *atys* and *torquatus* is as significant as Booth (by analogy) considered it. Dobroruka and Badalec (1966) expressed their dissatisfaction with this classification. They separated *atys* from *torquatus* and recognized *chrysogaster* and *agilis* as valid species, distinct from *galeritus*.

GENUS: Papio Muller, 1776

P. hamadryas Linnaeus, 1758

P. cynocephalus Linnaeus, 1766

The inclusion of Comopithecus is the synonymy of Papio is generally acknowledged. There is a question whether P. hamadryas and P. cynocephalus should be considered conspecific, and on this point the two co-authors are unable to agree. Kummer's data demonstrates a degree of interbreeding between the two forms, although it is not yet clear whether this is a narrow hybrid zone or a broad belt of intergradation; Kummer et al (this volume) are not in favor of making these two forms conspecific. The emerging picture is that all savanna baboons represent a single polytypic species with morphologically different subspecies interbreeding wherever they meet. With the inclusion of the hamadryas, the correct specific name of all these baboons would be Papio hamadryas. We would recommend this nomenclatorial adjustment unhesitatingly in a group familiar only to zoologists. However, baboons are used extensively in medical research and changing their commonly used names would be difficult and would cause undesirable bibliographical confusion. Amidst the existing confusion in baboon nomenclature, the name hamadryas has been unusually stable, referring to particular populations in Ethiopia and Arabia. To use this Latin name for all savanna baboons (while the common name "hamadryas" would undoubtedly continue to be used for just the north-eastern populations) would increase the confusion. Therefore one of us (R.W.T.) recommends continuance of the traditional usage of P. hamadryas and the use of P. cynocephalus for all other forms of savanna baboons, at least until the degree of intergradation between the two forms is better known and documented.

On the other hand we now have for savanna baboons a situation so far achieved for almost no other primate: namely, observation in the field of the passage from one form to another, with the formation of a hybrid population and the mutual invasion of the respective gene-pools (Kummer et al, this volume). Such a situation is usually inferred from museum material when field data are lacking; but in the present instance the existence of such a situation has actually been confirmed in the field, although it had not in fact previously been inferred on the strength of museum material. The second author (C.P.G.) would maintain that in such a clear-cut case it is the duty of the taxonomist to rearrange the classification in accordance with advances in our understanding of the interrelationships between existing forms, and that the biological reality must take precedence over matters of convenience.

P. sphinx Linnaeus, 1758

P. leucophaeus Cuvier, 1807

Jolly contends that drills and mandrills have overlapping ranges. Therefore they are retained as distinct species here. This differs from Buettner-Janusch's (1966) assessment. Since *Papio* (Muller, 1776) refers to a mandrill, since it is highly desirable to retain use of the name *Papio* for the savanna baboons, and since the forest and savanna baboons are a small monophyletic group, it seems most reasonable to include all in the same genus *Papio*. Gartlan's data and conclusions (this volume) strongly support this thesis.

GENUS: *Theropithecus* I. Geoffroy, 1843

T. gelada Rüppell, 1835

Buettner-Janusch (1966) synonymized *Theropithecus* with *Papio*. We retain it here to emphasize the behavioral and morphological differences between geladas and other baboons.

GENUS: Cercopithecus Linnaeus, 1758

C. aethiops Linnaeus, 1758

As discussed by Kuhn (1957) and Thorington (1969), Dandelot's (1959) morphological evidence strongly suggests that the C. aethiops group consists of interbreeding allopatric forms and thus should be considered a single polytypic species. Struhsaker and Gartlan (this volume) have adduced behavioral evidence supporting this thesis. Intergradation of sabaeus with the other forms seems not yet to have been clearly demonstrated, however.

C. cephus Linnaeus, 1766

C. ascanius Audebert, 1799

C. erythrogaster Gray, 1866

C. petaurista Schreber, 1774

Kuhn (1967) included these four species (plus C. erythrotis Waterhouse, 1938) in a single subspecies. Struhsaker's report to the Conference of a hybrid zone between the range of erythrotis and cephus, together with the ecological and ethological similarities, are the reasons for treating erythrotis as a subspecies of cephus. More documentation of this hybrid zone is desirable.

C. campbelli Waterhouse, 1838

C. mona Schreber, 1775

C. denti Thomas, 1907

C. wolfi Meyer, 1891

C. pogonias Bennett, 1833

The _mona_ subspecies, consisting of these five species, is treated similarly by Hill (1966) and Kuhn (1967), in accord with Booth (1955). The status of _C. denti_ is unclear; it may be a subspecies of _wolfi_ (Booth, 1955). Struhsaker's report (this volume) of some hybrids between _C. pogonias_ and _C. mona_ is interesting, in view of Booth's discussion of these species; it is noteworthy however that Struhsaker considers hybridization to occur solely in areas of habitat disruption by man, and that elsewhere there are wide areas of sympatry between the two.

C. hamlyni Pocock, 1907

C. neglectus Schlegel, 1876

C. lhoesti Sclater, 1899

C. mitis Wolf, 1822

C. nictitans Linnaeus, 1766

C. diana Linnaeus, 1758

C. dryas Schwarz, 1932

These seven names designate the other species recognized by Kuhn (1967) in his subgenus _Cercopithecus_. Some questions raised at the Conference were (a) do _mitis_ and _albogularis_ have sympatric subspecies, making it more desirable to consider them distinct species? (b) are _nictitans_ and _mitis_ better considered conspecific? (c) should _preussi_ be considered conspecific with _lhoesti_?

GENUS: _Miopithecus_ Geoffroy, 1842

M. talapoin Schreber, 1774

_M._ sp. nov.

Machado (1969) sets out in detail his opinion that the talapoins of Angola/Congo and of Gabon/Cameroun are specifically distinct, adding that, since all names thus far applied are referable to the Angola species, a new name must be found for the Gabon species. A full description and diagnosis are in course of preparation by Machado and Dandelot (see Machado, 1969).

On the basis of Verheyen's (1962) morphological work it would seem better to include the talapoin the the genus Cercopithecus. At the Conference it was concluded that the talapoin was behaviorally very different from the Cercopithecus species. This evaluation is the basis of the generic separation listed here. However, it might be more useful to represent the degree of difference of the talapoin by a subgeneric separation with Cercopithecus, as done by Kuhn (1967) and by Napier and Napier (1967).

GENUS: Allenopithecus Lang, 1923

A. nigroviridis Pocock, 1907

Since A. nigroviridis is morphologically more distinct than the talapoin, it should be separated from Cercopithecus at the same or higher taxonomic level. Again, this could reasonably be considered a subgenus of Cercopithecus. as by Kuhn (1967) and Napier and Napier (1967).

GENUS: Erythrocebus Trouessart, 1897

E. patas Schreber, 1774

There seems to be general agreement that the patas is most divergent from the Cercopithecus (sensu stricto) monkeys in its postcranial anatomy and behavior. Dandelot (1968) treated it as a subgenus of Cercopithecus, to emphasize its relationship to other members of this genus. Verheyen (1962) also considered it congeneric, on the basis of cranial similarities. If *Miopithecus* and *Allenopithecus* are treated as valid genera, certainly *Erythrocebus* should be.

SUBFAMILY: COLOBINAE BLYTH, 1875

The name "Colobinae" is predated by "Semnopithecinae" Owen, 1843, and by "Presbytinae" Gray, 1825. However we agree with Kuhn (1967) that Colobinae should be the name conserved, on the basis of Article 23b ii, iii or d ii.

GENUS: *Colobus* Illiger, 1811

SUBGENUS: *Colobus* Illiger, 1811

C. polykomos Zimmerman, 1780

C. satanas Waterhouse, 1838

C. angolensis Sclater, 1860

C. guereza Rüppell, 1835

Kuhn (1967) and Napier and Napier (1967) make *satanas* and *angolensis* conspecific with *polykomos*; both these are sympatric over wide areas with *guereza* (see map in Rahm's contribution to this volume) whereas *polykomos* is not. Behavioral and morphological data discussed during the Conference - especially by Rahm and Struhsaker - raise the question of the precise nature of the *polykomos*/*guereza* relationship. Accordingly, all four of Schwarz's old subspecies-groups are provisionally allowed specific rank. Dandelot (1968) also allows four species in the Black-and-White group.

SUBGENUS: *Piliocolobus* Rochebrune, 1886

C. badius Kerr, 1792

C. kirkii Gray, 1868

The clear-cut differences between the mainland representatives of this group, pointed out by Dandelot (1968) and independently studied by one of us (C. P. G.) make it questionable whether recognition of several species among what is usually called *C. badius* might not be more in line with current taxonomic practice. The well-marked differences in coat-color, at least between the Western races (*temmincki*, *badius* and *waldroni*) and the rest, are not of a clinal nature.

GENUS: *Procolobus* Rochebrune, 1886

P. verus Van Beneden, 1838

For reasons for restricting this genus to a single species, rather than including the *C. badius* group as does Kuhn (1967), see Groves (this volume).

GENUS: *Pygathrix* E. Geoffroy St. Hilaire, 1812

SUBGENUS: *Pygathrix* E. Geoffroy St. Hilaire, 1812

P. nemaeus Linnaeus, 1771

SUBGENUS: Rhinopithecus Milne-Edwards, 1872

P. roxellana Milne-Edwards, 1870

P. brelichi Thomas, 1903

P. avunculus Dollman, 1912

GENUS: Nasalis E.Geoffroy St.Hilaire, 1812

N. larvatus van Wurmb, 1781

N. concolor Miller, 1903

For the classification of this genus and the last, see Groves (this volume).

GENUS: Presbytis Eschscholtz, 1821

P. aygula Linnaeus, 1758

P. thomasi Collett, 1892

P. hosei Thomas, 1889

P. melalophos Raffles, 1821

Medway (this volume) discusses alternate interpretations of the distributions and morphology of these forms. We use his cautious interpretation here. His allocation of subspecies differs from that of Chasen (1940) and Kuhn (1967).

<u>P. frontata</u> Müller, 1838

<u>P. rubicunda</u> Müller, 1838

All modern authors recognize that these six species are closely related and form a monophyletic group (Pocock, 1934; Hill, 1939; Hooijer, 1962 etc.). Ideally they should be set apart as subgenus, however we use no subgenera at present because of the uncertainty surrounding the interrelationships of the other species, and the intermediate position of <u>P. potenziani</u> between the <u>aygula</u> and <u>cristata</u> groups

<u>P. potenziani</u> Bonaparte, 1856

<u>P. cristata</u> Raffles, 1821

<u>P. pileata</u> Blyth, 1843

<u>P. geei</u> Gee, 1956

There is a very close kinship between <u>geei</u> and <u>pileata</u>, of which it is probably only a subspecies as suggested by Oboussier and Maydell (1959). Because of the possibility of sympatry between them (Khajuria, 1956), we treat them as distinct species.

<u>P. obscura</u> Reid, 1837

<u>P. phayrei</u> Blyth, 1847

The distinctions between these two species are poorly defined, and they may prove conspecific.

<u>P. francoisi</u> Pousargues, 1898

As understood by Kuhn (1967) and others, this species includes a number of highly distinctive forms from a small area around the

North Vietnam-Laos-China border. Their status is poorly understood.

P. senex Erxleben, 1777

P. johnii Fischer, 1829

P. senex and *P. johnii* are probably conspecific

P. entellus Dufresne, 1797

FINAL COMMENT

We would like to emphasize the provisional nature of this classification. A Conference such as the one that this book reports, brings together people from a number of disciplines - traditional taxonomy, anatomy, biochemistry, paleontology, ethology, ecology - and forces each of the participants into an awareness of the other fields and gives him a wider perspective. This coming-together of scientists who are essentially approaching the same problem, but from different viewpoints, has its effect on the classification that is offered. In that a classification is, after all, a kind of one-dimensional phylogeny, the evolutionary history of the groups concerned is elucidated, and the interlocking elements of the total biological pattern is carried back into the past.

It is perfectly clear, at any rate, that the place of taxonomy as a focus and unifying field of biological data will in future become more and more firmly established. For this reason, as groups of organisms become better known, the traditional taxonomic schemes will need to be modified by new data derived, for example, from ethology or karyology. There is of course a danger that some workers will be overimpressed by some of these fields and will be tempted to reject taxonomic schemes derived from other fields, instead of modifying them. The unifying nature of taxonomy must be stressed again.

Conservation is an important concern of taxonomy. This was a topic much discussed at the 1969 Conference. The realization that systematic relationships can and should be studied in the field is an important step away from too great a reliance on the collection of large series for museums which has been so necessary in the past. While the collection of even quite large series need not be damaging to the survival prospects of a whole species or subspecies, it is hardly compatible with the aims of wild life protection; nor is it a good example to local hunters who are liable to imprisonment for "poaching" - for what they may well regard as legitimate reasons.

The preservation of an individual species does not ensure its survival when its habitat is destroyed. We are today witnessing a massive and horrifying destruction of the natural environment, both at home by pollution, and in the tropics by deforestation. In West Africa, in Thailand, in Malaya, a precious international heritage of primeval high forest is rapidly being cut down by timber companies bent on extracting the last cent of profit from the land before the chance is lost forever with the felling of the very last tree. This selfish waste, this insidious form of "neo-imperialism" (since in most cases the companies concerned are not indigenous), will exterminate many animals, primates among them, however many conservation regulations covering individual species may be in operation. The pitifully small areas set aside as Reserves or Sanctuaries must be enlarged, and controlled and managed in an intelligent manner if, for example, the Diana monkey, the Red Colobus, the Nilgiri langur or the Proboscis monkey are not to disappear for ever - for all our good intentions. This is the new role which taxonomy has to play; in its function as a focus for biological data, and in its emphasis on adaptation and the ecological setting of species, taxonomy has found a new role for itself in what may become one of the most fundamental questions of our time.

ACKNOWLEDGEMENTS

We wish to thank Drs C.O. Handley, Jr., H.-J. Kuhn, E.Mayr, and J.R. Napier for their criticisms of the manuscript. Our debt to the other conferees and to the Wenner-Gren Foundation is obvious and we thank them all. R.W.T. acknowledges support from N.I.H. grant FR 00168 to Harvard University.

REFERENCES

Booth, A.H. (1955). Speciation in the mona monkeys. J. of Mammalogy 35: 434-449

Booth, A.H. (1956). The Cercopithecidae of the Gold and Ivory coasts: geographic and systematic observations. Ann. Mag. Nat. Hist. 9(12): 476-480.

Buettner-Janusch, J. (1966). A problem in evolutionary systematics; nomenclature and classification of baboons, genus Papio. Folia primat. 4: 288-308.

Chasen, F.N. (1940). A handlist of Malaysian mammals. Bull. Raffles Mus. 15: 1-209.

Dandelot, P. (1959). Note sur la classification des cercopithèques du groupe aethiops. Mammalia 23: 357-368.

Dandelot, P. (1968). Preliminary Identification Manual for African Mammals. Primates: Anthropoidea. Smithsonian Institution, Washington, D.C.

Dobroruka, L.J. and J. Badalec. (1966). Zur Artbildung der Mangaben, Gattung Cercocebus. Rev. Zool. Bot. Afr. 73: 345-350.

Fooden, J. (1964). Rhesus and crab-eating macaques: intergradation in Thailand. Science 143: 363-365.

Fooden, J. (1967a). Identification of the stump-tailed monkey, Macaca speciosa I. Geoffroy, 1826. Folia primat. 5: 153-164.

Fooden, J. (1967b). Complementary specialization of male and female reproductive structures in the Bear Macaque, M. arctoides. Nature Lond. 214: 939-941

Fooden, J. (1969). Taxonomy and evolution of the monkeys of Celebes (Primates: Cercopithecidae). Bibliotheca primatologica No.10: 1-148.

Hill, W.C.O. (1939). An annotated systematic list of the leaf-monkeys. Ceylon J. Sci. (B) 21: 277-305.

Hill, W.C.O. (1966). Primates. Vol.6. Univ. Press: Edinburgh.

Hooijer, D.A. (1962). Quarternary langurs and macaques from the Malay Archipalago. Zool. Verh. Leiden 55: 1-64.

Khajuria, H. (1956). A new langur (Primates:Colobidae) from Goalpara District, Assam. Ann. Mag. Nat. Hist. 9(12): 86-88.

Kuhn, H.-J. (1967). Zur Systematik der Cercopithecidae. pp. 25-46. Progress in Primatology (Eds. Starck, Schneider, and Kuhn). Stuttgart: Gustav Fischer.

Machado, A. de Barros. (1969). Mamiferos de Angola ainda nao citados ou pouco conhecidos. Publ. cult. co. Diam. Ang. Lisboae No.46: 93-232.

Maydell, G.A. von and H. Oboussier. (1959). Zur Kenntnis des indischen Goldlangurs. Z. morph. Ökol. Tiere 48: 102-114.

Napier, J.R. and P.H. Napier. (1967). A Handbook of Living Primates. London and New York: Academic Press.

Pocock, R.I. (1934). The monkeys of the genera Pithecus (or Presbytis) and Pygathrix found to the east of the Bay of Bengal. Proc. zool. Soc. Lond. 895-961.

Simpson, G.G. (1954). The principles of classification and a classification of mammals. Bull. Amer. Mus. Nat. Hist. 85: 1-350.

Thorington, R.W., Jr. (1969). Comments on primate taxonomy and its use. Proc. 2nd Internat. Congr. Primat. Vol. 2 Basel: Karger.

Verheyen, W.N. (1962). Contribution à la craniologie comparée des Primates. Musée Roy. Afr. Centrale Tervuren, Belgique. Ser. 8. Sci. zool. 105: 1-156.

INDEX TO ANIMALS

*Indicates fossil form throughout.

N

INDEX TO AUTHORS